Pelican Books
Asimov's Guide to Science, Vol. 2
The Biological Sciences

Dr Isaac Asimov was born in Russia in 1920 and
received a PhD from Columbia University. A Professor
of Biochemistry at the School of Medicine, Boston
University, he has published many works, both fiction
and non-fiction. Asimov's chief contribution to science
fiction is said to have been his use of robots, from the
collection of stories, *I, Robot* (1950) to *The Caves of
Steel* (1954) and *The Naked Sun* (1957), both of which
employ a robot detective, as does *Asimov's Mysteries*
(1968), a collection which combines science fiction with
the mystery story. Under the pseudonym Paul French,
he has also written for children. Among his other works
are *The Foundation Trilogy* (1963), *Nine Tomorrows*
(1959), a collection of tales, and *The Universe* (1967).

Isaac Asimov

Asimov's Guide to Science

Volume 2
The Biological Sciences

Penguin Books

Penguin Books Ltd,
Harmondsworth, Middlesex, England
Penguin Books Australia Ltd,
Ringwood, Victoria, Australia
Penguin Books (N.Z.) Ltd,
182–190 Wairau Road, Auckland 10, New Zealand

First published by Basic Books, Inc., New York, 1972
Published in Pelican Books 1975
Copyright © Basic Books, Inc., 1960, 1965, 1972

Made and printed in Great Britain by
Hazell Watson & Viney Ltd, Aylesbury, Bucks
Set in Monotype Ehrhardt

Contents

Contents

1 The Molecule

Organic Matter

The term molecule (from a Latin word meaning 'small mass') originally meant the ultimate, indivisible unit of a substance, and in a sense it *is* an ultimate particle, because it cannot be broken down without losing its identity. To be sure, a molecule of sugar or of water can be divided into single atoms or groups, but then it is no longer sugar or water. Even a hydrogen molecule loses its characteristic chemical properties when it is broken down into its two component hydrogen atoms.

Just as the atom has furnished chief excitement in twentieth-century physics, so the molecule has been the subject of equally exciting discoveries in chemistry. Chemists have been able to work out detailed pictures of the structure of even very complex molecules, to identify the roles of specific molecules in living systems, to create elaborate new molecules, and to predict the behaviour of a molecule of a given structure with amazing accuracy.

By the mid twentieth century, the complex molecules that form the key units of living tissue, the proteins and nucleic acids, were being studied with all the techniques made possible by an advanced chemistry and physics. The two sciences, 'biochemistry' (the study of the chemical reactions going on in living tissue) and 'biophysics' (the study of the physical forces and phenomena involved in living processes), merged to form a brand new discipline – 'molecular biology'. Through the findings of molecular biology, modern science has in a single generation of effort all but wiped out the borderline between life and non-life.

Yet less than a century and a half ago, the structure of not even the simplest molecule was understood. About all the chemists of the early nineteenth century could do was to separate all matter into two great categories. They had long been aware (even in the days of the alchemists) that substances fell into two sharply distinct classes with respect to their response to heat. One group – for example, salt, lead, water – remained basically unchanged after being heated. Salt might glow red-hot when heated, lead might melt, water might vaporize – but when they were cooled back to the original temperature, they were restored to their original form, none the worse, apparently, for their experience. On the other hand, the second group of substances – for example, sugar, olive oil – were changed permanently by heat. Sugar became charred when heated and remained charred after it was cooled again; olive oil was vaporized and the vapour did not condense on cooling. Eventually the scientists noted that the heat-resisting substances generally came from the inanimate world of the air, ocean, and soil, while the combustible substances usually came from the world of life, either from living matter directly or from dead remains. In 1807, the Swedish chemist Jöns Jakob Berzelius named the combustible substances 'organic' (because they were derived, directly or indirectly, from the living organisms) and all the rest 'inorganic'.

Early chemistry focused mainly on the inorganic substances. It was the study of the behaviour of inorganic gases that led to the development of the atomic theory. Once that theory was established, it soon clarified the nature of inorganic molecules. Analysis showed that inorganic molecules generally consisted of a small number of different atoms in definite proportions. The water molecule contained two atoms of hydrogen and one of oxygen; the salt molecule contained one atom of sodium and one of chlorine; sulphuric acid contained two atoms of hydrogen, one of sulphur, and four of oxygen, and so on.

When the chemists began to analyse organic substances, the picture seemed quite different. Two substances might have exactly the same composition and yet show distinctly different properties. (For instance, ethyl alcohol is composed of two carbon atoms, one oxygen atom, and six hydrogen atoms; so is dimethyl ether – yet one is a liquid at room temperature while the other is a gas.) The organic molecules contained many more atoms than the simple inorganic ones, and there seemed to be no rhyme or reason in the way they were combined. Organic compounds simply could not be explained by the straightforward laws of chemistry that applied so beautifully to inorganic substances.

Berzelius decided that the chemistry of life was something apart which obeyed its own set of subtle rules. Only living tissue, he said, could make an organic compound. His point of view is an example of 'vitalism'.

Then in 1828 the German chemist Friedrich Wöhler, a student of Berzelius, produced an organic substance in the laboratory! He was heating a compound called ammonium cyanate, which was then generally considered inorganic. Wöhler was thunderstruck to discover that, on being heated, this material turned into a white substance identical in properties with 'urea', a component of urine. According to Berzelius's views, only the living kidney could form urea, and yet Wöhler had formed it from inorganic material merely by applying a little heat.

Wöhler repeated the experiment many times before he dared publish his discovery. When he finally did, Berzelius and others at first refused to believe it. But other chemists confirmed the results. Furthermore, they proceeded to synthesize many other organic compounds from inorganic precursors. The first to bring about the production of an organic compound from its elements was the German chemist Adolph Wilhelm Hermann Kolbe, who produced acetic acid in this fashion in 1845. It was this that really killed Berzelius's version of vitalism. More and more it became clear that the same chemical laws applied to inorganic and organic molecules alike. Eventually the distinction between organic and inorganic substances was given a simple definition: all substances containing carbon (with the possible exception of a few simple compounds such as carbon dioxide) are called organic; the rest are inorganic.

To deal with the complex new chemistry, chemists needed a simple shorthand for representing compounds, and fortunately Berzelius had already suggested a convenient, rational system of symbols. The elements were designated by abbreviations of their Latin names. Thus C would stand for carbon, O for oxygen, H for hydrogen, N for nitrogen, S for sulphur, P for phosphorus, and so on. Where two elements began with the same letter, a second letter was used to distinguish them: e.g., Ca for calcium, Cl for chlorine, Cd for cadmium, Co for cobalt, Cr for chromium, and so on. In only a comparatively few cases are the Latin or Latinized names (and initials) different from the English, thus: iron ('ferrum') is Fe; silver ('argentum') Ag; gold ('aurum') Au; copper ('cuprum') Cu; tin ('stannum') Sn; mercury ('hydrargyrum') Hg; antimony ('stibium') Sb; sodium ('natrium') Na; and potassium ('kalium') K.

With this system it is easy to symbolize the composition of a molecule. Water is written H_2O (thus indicating the molecule to consist of two hydrogen atoms and an oxygen atom); salt, NaCl; sulphuric acid, H_2SO_4, and so on. This is called the 'empirical formula' of a compound; it tells what the compound is made of but says nothing about its structure, that is, the manner in which the atoms of the molecule are interconnected.

Baron Justus von Liebig, a co-worker of Wöhler's, went on to work out the composition of a number of organic chemicals, thus applying 'chemical analysis' to the field of organic chemistry. He would carefully burn a small quantity of an organic substance and trap the gases formed (chiefly CO_2 and water vapour, H_2O) with appropriate chemicals. Then he would weigh the chemicals used to trap the combustion products to see how much weight had been added by the trapped products. From that weight he could determine the amount of carbon, hydrogen, and oxygen in the original substance. It was then an easy matter to calculate, from the atomic weights, the numbers of each type of atom in the molecule. In this way, for instance, he established that the molecule of ethyl alcohol had the formula C_2H_6O.

Liebig's method could not measure the nitrogen present in organic compounds, but the French chemist Jean Baptiste André Dumas devised a combustion method which did collect the gaseous nitrogen released from substances. He made use of his methods to analyse the gases of the atmosphere with unprecedented accuracy in 1841.

The methods of 'organic analysis' were made more and more delicate until veritable prodigies of refinement were reached in the 'micro-analytical' methods of the Austrian chemist Fritz Pregl. He devised techniques, in the early 1900s, for the accurate analysis of quantities of organic compounds barely visible to the naked eye and received the Nobel Prize for chemistry in 1923 in consequence.

Unfortunately, determining only the empirical formulas of organic compounds was not very helpful in elucidating their chemistry. In contrast to inorganic compounds, which usually consisted of two or three atoms or at most a dozen, the organic molecules were often huge. Liebig found that the formula of morphine was $C_{17}H_{19}O_3N$, and of strychnine, $C_{21}H_{22}O_2N_2$.

Chemists were pretty much at a loss to deal with such large molecules or make head or tail of their formulas. Wöhler and Liebig tried to group atoms into smaller collections called 'radicals' and to work out theories to show that various compounds were made up of specific radicals in different numbers and combinations. Some of the systems were most ingenious, but none really explained enough. It was particularly difficult to explain why two compounds with the same empirical formula, such as ethyl alcohol and dimethyl ether, should have different properties.

This phenomenon was first dragged into the light of day in the 1820s by Liebig and Wöhler. The former was studying a group of compounds called 'fulminates', the latter a group called 'isocyanates', and the two turned out to have identical empirical formulas – the elements were present in equal parts, so to speak. Berzelius, the chemical dictator of the day, was told of this and was reluctant to believe it until, in 1830, he discovered some examples for himself. He named such compounds, with different properties but with elements present in equal parts, 'isomers' (from Greek words meaning 'equal parts'). The structure of organic molecules was indeed a puzzle in those days.

The chemists, lost in the jungle of organic chemistry, began to see daylight in the 1850s when they noted that each atom could combine with only a certain number of other atoms. For instance, the hydrogen atom apparently could attach itself to only one atom: it could form hydrogen chloride, HCl, but never HCl_2. Likewise chlorine and sodium could each take only a single partner, so they formed NaCl. An oxygen atom, on the other hand, could take two atoms as partners – for instance, H_2O. Nitrogen could take on three: e.g., NH_3 (ammonia). And carbon could combine with as many as four: e.g., CCl_4 (carbon tetra-chloride).

In short, it looked as if each type of atom had a certain number of hooks by which it could hang on to other atoms. The English chemist Edward Frankland called these hooks 'valence' bonds, from a Latin word meaning 'power', to signify the combining powers of the elements.

The German chemist Friedrich August Kekulé von Stradonitz saw that if carbon were given a valence of 4 and if it were assumed that carbon atoms could use those valences, in part at least, to join up in chains, then a map could be drawn through the organic jungle. His technique was made more visual by the suggestion of a Scottish chemist, Archibald Scott Couper, that these combining forces between atoms ('bonds', as they are usually called) be pictured in the form of small dashes.

In 1861, Kekulé published a textbook with many examples of this system, which proved its convenience and value. The 'structural formula' became the hallmark of the organic chemist.

For instance, the methane (CH_4), ammonia (NH_3), and water (H_2O) molecules, respectively, could be pictured this way:

$$
\begin{array}{ccc}
\overset{\textstyle H}{\underset{\textstyle H}{H-\overset{|}{\underset{|}{C}}-H}} & \overset{\textstyle H}{H-\overset{|}{N}-H} & H-O-H
\end{array}
$$

Organic molecules could be represented as chains of carbon atoms with hydrogen atoms attached along the sides. Thus butane (C_4H_{10}) would have the structure:

$$
\begin{array}{c}
\text{H} \quad \text{H} \quad \text{H} \quad \text{H} \\
| \quad\; | \quad\; | \quad\; | \\
\text{H}-\text{C}-\text{C}-\text{C}-\text{C}-\text{H} \\
| \quad\; | \quad\; | \quad\; | \\
\text{H} \quad \text{H} \quad \text{H} \quad \text{H}
\end{array}
$$

Oxygen or nitrogen might enter the chain in the following manner, picturing the compounds methyl alcohol (CH_4O) and methylamine (CH_5N), respectively:

$$
\begin{array}{c}
\text{H} \\
| \\
\text{H}-\text{C}-\text{O}-\text{H} \\
| \\
\text{H}
\end{array}
\qquad\qquad
\begin{array}{c}
\text{H} \quad \text{H} \\
| \quad\; | \\
\text{H}-\text{C}-\text{N}-\text{H} \\
| \\
\text{H}
\end{array}
$$

An atom possessing more than one hook, such as carbon with its four, need not use each of them for a different atom: it might form a double or triple bond with one of its neighbours, as in ethylene (C_2H_4) or acetylene (C_2H_2):

$$
\begin{array}{c}
\text{H} \quad \text{H} \\
| \quad\; | \\
\text{H}-\text{C}=\text{C}-\text{H}
\end{array}
\qquad\qquad
\text{H}-\text{C}\equiv\text{C}-\text{H}
$$

Now it became easy to see how two molecules could have the same number of atoms of each element and still differ in properties. The two isomers must differ in the arrangement of those atoms. For instance, the structural formulas of ethyl alcohol and dimethyl ether, respectively, could be written:

$$
\begin{array}{c}
\text{H} \quad \text{H} \\
| \quad\; | \\
\text{H}-\text{C}-\text{C}-\text{O}-\text{H} \\
| \quad\; | \\
\text{H} \quad \text{H}
\end{array}
\qquad\qquad
\begin{array}{c}
\text{H} \qquad\quad \text{H} \\
| \qquad\quad | \\
\text{H}-\text{C}-\text{O}-\text{C}-\text{H} \\
| \qquad\quad | \\
\text{H} \qquad\quad \text{H}
\end{array}
$$

The greater the number of atoms in a molecule, the greater the number of possible arrangements and the greater the number of isomers. For instance, heptane, a molecule made up of seven carbon atoms and sixteen hydrogen atoms, can be arranged in nine different ways; in other words, there can be nine different heptanes, each with its own properties. These nine isomers resemble one another fairly closely, but it is only a family resemblance.

Chemists have prepared all nine of these isomers but have never found a tenth – good evidence in favour of the Kekulé system.

A compound containing forty carbon atoms and eighty-two hydrogen atoms could exist in some 62·5 billion arrangements, or isomers. And organic molecules of this size are by no means uncommon.

Only carbon atoms can hook to one another to form long chains. Other atoms do well if they can form a chain as long as half a dozen or so. That is why inorganic molecules are usually simple, and why they rarely have isomers. The greater complexity of the organic molecule introduces so many possibilities of isomerism that nearly 2 million organic compounds are known, new ones are being formed daily, and a virtually limitless number await discovery.

Structural formulas are now universally used as indispensable guides to the nature of organic molecules. As a short cut, chemists often write the formula of a molecule in terms of the groups of atoms ('radicals') that make it up, such as the methyl (CH_3) and methylene (CH_2) radicals. Thus the formula for butane can be written as $CH_3CH_2CH_2CH_3$.

The Details of Structure

In the latter half of the nineteenth century, chemists discovered a particularly subtle kind of isomerism which was to prove very important in the chemistry of life. The discovery emerged from the oddly asymmetrical effect that certain organic compounds had on rays of light passing through them.

A cross-section of a ray of ordinary light would show that the waves of which it consists undulate in all planes – up and down, from side to side, and obliquely. Such light is called 'unpolarized'. But when light passes through a crystal of the transparent substance called Iceland spar, for instance, it is refracted in such a way that the light emerges 'polarized'. It is as if the array of atoms in the crystal allows only certain planes of undulation to pass through (just as the palings of a fence might allow a person moving sideways to squeeze through but not one coming up to them broadside on). There are devices, such as the 'Nicol prism', invented by the Scottish physicist William Nicol in 1829, that let light through in only one plane. This has now been replaced, for most purposes, by materials such as Polaroid (crystals of a complex of quinine sulphate and iodine, lined up with axes parallel and embedded in nitrocellulose), first produced in the 1930s by Edwin Land.

Reflected light is often partly plane-polarized, as was first discovered in 1808 by the French physicist, Étienne Louis Malus. (He invented the term 'polarization' through the application of a remark of Newton's about poles of light particles, one occasion

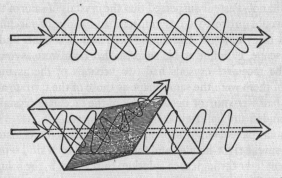

The polarization of light. The waves of light normally oscillate in all planes (*top*). The Nicol prism (*bottom*) lets through the oscillations in only one plane, reflecting away the others. The transmitted light is plane-polarized.

where Newton was wrong – but the name remains anyway.) The glare of reflected light from windows of buildings and cars, and even from paved highways, can therefore be cut to bearable levels by the use of Polaroid sunglasses.

Early in the nineteenth century the French physicist Jean Baptiste Biot had discovered that when plane-polarized light passed through quartz crystals, the plane of polarization was twisted. That is, the light went in undulating in one plane and came out undulating in a different plane. A substance that does this is said to display 'optical activity'. Some quartz crystals twisted the plane clockwise ('dextrorotation') and some counter-clockwise ('levorotation'). Biot found that certain organic compounds, such as camphor and tartaric acid, did the same thing. He thought it likely that some kind of asymmetry in the arrangement of the atoms in the molecules was responsible for the twisting of light. But for several decades this suggestion remained purely speculative.

In 1844, Louis Pasteur (only twenty-two at the time) took up this interesting question. He studied two substances: tartaric acid and racemic acid. Both had the same chemical composition, but tartaric acid rotated the plane of polarized light while racemic acid did not. Pasteur suspected that the crystals of salts of tartaric acid would prove to be asymmetric and those of racemic acid would be symmetric. Examining both sets of crystals under the microscope, he found to his surprise that both were asymmetric. But the racemate crystals had two versions of the asymmetry. Half of them were the same shape as those of the tartrate and the other half were mirror images. Half of the racemate crystals were left-handed and half right-handed, so to speak.

Pasteur painstakingly separated the left-handed racemate crystals from the right-handed and then dissolved each kind separately and sent light through each solution. Sure enough, the solution of the crystals possessing the same asymmetry as the tartrate crystals twisted the plane of polarized light just as the tartrate did, with the same specific rotation. Those crystals *were* tartrate. The other set twisted the plane of polarized light in the opposite direction, with the same amount of rotation. The reason the original racemate had shown no rotation of light, then, was that the two opposing tendencies cancelled each other.

Pasteur next reconverted the two separate types of racemate salt to acid again by adding hydrogen ions to the respective solutions. (A salt, by the way, is a compound in which some hydrogen ions of the acid molecule are replaced by other positively charged ions, such as those of sodium or potassium.) He found that each of these racemic acids was now optically active – one rotating polarized light in the same direction as tartaric acid did (for it *was* tartaric acid) and the other in the opposite direction.

Other pairs of such mirror-image compounds ('enantiomorphs', from Greek words meaning 'opposite shapes') were found. In 1863, the German chemist Johannes Wislicenus found that lactic acid (the acid of sour milk) formed such a pair. Furthermore, he showed the properties of the two forms to be identical *except* for the action on polarized light. This has turned out to be generally true of enantiomorphs.

So far, so good, but where did the asymmetry lie? What was there about the two molecules that made them mirror images of each other? Pasteur could not say. And although Biot, who had suggested the existence of molecular asymmetry, lived to be eighty-eight, he did not live long enough to see his intuition vindicated.

It was in 1874, twelve years after Biot's death, that the answer was finally presented. Two young chemists, a twenty-two-year-old Dutchman named Jacobus Hendricus Van 't Hoff and a twenty-seven-year-old Frenchman named Joseph Achille Le Bel, independently advanced a new theory of the carbon valence bonds that explained how mirror-image molecules could be constructed. (Later in his career, Van 't Hoff studied the behaviour of substances in solution and showed how the laws governing their behaviour resembled the laws governing the behaviour of gases. For this achievement he was the first man, in 1901, to be awarded the Nobel Prize in chemistry.)

Kekulé had drawn the four bonds of the carbon atom all in the same plane, not necessarily because this was the way they were actually arranged but because it was the convenient way of drawing them on a flat piece of paper. Van 't Hoff and Le Bel now suggested a three-dimensional model in which the bonds were directed in two mutually perpendicular planes, two in one plane and two in the other. A good way to picture this is to imagine the carbon atom as standing on any three of its bonds as legs, in which case the fourth bond points vertically upwards (see the drawing on p. 13). If you suppose the carbon atom to be at the centre of a tetrahedron (a four-sided geometrical figure with triangular sides), then the four bonds point to the four vertices of the figure. The model is therefore called the 'tetrahedral carbon atom'.

Now let us attach to these four bonds two hydrogen atoms, a chlorine atom, and a bromine atom. Regardless of which atom we attach to which bond, we will always come out with the same arrangement. Try it and see. With four toothpicks stuck into a marshmallow (the carbon atom) at the proper angles, you could represent the four bonds. Now suppose you stick two black olives (the hydrogen atoms), a green olive (chlorine), and a cherry (bromine) on the ends of the toothpicks in any order. Let us say that when you stand this on three legs with a black olive on the fourth pointing upwards, the order on the three standing legs in the clockwise direction is black olive, green olive, cherry. You might now switch the green olive and cherry so that the order runs black olive, cherry, green olive. But all you need to do to see the same order as before is to turn the structure over so that the black olive serving as one of the supporting legs sticks up in the air and the one that was in the air rests on the table. Now the order of the standing legs again is black olive, green olive, cherry.

In other words, when at least two of the four atoms (or groups of atoms) attached to carbon's four bonds are identical, only one structural arrangement is possible. (Obviously this is true when three or all four of the attachments are identical.)

But when all four of the attached atoms (or groups of atoms) are different, the situation changes. Now two different structural arrangements are possible – one the mirror image of the other. For instance, suppose you stick a cherry on the upward leg and a black olive, a green olive, and a cocktail onion on the three standing legs. If you then switch the black olive and green olive so that the clockwise order runs green olive, black olive, onion, there is no way you can turn the structure to make the order come out black olive, green olive, onion, as it was before you made the switch. Thus with four different attachments you can always form two different structures, mirror images of each other. Try it and see.

Van 't Hoff and Le Bel thus solved the mystery of the asymmetry of optically active substances. The mirror-image substances that rotated light in opposite directions were substances containing carbon atoms with four different atoms or groups of atoms attached to the bonds. One of the two possible arrangements of these four attachments rotated polarized light to the right; the other rotated it to the left.

More and more evidence beautifully supported Van 't Hoff's and Le Bel's tetrahedral model of the carbon atom, and, by 1885, their theory (thanks, in part, to the enthusiastic support of the respected Wislicenus) was universally accepted.

The notion of three-dimensional structure also was applied to atoms other than carbon. The German chemist Viktor Meyer applied it successfully to nitrogen, while the English chemist

The tetrahedral carbon atom.

William Jackson Pope applied it to sulphur, selenium, and tin. The German-Swiss chemist Alfred Werner added other elements and, indeed, beginning in the 1890s, worked out a 'coordination theory' in which the structure of complex inorganic substances was explained by careful consideration of the distribution of atoms and atom groupings about some central atom. For this work, Werner was awarded the Nobel Prize in chemistry for 1913.

The two racemic acids that Pasteur had isolated were named *d*-tartaric acid (for 'dextrorotatory') and *l*-tartaric acid (for 'levorotatory'), and mirror-image structural formulas were written for them. But which was which? Which was actually the right-handed and which the left-handed compound? There was no way of telling at the time.

To provide chemists with a reference, or standard of comparison, for distinguishing right-handed and left-handed substances, the German chemist Emil Fischer chose a simple compound called 'glyceraldehyde', a relative of the sugars, which were among the most thoroughly studied of the optically active compounds. He arbitrarily assigned left-handedness to one form which he named L-glyceraldehyde, and right-handedness to its mirror image, named D–glyceraldehyde. His structural formulas for them were:

$$
\begin{array}{ccc}
\text{CHO} & & \text{CHO} \\
| & & | \\
\text{H} - \text{C} - \text{OH} & & \text{HO} - \text{C} - \text{H} \\
| & & | \\
\text{CH}_2\text{OH} & & \text{CH}_2\text{OH}
\end{array}
$$

D-Glyceraldehyde   L-Glyceraldehyde

Any compound that could be shown by appropriate chemical methods (rather careful ones) to have a structure related to L-glyceraldehyde would be considered in the 'L-series' and would have the prefix 'L' attached to its name, regardless of whether it was levorotatory or dextrorotatory as far as polarized light was concerned. As it turned out, the levorotatory form of tartaric acid was found to belong to the D–series instead of the L–series. (Nowadays, a compound that falls in the D–series structurally but rotates light to the left has its name prefixed by 'D(—)'. Similarly, we have 'D(+)', 'L(—)', and 'L(+)'.)

This preoccupation with the minutiae of optical activity has turned out to be more than a matter of idle curiosity. As it happens, almost all the compounds occurring in living organisms contain asymmetric carbon atoms. And in every such case the organism makes use of only one of the two mirror-image forms of the compound. Furthermore, similar compounds generally fall in the same series. For instance, virtually all the simple sugars found in living tissue belong to the D–series, while virtually all the amino acids (the building blocks of proteins) belong to the L–series.

In 1955, a chemist named J. M. Bijvoet finally determined what structure tended to rotate polarized light to the left, and vice versa. It turned out that Fischer had, by chance, guessed right in naming the levorotatory and dextrorotatory forms.

For some years after the secure establishment of the Kekulé system of structural formulas, one compound with a rather simple molecule resisted formulation. That compound was benzene (discovered in 1825 by Faraday). Chemical evidence showed that it consisted of six carbon atoms and six hydrogen atoms. What happened to all the extra carbon bonds? (Six carbon atoms linked to one another by single bonds could hold 14 hydrogen atoms, and they do in the well-known compound called hexane, C_6H_{14}.) Evidently the carbon atoms in benzene were linked together by double or triple bonds. Thus benzene might have a structure such as $CH=C—CH=CH—CH=CH_2$. But the trouble was that the known compounds with that sort of structure had properties quite different from those of benzene. Besides, all the chemical evidence seemed to indicate that the benzene molecule was very symmetrical, and six carbons and six hydrogens could not be arranged in a chain in any reasonably symmetrical fashion.

In 1865, Kekulé himself came up with the answer. He related some years later that the vision of the benzene molecule came to him while he was riding on a bus and sunk in a reverie, half-asleep. In his dream, chains of carbon atoms seemed to come alive and dance before his eyes, and then suddenly one coiled on itself like a snake. Kekulé awoke from his reverie with a start and could have cried 'Eureka!' He had the solution: the benzene molecule was a ring.

Kekulé suggested that the six carbon atoms of the molecule were arranged as follows:

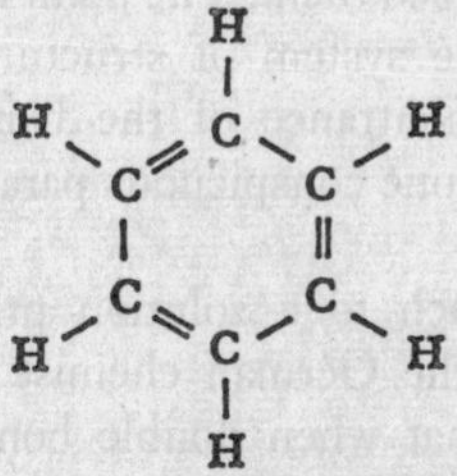

Here at last was the required symmetry. It explained, among other things, why the substitution of another atom for one of benzene's hydrogen atoms always yielded just one unvarying product. Since all the carbons in the ring were indistinguishable from one another in structural terms, no matter where you made the substitution for a hydrogen atom on the ring you would get the same product. Second, the ring structure showed that there were just three ways in which you could replace two hydrogen atoms on the ring: you could make the substitutions on two adjacent carbon atoms in the ring, on two separated by a single skip, or on two separated by a double skip. Sure enough, it was found that just three double substituted benzene isomers could be made.

Kekulé's blueprint of the benzene molecule, however, presented an awkward question. Generally, compounds with double bonds are more reactive, which is to say more unstable, than those with only single bonds. It is as if the extra bond were ready and more than willing to desert the attachment to the carbon atom and form a new attachment. Double-bonded compounds readily add on hydrogen or other atoms and can even be broken down without much difficulty. But the benzene ring is extraordinarily stable – more stable than carbon chains with only single bonds. (In fact, it is so stable and common in organic matter that molecules containing benzene rings make up an entire class of organic compounds, called 'aromatic', all the rest being lumped together as the 'aliphatic' compounds.) The benzene molecule resists taking on more hydrogen atoms and is hard to break down.

The nineteenth-century organic chemists could find no explanation for this queer stability of the double bonds in the benzene molecule, and it disturbed them. The point may seem a small one, but the whole Kekulé system of structural formulas was endangered by the recalcitrance of the benzene molecule. The failure to explain this one conspicuous paradox made all the rest uncertain.

The closest approach to a solution prior to the twentieth century was that of the German chemist Johannes Thiele. In 1899, he suggested that when double bonds and single bonds alternated, the nearer ends of a pair of double bonds somehow neutralized each other and cancelled each other's reactive nature. Consider, as an example, the compound 'butadiene', which contains, in simplest form, the case of two double bonds separated by a single bond ('conjugated double bonds'). Now if two atoms are added to the compound, they add on to the end carbons, as shown in the formula below. Such a view explained the non-reactivity of benzene, since the three double bonds of the benzene rings, being arranged in a ring, neutralize each other completely.

$$CH_2 = CH - CH = CH_2$$
......atoms add......
here

Some forty years later, a better answer was found by way of the new theory of chemical bonds that pictured atoms as linked together by sharing electrons.

The chemical bond, which Kekulé had drawn as a dash between atoms, came to be looked upon as representing a shared pair of electrons (see Vol. 1, pp. 251–4). Each atom that formed a combination with a partner shared one of its electrons with the partner, and the partner reciprocated by donating one of *its* electrons to the bond. Carbon, with four electrons in its outer shell, could form four attachments; hydrogen could donate its one electron to a bond with one other atom, and so on.

Now the question arose: How were the electrons shared? Obviously, two carbon atoms share the pair of electrons between them equally, because each atom has an equal hold on electrons. On the other hand, in a combination such as H_2O, the oxygen atom, which has a stronger hold on electrons than a hydrogen atom, takes possession of the greater share of the pair of electrons it has in common with each hydrogen atom. This means that the oxygen atom, by virtue of its excessive portion of electrons, has a slight excess of negative charge. By the same token, the hydrogen atom, suffering from an electron deficiency, has a slight excess of positive charge. A molecule containing an oxygen–hydrogen pair, such as water or ethyl alcohol, possesses a small concentration of negative charge in one part of the molecule and a small concentration of positive charge in another. It possesses two poles of charge, so to speak, and is called a 'polar molecule'.

This view of molecular structure was first proposed in 1912 by the Dutch chemist Peter Joseph Wilhelm Debye, who later pursued his research in the United States. He used an electric field to measure the amount of separation of poles of electric charge in a molecule. In such a field, polar molecules line themselves up with the negative ends pointing towards the positive pole and the positive ends towards the negative pole, and the ease with which this is done is the measure of the 'dipole moment' of the molecule. By the early 1930s, measurements of dipole moments had become routine, and in 1936, for this and other work, Debye was awarded the Nobel Prize in chemistry.

The new picture explained a number of things that earlier views of molecular structure could not. For instance, it explained some anomalies of the boiling points of substances. In general, the greater the molecular weight, the higher the boiling point. But this rule is commonly broken. Water, with a molecular weight of only 18, boils at 100°C, whereas propane with more than twice this molecular weight (44), boils at the much lower temperature of −42°C. Why should that be? The answer is that water is a polar molecule with a high dipole moment, while propane is 'non-polar' – it has no poles of charge. Polar molecules tend to orient themselves with the negative pole of one molecule adjacent to the positive pole of its neighbour. The resulting electrostatic attraction between neighbouring molecules makes it harder to tear the molecules apart, and so such substances have relatively high boiling points. This accounts for the fact that ethyl alcohol has a much higher boiling point (78°C) than its isomer dimethyl ether, which boils at −24°C, although both substances have the same molecular weight (46). Ethyl alcohol has a large dipole moment and dimethyl ether only a small one. Water has a dipole moment even larger than that of ethyl alcohol.

When de Broglie and Schrödinger formulated the new view of electrons not as sharply defined particles but as packets of waves (see Vol. 1, pp. 398–403), the idea of the chemical bond underwent a further change. In 1939, the American chemist Linus Pauling presented a quantum-mechanical concept of molecular bonds in a book entitled *The Nature of the Chemical Bond*. His theory finally explained, among other things, the paradox of the stability of the benzene molecule.

Pauling pictured the electrons that form a bond as 'resonating' between the atoms they join. He showed that under certain conditions it was necessary to view an electron as occupying any one of a number of positions (with varying probability). The electron, with its wave-like properties, might then best be presented as being spread out into a kind of blur, representing the weighted average of the individual probabilities of position. The more evenly the electron was spread out, the more stable was the compound. Such 'resonance stabilization' was most likely to occur when the molecule possessed conjugated bonds in one plane and when the existence of symmetry allowed a number of alternative positions for the electron (viewed as a particle). The benzene ring is planar and symmetrical, and Pauling showed that the bonds of the ring were not really double and single in alternation, but that the electrons were smeared out, so to speak, into an equal distribution which resulted in all the bonds being alike and all being stronger and less reactive than ordinary single bonds.

The resonance structures, though they explain chemical behaviour satisfactorily, are difficult to present in simple symbolism on paper. Therefore the old Kekulé structures, although now understood to represent only approximations of the actual electronic situation, are still universally used and will undoubtedly continue to be used through the foreseeable future.

Organic Synthesis

After Kolbe had produced acetic acid, there came in the 1850s a chemist who went systematically and methodically about the business of synthesizing organic substances in the laboratory. He was the Frenchman Pierre Eugène Marcelin Berthelot. He prepared a number of simple organic compounds from still simpler inorganic compounds such as carbon monoxide. Berthelot built his simple organic compounds up through increasing complexity until he finally had ethyl alcohol, among other things. It was 'synthetic ethyl alcohol', to be sure, but absolutely indistinguishable from the 'real thing', because it *was* the real thing.

Ethyl alcohol is an organic compound familiar to all and highly valued by most. No doubt the thought that the chemist could make ethyl alcohol from coal, air, and water (coal to supply the carbon, air the oxygen, and water the hydrogen), without the necessity of fruits or grain as a starting point, must have created enticing visions and endowed the chemist with a new kind of reputation as a miracle worker. At any rate, it put organic synthesis on the map.

For chemists, however, Berthelot did something even more significant. He began to form products that did not exist in nature. He took 'glycerol', a compound discovered by Scheele in 1778 and obtained from the breakdown of the fats of living organisms, and combined it with acids not known to occur naturally in fats (although they occurred naturally elsewhere). In this way he obtained fatty substances which were not quite like those that occurred in organisms.

Thus Berthelot laid the groundwork for a new kind of organic chemistry – the synthesis of molecules that nature could not supply. This meant the possible formation of a kind of 'synthetic' which might be a substitute – perhaps an inferior substitute – for some natural compound that was hard or impossible to get in the needed quantity. But it also meant the possibility of 'synthetics' which were improvements on anything in nature.

This notion of improving on nature in one fashion or another, rather than merely supplementing it, has grown to colossal proportions since Berthelot showed the way. The first fruits of the new outlook were in the field of dyes.

The beginnings of organic chemistry were in Germany. Wöhler and Liebig were both German, and other men of great ability followed them. Before the middle of the nineteenth century, there were no organic chemists in England even remotely comparable to those in Germany. In fact, English schools had so low an opinion of chemistry that they taught the subject only during the lunch recess, not expecting (or even perhaps desiring) many students to be interested. It is odd, therefore, that the first feat of synthesis with world-wide repercussions was actually carried through in England.

It came about in this way. In 1845, when the Royal College of Science in London finally decided to give a good course in chemistry, it imported a young German to do the teaching. He was August Wilhelm von Hofmann, only twenty-seven at the time, and he was hired at the suggestion of Queen Victoria's husband, the Prince Consort Albert (who was himself of German birth).

Hofmann was interested in a number of things, among them coal tar, which he had worked with on the occasion of his first research project under Liebig. Coal tar is a black, gummy material given off by coal when it is heated strongly in the absence of air. The tar is not an attractive material, but it is a valuable source of organic chemicals. In the 1840s, for instance, it served as a source of large quantities of reasonably pure benzene and of a nitrogen-containing compound called 'aniline', related to benzene, which Hofmann had been the first to obtain from coal tar.

About ten years after he arrived in England, Hofmann came across a seventeen-year-old boy studying chemistry at the college. His name was William Henry Perkin. Hofmann had a keen eye for talent and knew enthusiasm when he saw it. He took on the youngster as an assistant and set him to work on coal-tar compounds. Perkin's enthusiasm was tireless. He set up a laboratory in his home and worked there as well as at school.

Hofmann, who was also interested in medical applications of chemistry, mused aloud one day in 1856 on the possibility of synthesizing quinine, a natural substance used in the treatment of malaria. Now those were the days before structural formulas had come into their own. The only thing known about quinine was its composition, and no one at the time had any idea of just how complicated its structure was. (It was not till 1908 that the structure was correctly deduced.)

Blissfully ignorant of its complexity, Perkin, at the age of eighteen, tackled the problem of synthesizing quinine. He began with allyltoluidine, one of his coal-tar compounds. This molecule seemed to have about half the numbers of the various types of atoms that quinine had in its molecule. If he put two of these molecules together and added some missing oxygen atoms (say by mixing in some potassium dichromate, known to add oxygen atoms to chemicals with which it was mixed), Perkin thought he might get a molecule of quinine.

Naturally this approach got Perkin nowhere. He ended with a dirty, red-brown goo. Then he tried aniline in place of allyltoluidine and got a blackish goo. This time, though, it seemed to him that he caught a purplish glint in it. He added alcohol to the mess, and the colourless liquid turned a beautiful purple. At once Perkin thought of the possibility that he had discovered something that might be useful as a dye.

Dyes had always been greatly admired, and expensive, substances. There were only a handful of good dyes – dyes that stained fabric permanently and brilliantly and did not fade or wash out. There was dark blue indigo, from the indigo plant and the closely related 'woad' for which Britain was famous in early Roman times; there was 'Tyrian purple', from a snail (so-called because ancient Tyre grew rich on its manufacture – in the later Roman Empire the royal children were born in a room with hangings dyed with Tyrian purple, whence the phrase 'born to the purple'); and there was reddish alizarin, from the madder plant ('alizarin' came from Arabic words meaning 'the juice'). To these inheritances from ancient and medieval times later dyers had added a few tropical dyes and inorganic pigments (today used chiefly in paints).

This explains Perkin's excitement about the possibility that his purple substance might be a dye. At the suggestion of a friend, he sent a sample to a firm in Scotland which was interested in dyes, and quickly the answer came back that the purple compound had good properties. Could it be supplied cheaply? Perkin proceeded to patent the dye (there was considerable argument as to whether an eighteen-year-old could obtain a patent, but eventually he obtained it), to leave school, and to go into business.

His project was not easy. Perkin had to start from scratch, preparing his own starting materials from coal tar with equipment of his own design. Within six months, however, he was producing what he named 'aniline purple' – a compound not found in nature and superior to any natural dye in its colour range.

French dyers, who took to the new dye more quickly than did the more conservative English, named the colour 'mauve', from the mallow (Latin name 'malva'), and the dye itself came to be known as 'mauveine'. Quickly it became the rage (the period being sometimes referred to as the 'Mauve Decade'), and Perkin grew rich. At the age of twenty-three he was the world authority on dyes.

The dam had broken. A number of organic chemists, inspired by Perkin's astonishing success, went to work synthesizing dyes, and many succeeded. Hofmann himself turned to this new field, and, in 1858, he synthesized a red-purple dye which was later given the name 'magenta' by the French dyers (then, as now, arbiters of the world's fashions). The dye was named after the Italian city where the French defeated the Austrians in a battle in 1859.

Hofmann returned to Germany in 1865, carrying his new interest in dyes with him. He discovered a group of violet dyes still known as 'Hofmann's violets'. By the mid twentieth century, no less than 3,500 synthetic dyes were in commercial use.

Chemists also synthesized the natural dyestuffs in the laboratory. Karl Graebe of Germany and Perkin both synthesized alizarin in 1869 (Graebe applying for the patent one day sooner than Perkin), and in 1880 the German chemist Adolf von Baeyer worked out a method of synthesizing indigo. (For his work on dyes von Baeyer received the Nobel Prize in chemistry in 1905.)

Perkin retired from business in 1874, at the age of thirty-five, and returned to his first love, research. By 1875, he had managed to synthesize coumarin (a naturally occurring substance which has the pleasant odour of new-mown hay); this served as the beginning of the synthetic perfume industry.

Perkin alone could not maintain British supremacy against the great development of German organic chemistry, and, by the turn of the century, 'synthetics' became almost a German monopoly. It was a German chemist, Otto Wallach, who carried on the work on synthetic perfumes that Perkin had started. In 1910, Wallach was awarded the Nobel Prize in chemistry for his investigations. The Croatian chemist Leopold Ruzicka, teaching in Switzerland, first synthesized musk, an important component of perfumes. He shared the Nobel Prize in chemistry in 1938. However, during the First World War, Great Britain and the United States, shut off from the products of the German chemical laboratories, were forced to develop chemical industries of their own.

Achievements in synthetic organic chemistry could not have proceeded at anything better than a stumbling pace if chemists had had to depend upon fortunate accidents such as the one that had been seized upon by Perkin. Fortunately the structural formulas of Kekulé, presented three years after Perkin's discovery, made it possible to prepare blueprints, so to speak, of the organic molecule. No longer did chemists have to prepare quinine by sheer guesswork and hope; they had methods for attempting to scale the structural heights of the molecule step by step, with advance knowledge of where they were headed and what they might expect.

Chemists learned how to alter one group of atoms to another; to open up rings of atoms and to form rings from open chains; to split groups of atoms in two; and to add carbon atoms one by one to a chain. The specific method of doing a particular architectural task within the organic molecule is still often referred to by the name of the chemist who first described the details. For instance, Perkin discovered a method of adding a two-carbon atom group by heating certain substances with chemicals named acetic anhydride and sodium acetate. This is still called the 'Perkin reaction'. Perkin's teacher, Hofmann, discovered that a ring of atoms which included a nitrogen could be treated with a substance called methyl iodide in the presence of silver compound in such a way that the ring was eventually broken and the nitrogen atom removed. This is the 'Hofmann degradation'. In 1877, the French chemist Charles Friedel, working with the American chemist

James Mason Crafts, discovered a way of attaching a short carbon chain to a benzene ring by the use of heat and aluminium chloride. This is now known as the 'Friedel–Crafts reaction'.

In 1900, the French chemist Victor Grignard discovered that magnesium metal, properly used, could bring about a rather large variety of different jointings of carbon chains; he presented the discovery in his doctoral dissertation. For the development of these 'Grignard reactions' he shared in the Nobel Prize in chemistry in 1912. The French chemist Paul Sabatier, who shared it with him, had discovered (with J. B. Senderens) a method of using finely divided nickel to bring about the addition of hydrogen atoms in those places where a carbon chain possessed a double bond. This is the 'Sabatier–Senderens reduction'.

In 1928, the German chemists Otto Diels and Kurt Alder discovered a method of adding the two ends of a carbon chain to opposite ends of a double bond in another carbon chain, thus forming a ring of atoms. For the discovery of this 'Diels–Alder reaction', they shared the Nobel Prize for chemistry in 1950.

In other words, by noting the changes in the structural formulas of substances subjected to a variety of chemicals and conditions, organic chemists worked out a slowly growing set of ground rules on how to change one compound into another at will. It was not easy. Every compound and every change had its own peculiarities and difficulties. But the main paths were blazed, and the skilled organic chemist found them clear signs towards progress in what had formerly seemed a jungle.

Knowledge of the manner in which particular groups of atoms behaved could also be used to work out the structure of unknown compounds. For instance, when simple alcohols react with metallic sodium and liberate hydrogen, only the hydrogen linked to an oxygen atom is released, not the hydrogens linked to carbon atoms. On the other hand, some organic compounds will take on hydrogen atoms under appropriate conditions while others will not. It turns out that compounds that add hydrogen generally possess double or triple bonds and add the hydrogen at these bonds. From such information a whole new type of chemical analysis of organic compounds arose; the nature of the atom groupings was determined, rather than just the numbers and kinds of various atoms present. The liberation of hydrogen by the addition of sodium signified the presence of an oxygen-bound hydrogen atom in the compound; the acceptance of hydrogen meant the presence of double or triple bonds. If the molecule was too complicated for analysis as a whole, it could be broken down into simpler portions by well-defined methods; the structures of the simpler portions could be worked out and the original molecule deduced from those.

Using the structural formula as a tool and guide, chemists could work out the structure of some useful naturally occurring organic compound (analysis) and then set about duplicating it or something like it in the laboratory (synthesis). One result was that something which was rare, expensive or difficult to obtain in nature might become cheaply available in quantity in the laboratory. Or, as in the case of the coal-tar dyes, the laboratory might create something that fulfilled a need better than did similar substances found in nature.

One startling case of a deliberate improvement on nature involves cocaine, found in the leaves of the coca plant, which is native to Bolivia and Peru, but is now grown chiefly in Java. Like the compounds strychnine, morphine, and quinine, all mentioned earlier, cocaine is an example of an 'alkaloid', a nitrogen-containing plant product that, in small concentration, has profound physiological effects on man. Depending on the dose, alkaloids can cure or kill. The most famous alkaloid death of all times was that of Socrates, who was killed by 'coniine', an alkaloid in hemlock.

The molecular structure of the alkaloids is, in some cases, extraordinarily complicated, but that just sharpened chemical curiosity. The English chemist Robert Robinson tackled the alkaloids systematically. He worked out the structure of morphine (for all but one dubious atom) in 1925, and the structure of strychnine in 1946. He received the Nobel Prize for chemistry in 1947 as recognition of the value of his work.

Robinson had merely worked out the structure of alkaloids without using that structure as a guide to their synthesis. The

American chemist Robert Burns Woodward took care of that. With his American colleague William von Eggers Doering, he synthesized quinine in 1944. It was the wild goose chase after this particular compound by Perkin that had had such tremendous results. And, if you are curious, here is the structural formula of quinine:

```
                        OH
                        |
                        CH ──────── CH ─ N ─ CH₂
CH₃                     |           |
    O      CH      C         CH      CH₂
     C        C       C                |
        ‖                  CH          CH₂
     CH      C       CH                |
        CH      N       CH    CH₂ ─ CH ─ CH ─ CH = CH₂
```

No wonder it stumped Perkin.

That Woodward and von Doering solved the problem is not merely a tribute to their brilliance. They had at their disposal the new electronic theories of molecular structure and behaviour worked out by men such as Pauling. Woodward went on to synthesize a variety of complicated molecules which had, before his time, represented hopeless challenges. In 1954, for instance, he synthesized strychnine.

Long before the structure of the alkaloids had been worked out, however, some of them, notably cocaine, were of intense interest to medical men. The South American Indians, it had been discovered, would chew coca leaves, finding it an antidote to fatigue and a source of happiness-sensation. The Scottish physician Robert Christison introduced the plant to Europe. (This is not the only gift to medicine on the part of the witch doctors and herb-women of pre-scientific societies. There are also quinine and strychnine, already mentioned, as well as opium, digitalis, curare, atropine, strophanthidin, and reserpine. In addition, the smoking of tobacco, the chewing of betel nuts, the drinking of alcohol, and the taking of such drugs as marijuana and peyote are all inherited from primitive societies.)

Cocaine was not merely a general happiness-producer. Doctors discovered that it deadened the body, temporarily and locally, to sensations of pain. In 1884, the American physician Carl Koller discovered that cocaine could be used as a pain-deadener when added to the mucous membranes around the eye. Eye operations could then be performed without pain. Cocaine could also be used in dentistry, allowing teeth to be extracted without pain.

This fascinated doctors, for one of the great medical victories of the nineteenth century had been that over pain. In 1799, Humphry Davy had prepared the gas 'nitrous oxide' (N_2O) and studied its effects. He found that when it was inhaled it released inhibitions so that men breathing it would laugh, cry, or otherwise act foolishly. Its common name is 'laughing gas', for that reason.

In the early 1840s, an American scientist, Gardner Quincy Cotton, discovered that nitrous oxide deadened the sensation of pain, and, in 1844, an American dentist, Horace Wells, used it in dentistry. By that time, something better had entered the field.

The American surgeon Crawford Williamson Long in 1842 had used ether to put a patient to sleep during tooth extractions. In 1846, the American dentist William Thomas Green Morton conducted a surgical operation under ether at the Massachusetts General Hospital. Morton usually gets the credit for the discovery, because Long did not describe his feat in the medical journals until after Morton's public demonstration, and Wells's earliest public demonstrations with nitrous oxide had been only indifferent successes.

The American poet and physician Oliver Wendell Holmes suggested that pain-deadening compounds be called 'anaesthetics' (from Greek words meaning 'no feeling'). Some people at the time felt that anaesthetics were a sacrilegious attempt to avoid pain inflicted on mankind by God, but if anything was needed to make anaesthesia respectable, it was its use by the Scottish physician James Young Simpson for Queen Victoria during childbirth.

Anaesthesia had finally converted surgery from torture-chamber butchery to something that was at least humane and, with the addition of antiseptic conditions, even life-saving. For that reason, any further advance in anaesthesia was seized on with great interest. Cocaine's special interest was that it was a 'local anaesthetic', deadening pain in a specific area without inducing general unconsciousness and lack of sensation, as in the case of such 'general anaesthetics' as ether.

There are several drawbacks to cocaine, however. In the first place, it can induce troublesome side-effects and can even kill patients sensitive to it. Second, it can bring about addiction and has had to be used skimpily and with caution. (Cocaine is one of the dangerous 'dopes' or 'narcotics' that deaden not only pain but other unpleasant sensations and give the user the illusion of euphoria. The user may become so accustomed to this that he may require increasing doses and, despite the actual bad effect upon his body, become so dependent on the illusions it carries with it that he cannot stop using it without developing painful 'withdrawal symptoms'. Such 'drug addiction' for cocaine and other drugs of this sort is an important social problem. Up to twenty tons of cocaine are produced illegally each year and sold with tremendous profits to a few and tremendous misery to many, despite world-wide efforts to stop the traffic.) Third, the molecule is fragile, and heating cocaine to sterilize it of any bacteria leads to changes in the molecule that interfere with its anaesthetic effect.

The structure of the cocaine molecule is rather complicated:

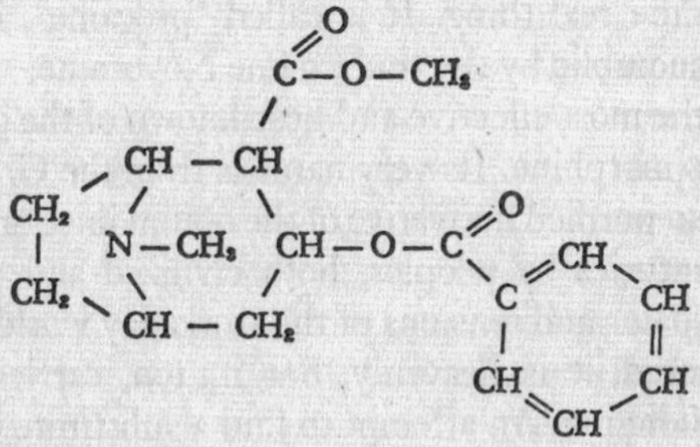

The double ring on the left is the fragile portion, and that is the difficult one to synthesize. (The synthesis of cocaine was not achieved until 1923, when the German chemist Richard Willstätter managed it.) However, it occurred to chemists that they might synthesize similar compounds in which the double ring was not closed. This would make the compound both easier to form and more stable. The synthetic substance might possess the anaesthetic properties of cocaine, perhaps without the undesirable side-effects.

For some twenty years, German chemists tackled the problem, turning out dozens of compounds, some of which were pretty good. The most successful modification was obtained in 1909, when a compound with the following formula was prepared:

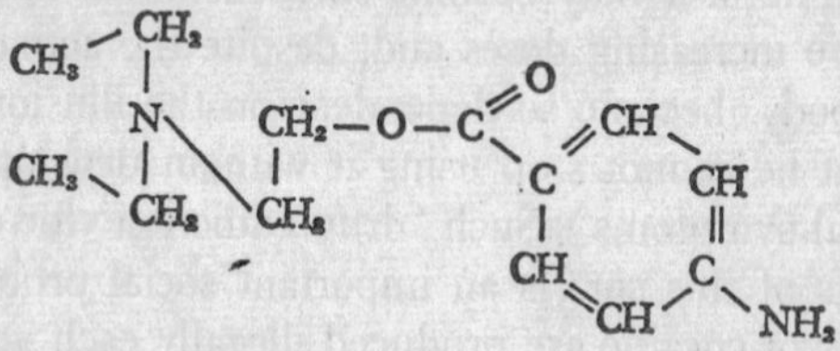

Compare this with the formula for cocaine and you will see the similarity, and also the important fact that the double ring no longer exists. This simpler molecule – stable, easy to synthesize, with good anaesthetic properties and very little in the way of side-effects – does not exist in nature. It is a 'synthetic substitute' far better than the real thing. It is called 'procaine', but is better known to the public by the trade name Novocaine.

Perhaps the most effective and best-known of the general pain-deadeners is morphine. Its very name is from the Greek word for 'sleep'. It is a purified derivative of the opium juice or 'laudanum' used for centuries by peoples, both civilized and primitive, to combat the pains and tensions of the workaday world. As a gift to the pain-racked, it is heavenly, but it, too, carries the deadly danger of addiction. An attempt to find a substitute backfired. In 1898, a synthetic derivative, 'diacetylmorphine', better known as 'heroin', was introduced in the belief that it would be safer. Instead, it turned out to be the most dangerous dope of all.

Less dangerous 'sedatives' (sleep-inducers) are chloral hydrate and, particularly, the barbiturates. The first example of this latter group was introduced in 1902, and they are now the most common constituents of 'sleeping pills'. Harmless enough when used properly, they can nevertheless induce addiction, and an over-dose can cause death. In fact, because death comes quietly as the end product of a gradually deepening sleep, barbiturate overdosage is a rather popular method of suicide, or attempted suicide.

The most common sedative, and the longest in use, is, of course, alcohol. Methods of fermenting fruit juice and grain were known in prehistoric times, as was distillation to produce stronger liquors than could be produced naturally. The value of light wines in areas where the water supply is nothing but a short cut to typhoid fever and cholera, and the social acceptance of drinking in moderation, make it difficult to treat alcohol as the drug it is, although it induces addiction as surely as morphine and through sheer quantity of use does much more harm. Legal prohibition of sale of liquor seems to be unhelpful; certainly the American experiment of 'Prohibition' (1920–33) was a disastrous failure. Nevertheless, alcoholism is more and more being treated as the disease it is rather than as a moral disgrace. The acute symptoms of alcoholism ('delirium tremens') are probably not so much due to the alcohol itself as to the vitamin deficiencies induced in those who eat little while drinking much.

Man now has at his disposal all sorts of synthetics of great potential use and misuse: explosives, poison gases, insecticides, weed-killers, antiseptics, disinfectants, detergents, drugs – almost no end of them, really. But synthesis is not merely the handmaiden of consumer needs. It can also be placed at the service of pure chemical research.

It often happens that a complex compound, produced either by living tissue or by the apparatus of the organic chemist, can only be assigned a tentative molecular structure, after all possible deductions have been drawn from the nature of the reactions it undergoes. In that case, a way out is to synthesize a compound by means of reactions designed to yield a molecular structure like the one that has been deduced. If the properties of the resulting compound are identical with the compound being investigated in the first place, the assigned structure becomes something in which a chemist can place his confidence.

An impressive case in point involves haemoglobin, the main component of the red blood cells and the pigment that gives the blood its red colour. In 1831, the French chemist L. R. LeCanu split haemoglobin into two parts, of which the smaller portion, called 'haem', made up 4 per cent of the mass of haemoglobin. Haem was found to have the empirical formula $C_{34}H_{32}O_4N_4Fe$. Such compounds as haem were known to occur in other vitally important substances, both in the plant and animal kingdoms, and so the structure of the molecule was a matter of great moment to biochemists. For nearly a century after LeCanu's isolation of haem, however, all that could be done was to break it down into smaller molecules. The iron atom (Fe) was easily removed, and what was left then broke up into pieces roughly a quarter the size of the original molecule. These fragments were found to be 'pyrroles' – molecules built on rings of five atoms, of which four are carbon and one nitrogen. Pyrrole itself has the following structure:

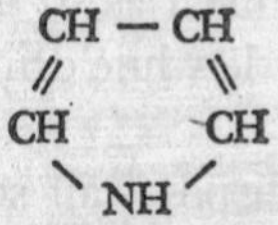

$$CH \!-\! CH$$
$$\!/\!/ \qquad \backslash\backslash$$
$$CH \qquad CH$$
$$\backslash \; NH \; /$$

The pyrroles actually obtained from haem possessed small groups of atoms containing one or two carbon atoms attached to the ring in place of one or more of the hydrogen atoms.

In the 1920s, the German chemist Hans Fischer tackled the problem further. Since the pyrroles were one quarter the size of the original haem, he decided to try to combine four pyrroles and see what he got. What he finally succeeded in getting was a four-ring compound which he called 'porphin' (from a Greek word meaning 'purple', because of its purple colour). Porphin would look like this:

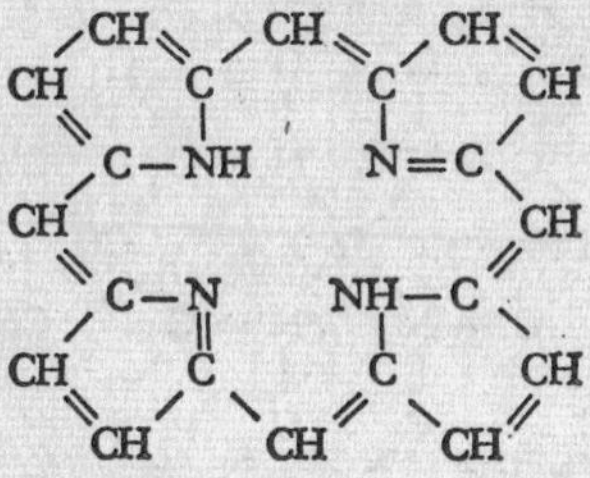

However, the pyrroles obtained from haem in the first place contained small 'side chains' attached to the ring. These remained in place when the pyrroles were joined to form porphin. The porphin with various side chains attached make up a family of compounds called the 'porphyrins'. It was obvious to Fischer upon comparing the properties of haem with those of the porphyrins he had synthesized that haem (minus its iron atom) was a porphyrin. But which one? No fewer than fifteen compounds could be formed from the various pyrroles obtained from haem, according to Fischer's reasoning, and any one of those fifteen might be haem itself.

A straightforward answer could be obtained by synthesizing all fifteen and testing the properties of each one. Fischer put his students to work preparing, by painstaking chemical reactions that allowed only a particular structure to be built up, each of the fifteen possibilities. As each different porphyrin was formed, he compared its properties with those of the natural porphyrin of haem.

In 1928, he discovered that the porphyrin numbered nine in his series was the one he was after. The natural variety of porphyrin is therefore called 'porphyrin IX' to this day. It was a simple procedure to convert porphyrin IX to haem by adding iron. Chemists at last felt confident that they knew the structure of that important compound. Here is the structure of haem, as worked out by Fischer:

```
                            CH₂
                       CH        CH₃
                    C     ‖    C
                  C     C   C     C
              CH₃   C — N     N — C    CH
                  CH      Fe        CH     ‖ CH₂
              CH₃   C — N     N = C    CH₃
                  C     C   C     C
                    C      CH     C
                      CH₃        CH₃
                      CH₃        CH₃
                      C=O        C=O
                      OH         OH
```

For his achievement Fischer was awarded the Nobel Prize in chemistry in 1930.

All the triumphs of synthetic organic chemistry through the nineteenth century and the first half of the twentieth century, great as they were, were won by means of the same processes used by the alchemists of ancient times – mixing and heating substances. Heat was the one sure way of adding energy to molecules and making them interact, but the interactions were usually random in nature and took place by way of briefly existent, unstable intermediates, whose nature could only be guessed at.

What chemists needed was a more refined, more direct method for producing energetic molecules – a method that would produce a group of molecules all moving at about the same speed in about the same direction. This would remove the random nature of interactions, for whatever one molecule would do, all would do. One way would be to accelerate ions in an electric field, much as sub-atomic particles are accelerated in cyclotrons.

In 1964, the German–American chemist Richard Leopold

Wolfgang accelerated ions and molecules to high energies and, by means of what might be called a 'chemical accelerator', produced ion speeds that heat would produce only at temperatures of from 10,000° to 100,000°C. Furthermore, the ions were all travelling in the same direction.

If the ions so accelerated are provided with a supply of electrons they can snatch up, they will be converted to neutral molecules which will still be travelling at great speeds. Such neutral beams were produced by the American chemist Leonard Wharton in 1969.

As to the brief intermediate stages of a chemical reaction, computers could help. It was necessary to work out the quantum-mechanical equations governing the state of the electrons in different atom-combinations and to work out the events that would take place on collision. In 1968, for instance, a computer guided by the Italian-American chemist Enrico Clementi 'collided' ammonia and hydrochloric acid on closed-circuit television to make ammonium chloride, with the computer working out the events that must take place. The computer indicated that the ammonium chloride which was formed could exist as a high-pressure gas at 700°C. This was not previously known, but was proved experimentally a few months later.

In the last decade, chemists have developed brand-new tools, both theoretically and experimentally. Intimate details of reactions not hitherto available will be known, and new products – unattainable before or at least attainable only in small lots – will be formed. We may be at the threshold of unexpected wonders.

Polymers and Plastics

When we consider molecules like those of haem and quinine, we are approaching a complexity with which even the modern chemist can cope only with great difficulty. The synthesis of such a compound requires so many steps and such a variety of procedures that we can hardly expect to produce it in quantity without the help of some living organism (other than the chemist). This is nothing about which to get an inferiority complex, however. Living tissue itself approaches the limit of its capacity at this level of complexity. Few molecules in nature are more complex than haem and quinine.

To be sure, there are natural substances composed of hundreds of thousands, even millions, of atoms, but these are not really individual molecules, constructed in one piece, so to speak. Rather, these large molecules are built up of units strung together like beads in a necklace. Living tissue usually synthesizes some small, fairly simple compound and then merely hooks the units together in chains. And that, as we shall see, the chemist also is capable of doing.

In living tissue this union of small molecules ('condensation') is usually accompanied by the over-all elimination of two hydrogen atoms and an oxygen atom (which combine to form a water molecule) at each point of junction. Invariably, the process can be reversed (both in the body and in the test tube): by the addition of water, the units of the chain can be loosened and separated. This reverse of condensation is called 'hydrolysis', from Greek words meaning 'loosening through water'. In the test tube the hydrolysis of these long chains can be hastened by a variety of methods, the most common being the addition of a certain amount of acid to the mixture.

The first investigation of the chemical structure of a large molecule dates back to 1812, when a Russian chemist, Gottlieb Sigismund Kirchhoff, found that boiling starch with acid produced a sugar identical in properties with glucose, the sugar obtained from grapes. In 1819 the French chemist Henri Braconnot also obtained glucose by boiling various plant products such as sawdust, linen, and bark, all of which contain a compound called 'cellulose'. It was easy to guess that both starch and cellulose were built of glucose units, but the details of the molecular structure of starch and cellulose had to await knowledge of the molecular structure of glucose. At first, before the days of structural formulas, all that was known of glucose was its empirical formula, $C_6H_{12}O_6$. This proportion suggested that there was one water molecule, H_2O, attached to each of the six carbon atoms. Hence glucose, and compounds similar to it in structure, were called 'carbohydrates' ('watered carbon').

The structural formula of glucose was worked out in 1886 by the German chemist Heinrich Kiliani. He showed that its molecule consisted of a chain of six carbon atoms, to which hydrogen atoms and oxygen–hydrogen groups were separately attached. There were no intact water combinations anywhere in the molecule.

Over the next decade or so the German chemist Emil Fischer studied glucose in detail and worked out the exact arrangement of the oxygen–hydrogen groups around the carbon atoms, four of which were asymmetric. There are sixteen possible arrangements of these groups, and therefore sixteen possible optical isomers, each with its own properties. Chemists have, indeed, made all sixteen, only a few of which actually occur in nature. It was as a result of his work on the optical activity of these sugars that Fischer suggested the establishment of the L-series and D-series of compounds. For putting carbohydrate chemistry on a firm structural foundation, Fischer received the Nobel Prize in chemistry in 1902.

Here are the structural formulas of glucose and of two other common sugars, fructose and galactose:

glucose   fructose   galactose

Once chemists knew the structure of the simple sugars, it was relatively easy to work out the manner in which they were built up into more complex compounds. For instance, a glucose molecule and a fructose can be condensed to the 'double-sugar' sucrose – the sugar we use at the table. Glucose and galactose combine to form lactose, which occurs in nature only in milk.

There is no reason why such condensations cannot continue indefinitely, and in starch and cellulose they do. Each consists of long chains of glucose units, condensed in a particular pattern.

The details of the pattern are important, because although both compounds are built up of the same unit, they are profoundly different. Starch in one form or another forms the major portion of humanity's diet, while cellulose is completely inedible. The difference in the pattern of condensation, as painstakingly worked out by chemists, is analogous to the following: Suppose a glucose molecule is viewed as either right-side-up (when it may be symbolized as 'u') or upside-down (symbolized as 'n'). The starch molecule can then be viewed as consisting of a string of glucose molecules after this fashion '. . . uuuuuuuuu . . .', while cellulose consists of '. . . unununun . . .' The body's digestive juices possess the ability to hydrolyse the 'uu' linkage of starch, breaking it up to glucose, which we can then absorb to obtain energy. Those same juices are helpless to touch the 'un' linkage of cellulose, and any cellulose we ingest travels through the alimentary canal and out.

There are certain micro-organisms that can digest cellulose, though none of the higher animals can. Some of these micro-organisms live in the intestinal tracts of ruminants and termites, for instance. It is thanks to these small helpers that cows can live on grass, and termites live on wood. The micro-organisms form glucose from cellulose in quantity, use what they need, and the host uses the overflow. The micro-organisms supply the processed food, while the host supplies the raw material and the living quarters. This form of cooperation between two forms of life for mutual benefit is called 'symbiosis', from Greek words meaning 'life together'.

Christopher Columbus discovered South American natives playing with balls of a hardened plant juice. Columbus and the other explorers who visited South America over the next two centuries were fascinated by these bouncy balls (obtained from the sap of trees in Brazil). Samples were brought back to Europe eventually as a curiosity. About 1770, Joseph Priestley (soon to discover oxygen) found that a lump of this bouncy material would rub out pencil marks, so he invented the uninspired name of 'rubber', still the English word for the substance. The British call it 'India rubber', because it came from the 'Indies' (the original name of Columbus's new world).

People eventually found other uses for rubber. In 1823, a Scotsman named Charles Macintosh patented garments made of a layer of rubber between two layers of cloth for use in rainy weather, and raincoats are still sometimes called 'mackintoshes' (with an added 'k').

The trouble with rubber as used in this way, however, was that in warm weather it became gummy and sticky, while in cold weather it was leathery and hard. A number of individuals tried to discover ways of treating rubber so as to remove these undesirable characteristics. Among them was an American named Charles Goodyear, who was innocent of chemistry but worked stubbornly along by trial and error. One day in 1839, he accidentally spilled a mixture of rubber and sulphur on a hot stove. He scraped it off as quickly as he could and found, to his amazement, that the heated rubber-sulphur mixture was dry even while it was still warm. He heated it and cooled it and found that he had a sample of rubber that did not turn gummy with heat or leathery with cold but remained soft and springy throughout.

This process of adding sulphur to rubber is now called 'vulcanization' (after Vulcan, the Roman god of fire). Goodyear's discovery founded the rubber industry. It is sad to have to report that Goodyear himself never reaped a reward despite this multi-million-dollar discovery. He spent his life fighting for patent rights and died deeply in debt.

Knowledge of the molecular structure of rubber dates back to 1879, when a French chemist, Gustave Bouchardat, heated rubber in the absence of air and obtained a liquid called 'isoprene'. Its molecule is composed of five carbon atoms and eight hydrogen atoms, arranged as follows:

$$CH_2 = \overset{\overset{\displaystyle CH_3}{\displaystyle |}}{C} - CH = CH_2$$

A second type of plant juice ('latex'), obtained from certain trees in south-east Asia, yields a substance called 'gutta percha'. This lacks the elasticity of rubber, but when it is heated in the absence of air it, too, yields isoprene.

Both rubber and gutta percha are made up of thousands of isoprene units. As in the case of starch and cellulose, the difference between them lies in the pattern of linkage. In rubber, the isoprene units are joined in the '. . . uuuuu . . .' fashion and in such a way that they form coils, which can straighten out when pulled, thus allowing stretching. In gutta percha, the units join in the '. . . unununununun . . .' fashion, and these form chains that are straighter to begin with and therefore much less stretchable.

A simple sugar molecule, such as glucose, is a 'monosaccharide' (Greek for 'one sugar'); sucrose and lactose are 'disaccharides' ('two sugars'); and starch and cellulose are 'polysaccharides' ('many sugars'). Because two isoprene molecules join to form a well-known type of compound called 'terpene' (obtained from turpentine), rubber and gutta percha are called 'polyterpenes'.

The general term for such compounds was invented by Berzelius (a great inventor of names and symbols) as far back as 1830. He called the basic unit a 'monomer' ('one part') and the large molecule a 'polymer' ('many parts'). Polymers consisting of many units (say more than a hundred) are now called 'high polymers'. Starch, cellulose, rubber, and gutta percha are all examples of high polymers.

Polymers are not clear-cut compounds but are complex mixtures of molecules of different sizes. The average molecular weight can be determined by several methods. One involves measurement of 'viscosity' (the ease or difficulty with which a liquid flows under a given pressure). The larger the molecule and the more elongated it is, the more it contributes to the 'internal friction' of a liquid and the more it makes it pour like molasses, rather than like water. The German chemist Hermann Staudinger worked out this method in 1930 as part of his general work on

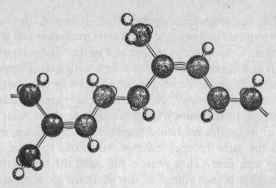

The gutta-percha molecule, a portion of which is shown here, is made up of thousands of isoprene units. The first five carbon atoms at the left (*black balls*) and the eight hydrogen atoms bonded to them make up an isoprene unit.

polymers, and in 1953 he was awarded the Nobel Prize for chemistry for his contribution towards the understanding of these giant molecules.

In 1913, two Japanese chemists discovered that natural fibres such as those of cellulose diffract X-rays, just as a crystal does. The fibres are not crystals in the ordinary sense, but they are 'microcrystalline' in character. That is, the long chains of units making up their molecules tend to run in parallel bundles for longer or shorter distances, here and there. Over the course of those parallel bundles, atoms are arranged in a repetitive order as they are in crystals, and X-rays striking those sections of the fibre are diffracted.

So polymers have come to be divided into two broad classes – crystalline and amorphous.

In a crystalline polymer, such as cellulose, the strength of the individual chains is increased by the fact that parallel neighbours are joined together by chemical bonds. The resulting fibres have considerable tensile strength. Starch is crystalline, too, but far less so than is cellulose. It therefore lacks the strength of cellulose or its capacity for fibre formation.

Rubber is an amorphous polymer. Since the individual chains

do not line up, cross-links do not occur. If heated, the various chains can vibrate independently and slide freely over and around one another. Consequently, rubber or a rubber-like polymer will grow soft and sticky and eventually melt with heat. (Stretching rubber straightens the chains and introduces a certain amount of microcrystalline character. Stretched rubber has considerable tensile strength, therefore.) Cellulose and starch, in which the individual molecules are bound together here and there, cannot undergo the same independence of vibration, so there is no softening with heat. They remain stiff until the temperature is high enough to induce vibrations that shake the molecule apart so that charring and smoke emission take place.

At temperatures below the gummy, sticky stage, amorphous polymers are often soft and springy. At still lower temperatures, however, they become hard and leathery, even glassy. Raw rubber is dry and elastic only over a rather narrow temperature range. The addition of sulphur to the extent of 5 to 8 per cent provides flexible sulphur links from chain to chain, which reduce the independence of the chains and thus prevent gumminess at moderate heat. They also increase the free play between the chains at moderately low temperatures; therefore the rubber does not harden. The addition of greater amounts of sulphur, up to 30 to 50 per cent, will bind the chains so tightly that the rubber grows hard. It is then known as 'hard rubber' or 'ebonite'.

(Even vulcanized rubber will turn glassy if the temperature is lowered sufficiently. An ordinary rubber ball, dipped in liquid air for a few moments, will shatter if thrown against a wall. This is a favourite demonstration in introductory chemistry courses.)

Various amorphous polymers show different physical properties at a given temperature. At room temperature natural rubber is elastic, various resins are glassy and solid, and chicle (from the sapodilla tree of South America) is soft and gummy (it is the chief ingredient of chewing gum).

Aside from our food, which is mainly made up of high polymers (meat, starch, and so on), probably the one polymer that man has depended on longest is cellulose. It is the major component of wood, which has been indispensable as a fuel and a construction material. Wood's cellulose is also used to make paper. In the pure fibrous forms of cotton and linen, cellulose has been man's most important textile material. And the organic chemists of the mid nineteenth century naturally turned to cellulose as a raw material for making other giant molecules.

One way of modifying cellulose is by attaching the 'nitrate group' of atoms (a nitrogen atom and three oxygen atoms) to the oxygen–hydrogen combinations ('hydroxyl groups') in the glucose units. When this was done, by treating cellulose with a mixture of nitric acid and sulphuric acid, an explosive of until-then unparalleled ferocity was created. The explosive was discovered by accident in 1846 by a German-born Swiss chemist named Christian Friedrich Schönbein (who, in 1839, had discovered ozone). He had spilled an acid mixture in the kitchen (where he was forbidden to experiment but where he had taken advantage of his wife's absence to do just that), and he snatched up his wife's cotton apron, so the story goes, to wipe up the mess. When he hung the apron over the fire to dry, it went poof, leaving nothing behind.

Schönbein recognized the potentialities at once, as can be told from the name he gave the compound, which in English translation is 'guncotton'. (It is also called 'nitrocellulose'.) Schönbein peddled the recipe to several governments. Ordinary gunpowder was so smoky that it blackened the gunners, fouled the cannon, which then had to be swabbed between shots, and raised such a pall of smoke that after the first volleys battles had to be fought by dead reckoning. War offices therefore leaped at the chance to use an explosive which was not only more powerful but also smokeless. Factories for the manufacture of guncotton began to spring up. And almost as fast as they sprang up, they blew up. Guncotton was too eager an explosive; it would not wait for the cannon. By the early 1860s, the abortive guncotton boom was over, figuratively as well as literally.

Later, however, methods were discovered for removing the small quantities of impurities that encouraged guncotton to explode. It then became reasonably safe to handle. The English chemist Dewar (of liquefied gas fame) and a co-worker, Frederick Augustus Abel, introduced the technique, in 1889, of mixing it with nitroglycerine, and adding Vaseline to the mixture to make it mouldable into cords (the mixture was called 'cordite'). That, finally, was a useful smokeless powder. The Spanish–American War of 1898 was the last of any consequence fought with ordinary gunpowder.

(The machine age added its bit to the horrors of gunnery, also. In the 1860s, the American inventor Richard Gatling produced the first 'machine gun' for the rapid firing of bullets, and this was improved by another American inventor, Hiram Stevens Maxim, in the 1880s. The 'Gatling gun' gave rise to the slang term 'gat' for gun. It and its descendant the 'Maxim gun' gave the unabashed imperialists of the late nineteenth century an unprecedented advantage over the 'lesser breeds', to use Rudyard Kipling's offensive phrase, of Africa and Asia. 'Whatever happens, we have got / The Maxim gun and they have not!' went a popular jingle.)

'Progress' of this sort continued in the twentieth century. The most important explosive in the First World War was 'trinitrotoluene', familiarly abbreviated as TNT. In the Second World War, an even more powerful explosive, 'cyclonite', came into use. Both contained the nitro group (NO_2) rather than the nitrate group (ONO_2). As lords of war, however, all chemical explosives gave way to nuclear bombs in 1945 (see Vol. 1, Chapter 9).

Nitroglycerine, by the way, was discovered in the same year as was guncotton. An Italian chemist named Ascanio Sobrero treated glycerol with a mixture of nitric acid and sulphuric acid and knew he had something when he nearly killed himself in the explosion that followed. Sobrero, lacking Schönbein's promotional impulses, felt nitroglycerine to be too dangerous a substance to deal with and virtually suppressed information about it. But within ten years a Swedish family, the Nobels, took to manufacturing it as a 'blasting oil' for use in mining and construction work. After a series of accidents, including one which took the life of a member of the family, Alfred Bernhard Nobel, the brother of the victim, discovered a method of mixing nitroglycerine with an absorbent earth called 'kieselguhr' or 'diatomaceous earth' (kieselguhr consists largely of the tiny skeletons of one-celled organisms called diatoms). The mixture consisted of three parts of nitroglycerine to one of kieselguhr, but such was the absorptive power of the latter that the mixture was virtually a dry powder. A stick of this impregnated earth (dynamite) could be dropped, hammered, even burned, without explosion. When set off by a percussion cap (electrically, and from a distance), it displayed all the shattering force of pure nitroglycerine.

Percussion caps contain sensitive explosives that detonate by heat or by mechanical shock and are therefore called 'detonators'. The strong shock of the detonation sets off the less sensitive dynamite. It might seem as though the danger were merely shifted from nitroglycerine to detonators, but it is not so bad as it sounds, since the detonator is only needed in tiny quantities. The detonators most used are mercury fulminate ($HgC_2N_2O_2$) and lead azide (PbN_6).

Sticks of dynamite eventually made it possible to carve the American West into railways, mines, highways, and dams at a rate unprecedented in history. Dynamite, and other explosives he discovered, made a millionaire of the lonely and unpopular Nobel (who found himself, against his humanitarian will, regarded as a 'merchant of death'). When he died in 1896, he left behind a fund out of which the famous Nobel Prizes, amounting to over 40,000 dollars each, were to be granted each year in five fields: chemistry, physics, medicine and physiology, literature, and peace. The first prizes were awarded on 10 December 1901, the fifth anniversary of his death, and these have now become the greatest honour any scientist can receive. (It is a pity that Nobel did not think to set up the category 'astronomy and earth sciences' so that such men as Shapley, Hubble, and others might have been properly rewarded for their work.)

Considering the nature of human society, explosives continued to take up a sizable fraction of the endeavour of great scientists. Since almost all explosives contain nitrogen, the chemistry of that element and its compounds was of key importance. (It is also, it must be admitted, of key importance to life as well.)

The German chemist Wilhelm Ostwald, who was interested in chemical theory rather than in explosives, studied the rates at which chemical reactions proceeded. He applied the mathematical principles associated with physics to chemistry, thus being one of the founders of 'physical chemistry'. Towards the turn of the century, he worked out new methods for converting ammonia (NH_3) to nitrogen oxides, which could then be used to manufacture explosives. For his theoretical work, particularly on catalysis, Ostwald received the Nobel Prize for chemistry in 1909.

The ultimate source of usable nitrogen was, in the early decades of the twentieth century, the nitrate deposits in the desert of northern Chile. During the First World War, these fields were placed out of reach of Germany by the British Navy. However, the German chemist Fritz Haber had devised a method by which the molecular nitrogen of the air could be combined with hydrogen under pressure, to form the ammonia needed for the Ostwald process. This 'Haber process' was improved by the German chemist Karl Bosch, who supervised the building of plants during the First World War for the manufacture of ammonia. Haber received the Nobel Prize for chemistry in 1918 and Bosch shared one in 1931. By the late 1960s, the United States alone was manufacturing 12 million tons of ammonia per year by the Haber process.

But let us return to modified cellulose. Clearly, it was the addition of the nitrate group that made for explosiveness. In guncotton all of the available hydroxyl groups were nitrated. What if only some of them were nitrated? Would they not be less explosive? Actually, such partly nitrated cellulose proved not to be explosive at all. However, it did burn very readily; the material was eventually named 'pyroxylin' (from Greek words meaning 'firewood').

Pyroxylin could be dissolved in mixtures of alcohol and ether. (This was discovered independently by the French scholar Louis Nicolas Ménard and an American medical student named J. Parkers Maynard – and an odd similarity in names that is.) When the alcohol and ether evaporated, the pyroxylin was left behind as a tough, transparent film, which was named 'collodion'. Its first use was as a coating over minor cuts and abrasions; it was called 'new skin'. However, the adventures of pyroxylin were only beginning. Much more lay ahead.

Pyroxylin itself is brittle in bulk. But the English chemist Alexander Parkes found that if it was dissolved in alcohol and ether and mixed with a substance such as camphor, the evaporation of the solvent left behind a hard solid that became soft and malleable when heated. It could then be modelled into some desired shape which it would retain when cooled and hardened. So nitrocellulose was transformed into the first artificial 'plastic', and the year in which this was done was 1865. Camphor, which introduced the plastic properties into an otherwise brittle substance, was the first 'plasticizer'.

What brought plastics to the attention of the public and made it more than a chemical curiosity was its dramatic introduction into the billiard room. Billiard balls were then made from ivory, a commodity which could be obtained only over an elephant's dead body – a point that naturally produced problems. In the early 1860s, a prize of 10,000 dollars was offered for the best substitute for ivory that would fulfil the billiard ball's manifold requirements of hardness, elasticity, resistance to heat and moisture, lack of grain, and so on. The American inventor John Wesley Hyatt was one of those who went out for the prize. He made no progress until he heard of Parkes' trick of plasticizing pyroxylin to a mouldable material that would set as a hard solid. Hyatt set about working out improved methods of manufacturing the material, using less of the expensive alcohol and ether and more in the way of heat and pressure. By 1869, Hyatt was turning out cheap billiard balls of this material, which he called 'celluloid'. It won him the prize.

Celluloid turned out to have significance away from the billiard table. It was versatile indeed. It could be moulded at the temperature of boiling water; it could be cut, drilled, and sawn at lower temperatures; it was strong and hard in bulk but could also be produced in the form of thin flexible films that served for shirt collars, babies' rattles, and so on. In the form of still thinner and more flexible films it could be used as a base for silver compounds in gelatin, and thus it became the first practical photographic film.

The one fault of celluloid was that, thanks to its nitrate groups, it had a tendency to burn with appalling quickness, particularly when in the form of thin film. It was the cause of a number of fire tragedies.

The substitution of acetate groups (CH_3COO-) for nitrate groups led to the formation of another kind of modified cellulose called 'cellulose acetate'. Properly plasticized, this has properties as good or almost as good as those of celluloid, plus the saving grace of being much less apt to burn. Cellulose acetate came into use just before the First World War, and after the war it completely replaced celluloid in the manufacture of photographic film and many other items.

Within half a century after the development of celluloid, chemists emancipated themselves from dependence on cellulose as the base for plastics. As early as 1872, Baeyer (who was later to synthesize indigo) had noticed that when phenols and aldehydes were heated together, a gooey, resinous mass resulted. Since he was interested only in the small molecules he could isolate from the reaction, he ignored this mess at the bottom of the flask (as nineteenth-century organic chemists typically tended to do when goo fouled up their glassware). Thirty-seven years later, the Belgian-born American chemist Leo Hendrik Baekeland, experimenting with formaldehyde, found that under certain conditions the reaction would yield a resin that on continued heating under pressure became first a soft solid, then a hard, insoluble substance. This resin could be moulded while soft and then be allowed to set into a hard, permanent shape. Or, once hard, it could be powdered, poured into a mould and set into one piece by heat and pressure. Very complex forms could be cast easily and quickly. Furthermore, the product was inert and impervious to most environmental vicissitudes.

Baekeland named his product 'Bakelite', after his own name. Bakelite belongs to the class of 'thermosetting plastics', which, once they set on cooling, cannot be softened again by heating (though, of course, they can be destroyed by intense heat). Materials such as the cellulose derivatives, which can be softened again and again, are called 'thermoplastics'. Bakelite has numerous uses – as an insulator, an adhesive, a laminating agent, and so on. Although the oldest of the thermosetting plastics, it is still the most used.

Bakelite was the first production, in the laboratory, of a useful high polymer from small molecules. For the first time the chemist had taken over this particular task completely. It does not, of course, represent synthesis in the sense of the synthesis of haem or quinine, where chemists must place every last atom into just the proper position, almost one at a time. Instead, the production of high polymers requires merely that the small units of which they are composed be mixed under the proper conditions. A reaction is then set up in which the units form a chain automatically, without the specific point-to-point intervention of the chemist. The chemist can, however, alter the nature of the chain indirectly by varying the starting materials or the proportions among them, or by the addition of small quantities of acids, alkalies, or various substances that act as 'catalysts' and tend to guide the precise nature of the reaction.

With the success of Bakelite, chemists naturally turned to other possible starting materials in search of more synthetic high polymers that might be useful plastics. And, as time went on, they succeeded many times over.

British chemists discovered in the 1930s, for instance, that the gas ethylene ($CH_2 = CH_2$), under heat and pressure, would form very long chains. One of the two bonds in the double bond between the carbon atoms opens up and attaches itself to a neighbouring molecule. With this happening over and over again, the result is a long-chain molecule called 'polythene' in Britain and 'polyethylene' in the United States.

The paraffin-wax molecule is a long chain made up of the same units, but the molecule of polythene is even longer. Polythene is therefore like wax, but more so. It has the cloudy whiteness of wax, the slippery feel, the electrical insulating properties, the waterproofness, and the lightness (it is about the only plastic that will float on water). It is, however, at its best, much tougher than wax and much more flexible.

As it was first manufactured, polythene required dangerous pressures, and the product had a rather low melting point – just above the boiling point of water. It softened to uselessness at temperatures below the melting point. Apparently this was due to the fact that the carbon chain had branches which prevented the molecules from forming close-packed, crystalline arrays. In 1953, a German chemist named Karl Ziegler found a way to produce unbranched polythene chains, and without the need for high pressures. The result was a new variety of polythene, tougher and stronger than the old, and capable of withstanding boiling-water temperatures without softening too much. Ziegler accomplished this by using a new type of catalyst – a resin with ions of metals such as aluminium or titanium attached to negatively charged groups along the chain.

On hearing of Ziegler's development of metal-organic catalysts for polymer formation, the Italian chemist Giulio Natta began applying the technique to propylene (ethylene to which a small one-carbon methyl group, CH_3-, was attached). Within ten weeks, he had found that in the resultant polymer all the methyl groups face in the same direction, rather than (as was usual in polymer formation before that time) facing, in random fashion, in either direction. Such 'isotactic polymers' (the name was proposed by Mrs Natta) proved to have useful properties, and these can now be manufactured virtually at will. Chemists can design polymers, in other words, with greater precision than ever before. For their work in this field, Ziegler and Natta shared the 1963 Nobel Prize for chemistry.

The atomic-bomb project contributed another useful high polymer in the form of an odd relative of polythene. In the separation of uranium 235 from natural uranium, the nuclear physicists had to combine the uranium with fluorine in the gaseous compound uranium hexafluoride. Fluorine is the most active of all substances and will attack almost anything. Looking for lubricants and seals for their vessels that would be impervious to attack by fluorine, the physicists resorted to 'fluorocarbons' –

substances in which the carbon was already combined with fluorine (replacing hydrogen).

Until then, fluorocarbons had been only laboratory curiosities. The first (and simplest) of this type of molecule, 'carbon tetrafluoride' (CF_4), had only been obtained in pure form in 1926. The chemistry of these interesting substances was now pursued intensively. Among the fluorocarbons studied was 'tetrafluoroethylene' ($CF_2 = CF_2$), which had first been synthesized in 1933 and is, as you see, ethylene with its four hydrogens replaced by four fluorines. It was bound to occur to someone that tetrafluoroethylene might polymerize as ethylene itself did. After the war, Du Pont chemists produced a long-chain polymer which was as monotonously $CF_2CF_2CF_2$ ... as polythene was $CH_2CH_2CH_2$ ... Its trade name is Teflon, the 'tefl' being an abbreviation of 'tetrafluoro-'.

Teflon is like polythene, only more so. The carbon–fluorine bonds are stronger than the carbon–hydrogen bonds and offer even less opportunity for the interference of the environment. Teflon is insoluble in everything, unwettable by anything, an extremely good electrical insulator, and considerably more resistant to heat than is even the new and improved polythene. Teflon's best-known application, so far as the housewife is concerned, is as a coating upon frying pans, thus enabling food to be fried without fat, since fat will not stick to the standoffish fluorocarbon polymer.

An interesting compound that is not quite a fluorocarbon is Freon (CF_2Cl_2), introduced in 1932 as refrigerant. It is more expensive than the ammonia or sulphur dioxide used in large-scale freezers, but, on the other hand, Freon is non-odorous, non-toxic, and non-flammable, so that accidental leakage introduces a minimum of danger. It is through Freon that room air conditioners have become so characteristic a part of the American scene since the Second World War.

Plastic properties do not, of course, belong solely to the organic world. One of the most ancient of all plastic substances is glass. The large molecules of glass are essentially chains of silicon and oxygen atoms; that is, —Si—O—Si—O—Si—O—Si—, and so on indefinitely. Each silicon atom in the chain has two unoccupied bonds to which other groups can be added. The silicon atom, like the carbon atom, has four valence bonds. The silicon-silicon bond, however, is weaker than the carbon–carbon bond, so that only short silicon chains can be formed, and those (in compounds called 'silanes') are unstable. The silicon–oxygen bond is a strong one, however, and such chains are even more stable than those of carbon. In fact, since the earth's crust is half oxygen and a quarter silicon, the solid ground we stand upon may be viewed as essentially a silicon–oxygen chain.

Although the beauties and usefulness of glass (a kind of sand, made transparent) are infinite, it possesses the great disadvantage of being breakable. And in the process of breaking, it produces hard, sharp pieces which can be dangerous, even deadly. With untreated glass in the windscreen of a car, a crash may convert the car into a shrapnel bomb.

Glass can be prepared, however, as a double sheet between which is placed a thin layer of a transparent polymer, which hardens and acts as an adhesive. This is 'safety glass', for when it is shattered, even into powder, each piece is held firmly in place by the polymer. None goes flying out on death-dealing missions. Originally, as far back as 1905, collodion was used as the binder, but nowadays that has been replaced for the most part by polymers built of small molecules such as vinyl chloride. (Vinyl chloride is like ethylene, except that one of the hydrogen atoms is replaced by a chlorine atom.) The 'vinyl resin' is not discoloured by light, so safety glass can be trusted not to develop a yellowish cast with time.

Then there are the transparent plastics that can completely replace glass, at least in some applications. In the middle 1930s, Du Pont polymerized a small molecule called methyl methacrylate and cast the polymer that resulted (a 'polyacrylic plastic') into clear, transparent sheets. The trade names of these products are Plexiglas and Lucite. Such 'organic glass' is lighter than ordinary glass, more easily moulded, less brittle, and simply snaps instead of shattering when it does break. During the Second World War, moulded transparent plastic sheets came into important use as windows and transparent domes in aeroplanes, where lightness and non-brittleness are particularly useful. To be sure, the poly-acrylic plastics have their disadvantages. They are affected by organic solvents, are more easily softened by heat than glass is, and are easily scratched. Polyacrylic plastics used in the wind-screens of cars, for instance, would quickly scratch under the impact of dust particles and become dangerously hazy. Conse-quently, glass is not likely ever to be replaced entirely. In fact, it is actually developing new versatility. Glass fibres have been spun into textile material that has all the flexibility of organic fibres and the inestimable further advantage of being absolutely fire-proof.

In addition to glass substitutes, there is also what might be called a glass compromise. As I said, each silicon atom in a silicon–oxygen chain has two spare bonds for attachment to other atoms. In glass those other atoms are oxygen atoms, but they need not be. What if carbon-containing groups are attached in-stead of oxygen? You will then have an inorganic chain with organic offshoots, so to speak – a compromise between an organic and an inorganic material. As long ago as 1908, the English chemist Frederic Stanley Kipping formed such compounds, and they have come to be known as 'silicones'.

During the Second World War, long-chain 'silicone resins' came into prominence. Such silicones are essentially more resis-tant to heat than purely organic polymers. By varying the length of the chain and the nature of the side chains, a list of desirable properties not possessed by glass itself can be obtained. For in-stance, some silicones are liquid at room temperature and change very little in viscosity over large ranges of temperature. (That is, they do not thin out with heat or thicken with cold.) This is a particularly useful property for a hydraulic fluid – the type of fluid used to lower landing gear on aeroplanes, for instance. Other silicones form soft, putty-like sealers that do not harden or crack at the low temperatures of the stratosphere and are remarkably water-repellent. Still other silicones serve as acid-resistant lubricants, and so on.

By the late 1960s, plastics of all sorts were being used at the rate of over 7 million tons a year, creating a serious problem as far as waste-disposal is concerned.

A possible polymer, utterly unexpected and of potentially fascinating theoretical implications, was announced in 1962. In that year the Soviet physicist Boris Vladimirovich Deryagin reported that water in very thin tubes seemed to have most peculiar properties. Chemists generally were sceptical, but eventually investigators in the United States confirmed Deryagin's findings. What seemed to happen was that under constricted conditions, water molecules lined up in orderly fashion, with the atoms approaching each other more closely than under ordinary conditions. The structure resembles a polymer composed of H_2O units, and the expression 'polywater' came to be used for it.

Polywater was 1·4 times as dense as ordinary water, could be heated to 500°C before being made to boil, and froze to a glassy ice only at −40°C. What gave it particular interest to biologists was the speculation that polywater might exist in the constricted confines of the cell interior and that its properties might be a key to some life processes.

However, reports soon began to filter out of chemistry laboratories that polywater might be ordinary water that had dissolved sodium silicate out of glass, or might be contaminated with perspiration. In short, polywater may only be impure water. The weight of the evidence seems to be shifting in the direction of the negative, so that polywater, after a brief and exciting life, may be dismissed – but the controversy, at the time of writing, is not yet quite over.

Fibres

In the story of organic synthesis, a particularly interesting chapter is that of the synthetic fibres. The first artificial fibres (like the first bulk plastics) were made from cellulose as the starting material. Naturally, the chemists began with cellulose nitrate, since it was available in reasonable quantity. In 1884, Hilaire

Bernigaud de Chardonnet, a French chemist, dissolved cellulose nitrate in a mixture of alcohol and ether and forced the resulting thick solution through small holes. As the solution sprayed out, the alcohol and ether evaporated, leaving behind the cellulose nitrate as a thin thread of collodion. (This is essentially the manner in which spiders and silkworms spin their threads. They eject a liquid through tiny orifices and this becomes a solid fibre on exposure to air.) The cellulose–nitrate fibres were too flammable for use, but the nitrate groups could be removed by appropriate chemical treatment, and the result was a glossy cellulose thread that resembled silk.

De Chardonnet's process was expensive, of course, what with nitrate groups being first put on and then taken off, to say nothing of the dangerous interlude while they were in place and of the fact that the alcohol–ether mixture used as solvent was also dangerously flammable. In 1892 methods were discovered for dissolving cellulose itself. The English chemist Charles Frederick Cross, for instance, dissolved it in carbon disulphide and formed a thread from the resulting viscous solution (named 'viscose'). The trouble was that carbon disulphide is flammable, toxic, and evil-smelling. In 1903, a competing process employing acetic acid as part of the solvent, and forming a substance called cellulose acetate, came into use.

These artificial fibres were first called 'artificial silk', but were later named 'rayon' because their glossiness reflected rays of light. The two chief varieties of rayon are usually distinguished as 'viscose rayon' and 'acetate rayon'.

Viscose, by the way, can be squirted through a slit to form a thin, flexible, waterproof, transparent sheet – 'cellophane' – a process invented in 1908 by a French chemist, Jacques Edwin Brandenberger. Some synthetic polymers also can be extruded through a slit for the same purpose. Vinyl resins, for instance, yielded the covering material known as Saran.

It was in the 1930s that the first completely synthetic fibre was born.

Let me begin by saying a little about silk. Silk is an animal product made by certain caterpillars which are exacting in their requirements for food and care. The fibre must be tediously un-ravelled from their cocoons. For these reasons, silk is expensive and cannot be turned out on a mass-production basis. It was first produced in China more than 2,000 years ago, and the secret of its preparation was jealously guarded by the Chinese, so that it could be kept a lucrative monopoly for export. However, secrets cannot be kept for ever, despite all security measures. The secret spread to Korea, Japan, and India. Ancient Rome received silk by the long overland route across Asia, with middlemen levying tolls every step of the way; thus the fibre was beyond the reach of anyone except the most wealthy. In A.D. 550 silkworm eggs were smuggled into Constantinople, and silk production in Europe got its start. Nevertheless, silk has always remained more or less a luxury item. Moreover, until recently there was no good sub-stitute for it. Rayon could imitate its glossiness but not its sheer-ness or strength.

After the First World War, when silk stockings became an in-dispensable item of the feminine wardrobe, the pressure for greater supplies of silk or of some adequate substitute became very strong. This was particularly true in the United States, where silk was used in greatest quantity and where relations with the chief supplier, Japan, were steadily deteriorating. Chemists dreamed of somehow making a fibre that could compare with it.

Silk is a protein. Its molecule is built up of monomers called 'amino acids', which in turn contain 'amino' ($-NH_2$) and 'carboxyl' ($-COOH$) groups. The two groups are joined by a carbon atom between them; labelling the amino group a and the carboxyl group c, and symbolizing the intervening carbon by a hyphen, we can write an amino acid like this: a - c. These amino acids polymerize in head-to-tail fashion; that is, the amino group of one condenses with the carboxyl group of the next. Thus the structure of the silk molecule runs like this:

. . .a - c.a - c.a - c.a - c . . .

In the 1930s, a Du Pont chemist named Wallace Hume Carothers was investigating molecules containing amine groups and carboxyl groups in the hope of discovering a good method of making them condense in such a way as to form molecules with large rings. (Such molecules are of importance in perfumery.) Instead, he found them condensing to form long-chain molecules.

Carothers had already suspected that long chains might be possible and he was not caught napping. He lost little time in following up this development. He eventually formed fibres from adipic acid and hexamethylenediamine. The adipic acid molecule contains two carboxyl groups separated by four carbon atoms, so it can be symbolized as: c - - - - c. Hexamethylenediamine consists of two amine groups separated by six carbon atoms, thus: a - - - - - - a. When Carothers mixed the two substances together, they condensed to form a polymer like this:

. . . a - - - - - - a . c - - - - c . a - - - - - - a . c - - - - c . a - - - - - - a . . . .

The points at which condensation took place had the c.a configuration found in silk, you will notice.

At first the fibres produced were not much good. They were too weak. Carothers decided that the trouble lay in the presence of the water produced in the condensation process. The water set up a counteracting hydrolysis reaction which prevented polymerization from going very far. Carothers found a cure for this: he arranged to carry on the polymerization under low pressure, so that the water vaporized and was easily removed by letting it condense on a cooled glass surface held close to the reacting liquid and so slanted as to carry the water away (a 'molecular still'). Now the polymerization could continue indefinitely. It formed nice long, straight chains, and in 1935 Carothers finally had the basis for a dream fibre.

The polymer formed from adipic acid and hexamethylenediamine was melted and extruded through holes. It was then stretched so that the fibres would lie side by side in crystalline bundles. The result was a glossy, silk-like thread that could be used to weave a fabric as sheer and beautiful as silk, and even stronger. This first of the completely synthetic fibres was named 'nylon'. Carothers did not live to see his discovery come to fruition, however. He died in 1937.

Du Pont announced the existence of the synthetic fibre in 1938 and began producing it commercially in 1939. During the Second

World War, the United States Armed Forces took all the production of nylon for parachutes and for a hundred other purposes. But after the war nylon completely replaced silk for hosiery; indeed, women's stockings are now called 'nylons'.

Nylons opened the way to the production of many other synthetic fibres. Acrylonitrile, or vinyl cyanide ($CH_2{=}CHCN$), can be made to polymerize into a long chain like that of polythene but with cyanide groups (completely non-poisonous in this case) attached to every other carbon. The result, introduced in 1950, is 'Orlon'. If vinyl chloride ($CH_2{=}CHCl$) is added, so that the eventual chain contains chlorine atoms as well as cyanide groups, 'Dynel' results. Or the addition of acetate groups, through the use of vinyl acetate ($CH_2{=}CHOOCCH_3$), produces 'Acrilan'.

The British in 1941 made a 'polyester' fibre, in which the carboxyl group of one monomer condenses with the hydroxyl group of another. The result is the usual long chain of carbon atoms, broken in this case by the periodic insertion of an oxygen in the chain. The British call it 'Terylene', but in the United States it has appeared under the name of 'Dacron'.

These new synthetic fibres are more water-repellent than most of the natural fibres; thus they resist dampness and are not easily stained. They are not subject to destruction by moths or beetles. Some are crease-resistant and can be used to prepare 'wash-and-wear' fabrics.

Rubbers

It is a bit startling to realize that man has been riding on rubber wheels for only about a hundred years. For thousands of years he had ridden on wooden or metal rims. When Goodyear's discovery made vulcanized rubber available, it occurred to a number of people that rubber rather than metal might be wrapped around wheels. In 1845, a British engineer, Robert William Thomson, went one better: he patented a device consisting of an inflated rubber tube that would fit over a wheel. By 1890, 'tyres' were routinely used for bicycles, and, in 1895, they were placed on horseless carriages.

Amazingly enough, rubber, though a soft, relatively weak substance, proved to be much more resistant to abrasion than wood or metal. This durability, coupled with its shock-absorbing qualities and the air-cushioning idea, introduced man to unprecedented riding comfort.

As the car increased in importance, the demand for rubber for tyres grew astronomical. In half a century, the world production of rubber increased 42-fold. You can judge the quantity of rubber in use for tyres today when I tell you that, in the United States, they leave no less than 200,000 tons of abraded rubber on the highways each year, in spite of the relatively small amount abraded from the tyres of an individual car.

The increasing demand for rubber introduced a certain insecurity in the war resources of many nations. As war was mechanized, armies and supplies began to move on rubber and rubber could be obtained in significant quantity only from the Malayan peninsula, far removed from the 'civilized' nations most apt to engage in 'civilized' warfare. (The Malayan peninsula is not the natural habitat of the rubber tree. The tree was transplanted there, with great success, from Brazil, where the original rubber supply steadily diminished.) The supply of the United States was cut off at the beginning of its entry into the Second World War when the Japanese overran Malaya. American apprehensions in this respect were responsible for the fact that the very first object rationed during the war emergency, even before the attack on Pearl Harbor, was rubber tyres.

Even in the First World War, when mechanization was just beginning, Germany was hampered by being cut off from rubber supplies by Allied sea power.

By the time of the First World War, then, there was reason to consider the possibility of constructing a synthetic rubber. The natural starting material for such a synthetic rubber was isoprene, the building block of natural rubber. As far back as 1880, chemists had noted that isoprene, on standing, tended to become gummy, and, if acidified, would set into a rubber-like material.

Kaiser Wilhelm II eventually had the tyres of his official car made of such material, as a kind of advertisement of Germany's chemical virtuosity.

However, there were two catches to the use of isoprene as the starting material for synthesizing rubber. First, the only major source of isoprene was rubber itself. Second, when isoprene polymerizes, it is most likely to do so in a completely random manner. The rubber chain possesses all the isoprene units oriented in the same fashion: – – – uuuuuuuuu – – –. The gutta percha chain has them oriented in strict alternation: – – – – unununununun – – – –. When isoprene is polymerized in the laboratory under ordinary conditions, however, the *u*'s and *n*'s are mixed randomly, forming a material which is neither rubber nor gutta percha. Lacking the flexibility and resilience of rubber, it is useless for car tyres (except possibly for imperial cars used on state occasions).

Eventually, catalysts like those that Ziegler introduced in 1953 for manufacturing polythene made it possible to polymerize isoprene to a product almost identical with natural rubber, but by that time many useful synthetic rubbers, very different chemically from natural rubber, had been developed.

The first efforts, naturally, concentrated on attempts to form polymers from readily available compounds resembling isoprene. For instance, during the First World War, under the pinch of the rubber famine, Germany made use of dimethylbutadiene:

$$CH_2 = C - C = CH_2$$
$$| |$$
$$CH_3\ CH_3$$

Dimethylbutadiene differs from isoprene (see p. 39) only in containing a methyl group (CH_3) on both middle carbons of the four-carbon chain instead of on only one of them. The polymer built of dimethylbutadiene, called 'methyl rubber', could be formed cheaply and in quantity. Germany produced about 2,500 tons of it during the First World War. While it did not stand up well under stress, it was nonetheless the first of the usable synthetic rubbers.

About 1930, both Germany and the Soviet Union tried a new tack. They used as the monomer, butadiene, which has no methyl group at all:

$$CH_2 = CH - CH = CH_2$$

With sodium metal as a catalyst, they formed a polymer called 'Buna' (from '*bu*tadiene' and *Na* for sodium).

Buna rubber was a synthetic rubber which could be considered satisfactory in a pinch. It was improved by the addition of other monomers, alternating with butadiene at intervals in the chain. The most successful addition was 'styrene', a compound resembling ethylene but with a benzene ring attached to one of the carbon atoms. This product was called Buna S. Its properties were very similar to those of natural rubber, and, in fact, thanks to it, Germany's armed forces suffered no serious rubber shortage in the Second World War. The Soviet Union also supplied itself with rubber in the same way. The raw materials could be obtained from coal or petroleum.

The United States was later in developing synthetic rubber in commercial quantities, perhaps because it was in no danger of a rubber famine before 1941. But after Pearl Harbor it took up synthetic rubber with a vengeance. It began to produce buna rubber and another type of synthetic rubber called 'neoprene', built up of 'chloroprene':

$$CH_2 = C - CH = CH_2$$
$$|$$
$$Cl$$

This molecule, as you see, resembles isoprene except for the substitution of a chlorine atom for the methyl group.

The chlorine atoms, attached at intervals to the polymer chain, confer upon neoprene certain resistances that natural rubber does not have. For instance, it is more resistant to organic solvents such as petrol: it does not soften and swell nearly as much as would natural rubber. Thus neoprene is actually preferable to rubber for such uses as petrol hoses. Neoprene first clearly demonstrated that in the field of synthetic rubbers, as in many other fields, the product of the test tube need not be a mere substitute for nature, but could be an improvement.

Amorphous polymers with no chemical resemblance to natural rubber but with rubbery qualities have now been produced, and they offer a whole constellation of desirable properties. Since they are not actually rubbers, they are called 'elastomers' (an abbreviation of 'elastic polymer').

The first rubber-unlike elastomer had been discovered in 1918. This was a 'polysulphide rubber'; its molecule was a chain composed of pairs of carbon atoms alternating with groups of four sulphur atoms. The substance was given the name 'Thiokol', the prefix coming from the Greek word for sulphur. The odour involved in its preparation held it in abeyance for a long time, but eventually it was put into commercial production.

Elastomers have also been formed from acrylic monomers, fluorocarbons, and silicones. Here, as in almost every field he touches, the organic chemist works as an artist, using materials to create new forms and improve upon nature.

2 The Proteins

Key Molecules of Life

Early in their study of living matter, chemists noticed that there was a large group of substances that behaved in a peculiar manner. Heating changed these substances from the liquid to the solid state, instead of the other way round. The white of eggs, a substance in milk (casein), and a component of the blood (globulin) were among the things that showed this property. In 1777, the French chemist Pierre Joseph Macquer put all the substances that coagulated on heating into a special class that he called 'albuminous', after 'albumen', the name the Roman encyclopedist Pliny had given to egg white.

When the nineteenth-century organic chemists undertook to analyse the albuminous substances, they found these compounds considerably more complicated than other organic molecules. In 1839, the Dutch chemist Gerardus Johannes Mulder worked out a basic formula, $C_{40}H_{62}O_{12}N_{10}$, which he thought the albuminous substances had in common. He believed that the various albuminous compounds were formed by the addition of small sulphur-containing groups or phosphorus-containing groups to this central formula. Mulder named his root formula 'protein' (a word suggested to him by the inveterate word-coiner Berzelius), from a Greek word meaning 'of first importance'. Presumably the term was merely meant to signify that this core formula was of first importance in determining the structure of the albuminous substances, but as things turned out, it proved to be a very apt word for the substances themselves. The 'proteins', as they came to be known, were soon found to be of key importance to life.

Within a decade after Mulder's work, the great German organic chemist Justus von Liebig had established that proteins were even more essential for life than carbohydrates or fats; they supplied not only carbon, hydrogen, and oxygen, but also nitrogen, sulphur, and often phosphorus, which were absent from fats and carbohydrates.

The attempts of Mulder and others to work out complete empirical formulas for proteins were doomed to failure at the time they were made. The protein molecule was far too complicated to be analysed by the methods available. However, a start had already been made on another line of attack that was eventually to reveal, not only the composition, but also the structure of proteins. Chemists had begun to learn something about the building blocks of which they were made.

In 1820, Henri Braconnot, having succeeded in breaking down cellulose into its glucose units by heating the cellulose in acid (see p. 36), decided to try the same treatment with gelatin, an albuminous substance. The treatment yielded a sweet, crystalline substance. Despite Braconnot's first suspicions, this turned out to be not a sugar, but a nitrogen-containing compound, for ammonia (NH_3) could be obtained from it. Nitrogen-containing substances are conventionally given names ending in '-ine', and the compound isolated by Braconnot is now called 'glycine', from the Greek word for 'sweet'.

Shortly afterwards Braconnot obtained a white crystalline substance by heating muscle tissue with acid. He named this one 'leucine', from the Greek word for 'white'.

Eventually, when the structural formulas of glycine and leucine were worked out, they were found to have a basic resemblance:

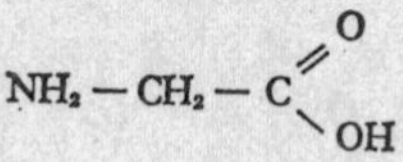

$$CH_3 \quad CH_3$$
$$CH$$
$$|$$
$$CH_2$$
$$|$$
$$NH_2 - CH_2 - C \overset{O}{\underset{OH}{\diagup}} \qquad NH_2 - CH - C \overset{O}{\underset{OH}{\diagup}}$$

glycine 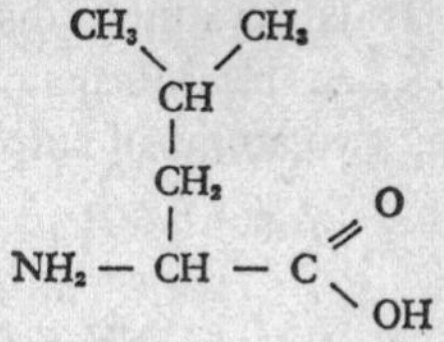   leucine

Each compound, as you see, has at its ends an amine group (NH_2) and a carboxyl group (COOH). Because the carboxyl group gives acid properties to any molecule that contains it, molecules of this kind were named 'amino acids'. Those that have the amine group and carboxyl group linked together by a single carbon atom between them, as both these molecules have, are called 'alpha-amino acids'.

As time went on, chemists isolated other amino acids from proteins. For instance, Liebig obtained one from the protein of milk (casein), which he called 'tyrosine' (from the Greek word for 'cheese'; casein itself comes from the Latin word for 'cheese').

The differences among the various alpha-amino acids lie entirely in the nature of the atom grouping attached to that single carbon atom between the amine and the carboxyl groups. Glycine, the simplest of all the amino acids, has only a pair of hydrogen atoms attached there. The others all possess a carbon-containing 'side chain' attached to that carbon atom.

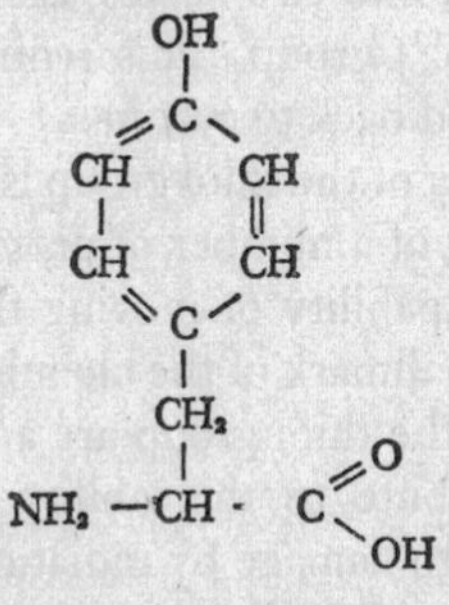

$$OH$$
$$|$$
$$C$$
$$CH \diagup \quad \diagdown CH$$
$$|$$
$$CH \diagdown \quad \diagup CH$$
$$C$$
$$|$$
$$CH_2$$
$$|$$
$$NH_2 - CH \cdot \quad C \overset{O}{\underset{OH}{\diagup}}$$

I shall give the formula of just one more amino acid, which will be useful in connection with matters to be discussed later in the chapter. It is 'cystine', discovered in 1899 by the German chemist K. A. H. Mörner. This is a double-headed molecule containing two atoms of sulphur:

$$NH_2 - CH - C \begin{smallmatrix} O \\ \\ OH \end{smallmatrix}$$
$$| \\ CH_2 \\ | \\ S \\ | \\ S \\ | \\ CH_2 \\ |$$
$$NH_2 - CH - C \begin{smallmatrix} O \\ \\ OH \end{smallmatrix}$$

Actually, cystine had first been isolated in 1810 by the English chemist William Hyde Wollaston from a bladder stone, and it had been named cystine from the Greek word for 'bladder' in consequence. What Mörner did was to show that this century-old compound was a component of protein as well as the substance in bladder stones.

Cystine is easily 'reduced' (a term that, chemically, is the opposite of 'oxidized'). This means that it will easily add on two hydrogen atoms, which fall into place at the S—S bond. The molecule then divides into two halves, each containing an —SH ('mercaptan' or 'thiol') group. This reduced half is 'cysteine', and it is easily oxidized back to cystine.

The general fragility of the thiol group is such that it is important to the functioning of a number of protein molecules. (A delicate balance and a capability of moving this way or that under slight impulse is the hallmark of the chemicals most important to life; the members of the thiol group are among the atomic combinations that contribute to this ability.) The thiol group is particularly sensitive to damage by radiation and the administration of cysteine either immediately before or immediately after exposure to radiation protects somewhat against radiation sickness. The injected cysteine undergoes the chemical changes to which, otherwise, important cellular components might be exposed. This is a very slight ray of hope in connection with an extremely dark cloud of worry.

Altogether, nineteen important amino acids (that is, occurring in most proteins) have now been identified. The last of these was discovered in 1935 by the American chemist William Cumming Rose. It is unlikely that any other common ones remain to be found.

By the end of the nineteenth century, biochemists had become certain that proteins were giant molecules built up of amino acids, just as cellulose was constructed of glucose and rubber of isoprene units. But there was this important difference: whereas cellulose and rubber were made with just one kind of building block, a protein was built from a number of different amino acids. That meant that working out its structure would pose special and subtle problems.

The first problem was to find out just how the amino acids were joined together in the protein chain molecule. Emil Fischer made a start on the problem by linking amino acids together in chains, in such a way that the carboxyl group of one amino acid was always joined to the amino group of the next. In 1901, he achieved his first such condensation, linking one glycine molecule to another with the elimination of a molecule of water:

$$NH_2 - CH_2 - C \overset{O}{\diagup} OH + NH_2 - CH_2 - C \overset{\diagup O}{\diagdown OH}$$

$$\downarrow$$

$$NH_2 - CH_2 - C \overset{O}{\diagup} NH - CH_2 - C \overset{\diagup O}{\diagdown OH} + H_2O$$

This is the simplest condensation possible. By 1907, Fischer had synthesized a chain made up of 18 amino acids, 15 of them glycine and the remaining 3 leucine. This molecule did not show any of the obvious properties of proteins, but Fischer felt that was only because the chain was not long enough. He called his synthetic chains 'peptides', from a Greek word meaning 'digest', because he believed that proteins broke down into such groups when they were digested. Fischer named the combination of the carboxyl's carbon with the amine group a 'peptide link'.

In 1932, the German biochemist Max Bergmann (a pupil of Fischer's) devised a method of building up peptides from various amino acids. Using Bergmann's method, the Polish-American biochemist Joseph Stewart Fruton prepared peptides that could be broken down into smaller fragments by digestive juices. Since there was good reason to believe that digestive juices would hydro-lyse (split by the addition of water) only one kind of molecular bond, this meant that the bond between the amino acids in the synthetic peptides must be of the same kind as the one joining amino acids in true proteins. The demonstration laid to rest any lingering doubts as to the validity of Fischer's peptide theory of protein structure.

Still, the synthetic peptides of the early decades of the twentieth century were very small and nothing like proteins in their proper-ties. Fischer had made one consisting of eighteen amino acids, as I have said; in 1916, the Swiss chemist Emil Abderhalden went one better by preparing a peptide with nineteen amino acids, but that held the record for thirty years. And chemists knew that such a peptide must be a tiny fragment indeed compared with the size of a protein molecule, because the molecule weights of proteins were enormous.

Consider, for instance, haemoglobin, a protein of the blood. Haemoglobin contains iron, making up just 0·34 per cent of the weight of the molecule. Chemical evidence indicates that the haemoglobin molecule has four atoms of iron, so the total mole-cular weight must be about 67,000; four atoms of iron, with a total weight of $4 \times 55 \cdot 85$, would come to 0·34 per cent of such a molecular weight. Consequently, haemoglobin must contain about 550 amino acids (the average molecular weight of the amino acids being about 120). Compare that with Abderhalden's puny nineteen. And haemoglobin is only an average-sized protein.

The best measurement of the molecular weights of proteins has been obtained by whirling them in a centrifuge, a spinning device that pushes particles outwards from the centre by centrifugal force. When the centrifugal force is more intense than the earth's gravitational force, particles suspended in a liquid will settle outwards away from the centre at a faster rate than they would settle downwards under gravity. For instance, red blood corpuscles will settle out quickly in such a centrifuge, and fresh milk will separate into two fractions, the fatty cream and the denser skim milk. These particular separations will take place slowly under ordinary gravitational forces, but centrifugation speeds them up.

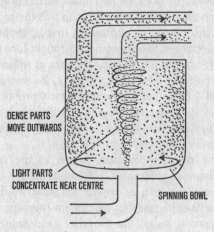

Principle of the centrifuge.

Protein molecules, though very large for molecules, are not heavy enough to settle out of solution under gravity; nor will they settle out rapidly in an ordinary centrifuge. But in the 1920s, the Swedish chemist Theodor Svedberg developed an 'ultra-centrifuge' capable of separating molecules according to their weight. This high-speed device whirls at more than 10,000 revolutions per second and produces centrifugal forces up to 900,000 times as intense as the gravitational force at the earth's surface. For his contributions to the study of suspensions, Svedberg received the Nobel Prize in chemistry in 1926.

With the ultra-centrifuge, chemists were able to determine the molecular weights of a number of proteins on the basis of their rate of sedimentation (measured in 'svedbergs' in honour of the chemist). Small proteins turned out to have molecular weights of only a few thousand and to contain perhaps not more than fifty amino acids (still decidedly more than nineteen). Other proteins have molecular weights in the hundreds of thousands and even in the millions, which means that they must consist of thousands or tens of thousands of amino acids. The possession of such large molecules put proteins into a class of substances that have only been studied systematically from the mid nineteenth century onwards.

The Scottish chemist Thomas Graham was the pioneer in this field through his interest in 'diffusion', that is, in the manner in which the molecules of two substances, brought into contact, will intermingle. He began by studying the rate of diffusion of gases through tiny holes or fine tubes. By 1831, he was able to show that the rate of diffusion of a gas was inversely proportional to the square root of its molecular weight ('Graham's law'). (It was through the operation of Graham's law that uranium 235 was separated from uranium 238, by the way.)

In following decades, Graham passed to the study of the diffusion of dissolved substances. He found that solutions of such compounds as salt, sugar, or copper sulphate would find their way through a blocking sheet of parchment (presumably containing sub-microscopic holes). On the other hand, solutions of such materials as gum arabic, glue, and gelatin would not. Clearly, the giant molecules of the latter group of substances would not fit through the holes in the parchment.

Graham called materials that could pass through parchment (and that happened to be easily obtained in crystalline form) 'crystalloids'. Those that did not, such as glue (in Greek, 'kolla'), he called 'colloids'. The study of giant molecules (or giant aggregates of atoms, even where these did not form distinct molecules) thus came to be known as 'colloid chemistry'. Because proteins and other key molecules in living tissue are of giant size, colloid chemistry is of particular importance to 'biochemistry' (the study of the chemical reactions proceeding in living tissue).

Advantage can be taken of the giant size of protein molecules in a number of ways. Suppose that pure water is on one side of a sheet of parchment and a colloidal solution of protein on the other. The protein molecules cannot pass through the parchment; moreover, they block the passage of some of the water molecules, which might otherwise move through. For this reason, water moves more readily into the colloidal portion of the system than out of it. Fluid builds up on the side of the protein solution and sets up an 'osmotic pressure'.

In 1877, the German botanist Wilhelm Pfeffer showed how one could measure this osmotic pressure and from that determine the molecular weight of a giant molecule. It was the first reasonably good method for estimating the size of such molecules.

Again, protein solutions could be placed in bags made of 'semi-permeable membranes' (membranes with pores large enough to permit the passage of small, but not of large, molecules). If these were placed in running water, small molecules and ions would pass through the membrane and be washed away, while the large protein molecule would remain behind. This process of 'dialysis' is the simplest method of purifying protein solutions.

Molecules of colloidal size are large enough to scatter light; small molecules cannot. Furthermore, light of short wavelength is more efficiently scattered than that of long wavelength. The first to note this effect, in 1869, was the Irish physicist John Tyndall; in consequence, it is called the 'Tyndall effect'. The blue of the sky is explained now by the scattering effect of dust particles in the atmosphere upon the short-wave sunlight. At sunset, when light passes through a greater thickness of atmosphere rendered particularly dusty by the activity of the day, enough light is scattered to leave only the red and orange, thus accounting for the beautiful ruddy colour of sunsets.

Light passing through a colloidal solution is scattered so that it can be seen as a visible cone of illumination when viewed from the side. Solutions of crystalloidal substances do not show such a visible cone of light when illuminated and are 'optically clear'. In 1902, the Austro-German chemist Richard Adolf Zsigmondy took advantage of this observation to devise an 'ultramicroscope', which viewed a colloidal solution at right angles, with individual particles (too small to be seen in an ordinary microscope) showing up as bright dots of light. For his endeavour, he received the Nobel Prize for chemistry in 1925.

The protein chemists naturally were eager to synthesize long, 'polypeptide' chains, with the hope of producing proteins. But the methods of Fischer and Bergmann allowed only one amino acid to be added at a time – a procedure that seemed at the time to be completely impractical. What was needed was a procedure that would cause amino acids to join up in a kind of chain reaction, such as Baekeland had used in forming his high-polymer plastics. In 1947, both the Israeli chemist E. Katchalski and the Harvard chemist Robert Woodward (who had synthesized quinine) reported success in producing polypeptides through chain-reaction polymerization. Their starting material was a slightly modified amino acid. (The modification eliminated itself neatly during the reaction.) From this beginning, they built up synthetic polypeptides consisting of as many as a hundred or even a thousand amino acids.

These chains are usually composed of only one kind of amino acid, such as glycine or tyrosine, and are therefore called 'polyglycine' or 'polytyrosine'. It is also possible, by beginning with a mixture of two modified amino acids, to form a polypeptide containing two different amino acids in the chain. But these synthetic constructions resemble only the very simplest kind of protein: for example, 'fibroin', the protein in silk.

Some proteins are as fibrous and crystalline as cellulose or nylon. Examples are fibroin; keratin, the protein in hair and skin; and collagen, the protein in tendons and in connective tissue. The German physicist R. O. Herzog proved the crystallinity of these substances by showing that they diffracted X-rays. Another German physicist, R. Brill, analysed the pattern of the diffraction and determined the spacing of the atoms in the polypeptide chain. The British biochemist William Thomas Astbury and others in the 1930s obtained further information about the structure of the chain by means of X-ray diffraction. They were able to calculate with reasonable precision the distances between adjacent atoms and the angles at which adjacent bonds were set. And they learned that the chain of fibroin was fully extended: that is, the atoms were in as nearly a straight line as the angles of the bonds between them would permit.

This full extension of the polypeptide chain is the simplest possible arrangement. It is called the 'beta configuration'. When hair is stretched, its keratin molecule, like that of fibroin, takes up this configuration. (If hair is moistened, it can be stretched up to three times its original length.) But in its ordinary, unstretched state, keratin shows a more complicated arrangement, called the 'alpha configuration'.

In 1951, Linus Pauling and Robert Brainard Corey of the California Institute of Technology suggested that, in the alpha configuration, polypeptide chains took a helical shape (a shape like that of a spiral staircase). After building various models to see how the structure would arrange itself if all the bonds between atoms lay in their natural directions without strain, they decided that each turn of the helix would have the length of 3·6 amino acids, or 5·4 angstrom units.

What enables a helix to hold its structure? Pauling suggested that the agent is the so-called 'hydrogen bond'. As we have seen, when a hydrogen atom is attached to an oxygen or a nitrogen atom, the latter holds the major share of the bonding electrons, so that the hydrogen atom has a slight positive charge and the oxygen or nitrogen a slight negative charge. In the helix, it appears, a hydrogen atom periodically occurs close to an oxygen or nitrogen atom on the turn of the helix immediately above or below it. The slightly positive hydrogen atom is attracted to its slightly negative neighbour. This attraction has only one twentieth of the force of an ordinary chemical bond, but it is strong enough to hold the helix in place. However, a pull on the fibre easily uncoils the helix and thereby stretches the fibre.

We have considered so far only the 'backbone' of the protein molecule – the chain that runs . . . CCNCCNCCNCCN . . . But the various side chains of the amino acids also play an important part in protein structure.

All the amino acids except glycine have at least one asymmetric carbon atom – the one between the carboxyl group and the amine group. Thus each could exist in two optically active isomers. The general formulas of the two isomers are:

$$
\begin{array}{cc}
\overset{\displaystyle O}{\overset{\displaystyle \parallel}{C}} - OH & \overset{\displaystyle O}{\overset{\displaystyle \parallel}{C}} - OH \\
\mid & \mid \\
H - C - NH_2 & NH_2 - C - H \\
\mid & \mid \\
\text{side chain} & \text{side chain}
\end{array}
$$

However, it seems quite certain from both chemical and X-ray analysis that polypeptide chains are made up only of L-amino acids. In this situation, the side chains stick out alternately on one side of the backbone and then the other. A chain composed of a mixture of both isomers would not be stable, because, whenever an L-amino and a D-amino acid were next to each other, there would be two side chains sticking out on the same side, which would crowd them and strain the bonds.

The side chains are important factors in holding neighbouring peptide chains together. Wherever a negatively charged side chain on one chain is near a positively charged side chain on its neighbour, they will form an electrostatic link. The side chains also provide hydrogen bonds that can serve as links. And the double-headed amino acid cystine (see p. 66) can insert one of its amine–carboxyl sequences in one chain and the other in the next. The two chains are then tied together by the two sulphur atoms in the side chain (the 'disulphide link'). The binding together of polypeptide chains accounts for the strength of protein fibres. It explains the remarkable toughness of the apparently fragile spider's web and the fact that keratin can form structures as hard as fingernails, tiger claws, alligator scales, and rhinoceros horns.

All this nicely describes the structure of protein fibres. What about proteins in solution? What sort of structure do they have?

They certainly possess a definite structure, but it is extremely delicate, for gentle heating or stirring of the solution or the addition of a bit of acid or alkali or any of a number of other environmental stresses will 'denature' a dissolved protein. That is, the protein loses its ability to perform its natural functions, and many of its properties change. Furthermore, denaturation usually is irreversible; for instance, a hard-boiled egg can never be un-hard-boiled again.

It seems certain that denaturation involves the loss of some specific configuration of the polypeptide backbone. Just what feature of the structure is destroyed? X-ray diffraction will not help us when proteins are in solution, but other techniques are available.

In 1928, for instance, the Indian physicist Chandrasekhara Venkata Raman found that light scattered by molecules in solution was, to some extent, altered in wavelength. From the nature of the alteration, deductions could be made as to the structure of the molecule. For this discovery of the 'Raman effect', Raman received the 1930 Nobel Prize for physics. (The altered wavelengths of light are usually referred to as the 'Raman spectrum' of the molecule doing the scattering.)

Another delicate technique was developed twenty years later, one that was based on the fact that atomic nuclei possess magnetic properties. Molecules exposed to a high-intensity magnetic field will absorb certain frequencies of radio waves. From such absorption, referred to as 'nuclear magnetic resonance' and frequently abbreviated as NMR, information concerning the bonds between atoms can be deduced. In particular, NMR techniques can locate the position of the small hydrogen atoms within molecules, something X-ray diffraction cannot do. NMR techniques were worked out in 1946 by two teams, working independently, one under E. M. Purcell (later to be the first to detect the radio waves emitted by the neutral hydrogen atom in space; see Vol. 1, p. 95) and the other under the Swiss-American physicist Felix Bloch. Purcell and Bloch shared the Nobel Prize for physics in 1952 for this feat.

To return, then, to the question of the denaturation of proteins in solution. The American chemists Paul Mead Doty and Elkan Rogers Blout used light-scattering techniques on solutions of synthetic polypeptides and found that they had a helical structure. By changing the acidity of the solution, they could break down the helices into randomly curved coils; by readjusting the acidity, they could restore the helices. And they showed that the conversion of the helices to random coils reduced the amount of the solution's optical activity. It was even possible to show which way a protein helix is twisted: it runs in the direction of a right-handed screw thread.

All this suggests that the denaturation of a protein involves the destruction of its helical structure.

Amino Acids in the Chain

What I have described so far represents an over-all look at the structure of the protein molecule – the general shape of the chain. What about the details of its construction? For instance, how many amino acids of each kind are there in a given protein molecule?

We might break down a protein molecule into its amino acids (by heating it with acid) and then determine how much of each amino acid is present in the mixture. Unfortunately, some of the amino acids resemble each other chemically so closely that it is almost impossible to get clear-cut separations by ordinary chemical methods. The amino acids can, however, be separated neatly by chromatography (see p. 78). In 1941, the British bio-chemists Archer John Porter Martin and Richard Laurence Millington Synge pioneered the application of chromatography to this purpose. They introduced the use of starch as the packing material in the column. In 1948, the American biochemists Stanford Moore and William Howard Stein brought the starch chromatography of amino acids to a high pitch of efficiency.

After the mixture of amino acids has been poured into the starch column and all the amino acids have attached themselves to the starch particles, they are slowly washed down the column with fresh solvent. Each amino acid moves down the column at its own characteristic rate. As each emerges at the bottom separately, the drops of solution of that amino acid are caught in a container. The solution in each container is then treated with a chemical that turns the amino acid into a coloured product. The intensity of the colour is a measure of the amount of the particular amino acid present. This colour intensity is measured by an instrument called a 'spectrophotometer', which indicates the intensity by means of the amount of light of that particular wavelength that is absorbed.

(Spectrophotometers can be used for other kinds of chemical

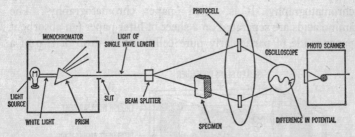

A spectrophotometer. The beam of light is split into two so that one beam passes through the specimen being analysed and the other goes directly to the photocell. Since the weakened beam that has passed through the specimen liberates fewer electrons in the photocell than the unabsorbed beam does, the two beams create a difference in potential that measures the amount of absorption of the light by the specimen.

analysis, by the way. If light of successively increased wavelength is sent through a solution, the amount of absorption changes smoothly, rising to maxima at some wavelengths and falling to minima at others. The result is an 'absorption spectrum'. A given atomic group has its own characteristic absorption peak or peaks. This is especially true in the region of the infra-red, as was first shown by the American physicist William Weber Coblentz shortly after 1900. His instruments were too crude to make the technique practical then, but since the Second World War the 'infra-red spectrophotometer', designed to scan, automatically, the spectrum from two to forty microns and to record the results has come into increasing use for analysis of the structure of complex compounds. 'Optical methods' of chemical analysis, involving radio-wave absorption, light absorption, light scattering, and so on, are extremely delicate and non-destructive – the sample survives the inspection, in other words – and are completely replacing the classical analytical methods of Liebig, Dumas, and Pregl that were mentioned in the previous chapter.)

The measurement of amino acids with starch chromatography is quite satisfactory, but by the time this procedure was developed, Martin and Synge had worked out a simpler method of chromatography. It is called 'paper chromatography'. The amino acids are separated on a sheet of filter paper (an absorbent paper made of particularly pure cellulose). A drop or two of a

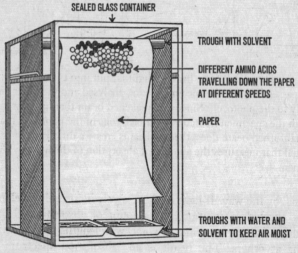

Paper chromatography.

mixture of amino acids is deposited near a corner of the sheet, and this edge of the sheet is then dipped into a solvent, such as butyl alcohol. The solvent slowly creeps up the paper through capillary action. (Dip the corner of a blotter into water and see it happen yourself.) The solvent picks up the molecules in the deposited drop and sweeps them along the paper. As in column chromatography, each amino acid moves up the paper at a characteristic rate. After a while the amino acids in the mixture become separated in a series of spots on the sheet. Some of the spots may contain two or three amino acids. To separate these, the filter paper, after being dried, is turned around ninety degrees from its first position, and the new edge is now dipped into a second solvent which will deposit the components in separate spots. Finally, the whole sheet, after once again being dried, is washed with chemicals that cause the patches of amino acids to show up as coloured or darkened spots. It is a dramatic thing to see: all the amino acids, originally mixed in a single solution, are now spread out over the length and breadth of the paper in a mosaic of colourful spots. Experienced biochemists can identify each amino acid by the spot it occupies, and thus they can read the composition of the original protein almost at a glance. By dissolving a spot they can even measure how much of that particular amino acid was present in the protein.

(Martin, along with A. T. James, applied the principles of this technique to the separation of gases in 1952. Mixtures of gases or vapours may be passed through a liquid solvent or over an adsorbing solid by means of a current of inert 'carrier gas', such as nitrogen or helium. The mixture is pushed through and emerges at the other end separated. Such 'gas chromatography' is particularly useful because of the speed of its separations and the great delicacy with which it can detect trace impurities.)

Chromatographic analysis yielded accurate estimates of the amino-acid contents of various proteins. For instance, the molecule of a blood protein called 'serum albumin' was found to contain 15 glycines, 45 valines, 58 leucines, 9 isoleucines, 31 prolines, 33 phenylalanines, 18 tyrosines, 1 tryptophan, 22 serines, 27 threonines, 32 cystines, 4 cysteines, 6 methionines, 25 arginines, 16 histidines, 58 lysines, 46 aspartic acids, and 80 glutamic acids – a total of 526 amino acids of 18 different types built into a protein with a molecular weight of about 69,000. (In addition to these eighteen, there is one other common amino acid – alanine.)

The German-American biochemist Erwin Brand suggested a system of symbols for the amino acids which is now in general use. To avoid confusion with the symbols of the elements, he designated each amino acid by the first three letters of its name, instead of just the initial. There are a few special variations: cystine is symbolized CyS, to show that its two halves are usually incorporated in two different chains; cysteine is CySH, to distinguish it from cystine; and isoleucine is Ileu rather than Iso, for 'iso' is the prefix of many chemical names.

In this shorthand, the formula of serum albumin can be written: $Gly_{15}Val_{45}Leu_{58}Ileu_9Pro_{31}Phe_{33}Tyr_{18}Try_1Ser_{22}Thr_{27}CyS_{32}$-$CySH_4Met_6Arg_{25}His_{16}Lys_{58}Asp_{46}Glu_{80}$. This is, you will admit, more concise, though certainly nothing to be rattled off.

Discovering the empirical formula of a protein was only half the battle – in fact, much less than half. Now came the far more difficult task of deciphering the structure of a protein molecule. There was every reason to believe that the properties of every protein depended on exactly how – in what order – all those amino acids are arranged in the molecular chain. This presents the biochemist with a staggering problem. The number of possible arrangements in which nineteen amino acids can be placed in a chain (even assuming that only one of each is used) comes to nearly 120 thousand billion. If you find this hard to believe, try multiplying out 19 times 18 times 17 times 16, and so on, which is the way the number of possible arrangements is calculated. And if you do not trust the arithmetic, get nineteen counters, number them 1 to 19, and see in how many different orders you can arrange them. I guarantee you will not continue the game long.

When you have a protein of the size of serum albumin, composed of more than 500 amino acids, the number of possible arrangements comes out to something like 10^{600} – that is, 1 followed by 600 noughts. This is a completely fantastic number – far more than the number of sub-atomic particles in the entire known universe, or, for that matter, far more than the universe could hold if it were packed solid with such particles.

Nevertheless, although the task of finding out which one of all those possible arrangements a serum albumin molecule actually possesses may seem hopeless, this sort of problem has actually been tackled and solved.

In 1945, the British biochemist Frederick Sanger set out to determine the order of amino acids in a peptide chain. He started by trying to identify the amino acid at one end of the chain – the amine end.

Obviously, the amine group of this end amino acid (called the 'N-terminal amino acid') is free: that is, not attached to another amino acid. Sanger made use of a chemical that combines with a free amine group but not with an amine group that is bound to a carboxyl group. This produces a DNP (dinitrophenyl) derivative of the peptide chain. With DNP he could label the N-terminal amino acid, and since the bond holding this combination together is stronger than those linking the amino acids in the chain, he could break up the chain into its individual amino acids and isolate the one with the DNP label. As it happens, the DNP group has a yellow colour, so this particular amino acid, with its DNP label, shows up as a yellow spot on a paper chromatogram.

Thus, Sanger was able to separate and identify the amino acid at the amine end of a peptide chain. In a similar way, he identified the amino acid at the other end of the chain – the one with a free carboxyl group, called the 'C-terminal amino acid'. He was also able to peel off a few other amino acids one by one and identify the 'end sequence' of a peptide chain in several cases.

Now Sanger proceeded to attack the peptide chain all along its length. He worked with insulin, a protein that has the merit of being very important to the functioning of the body and which has the added virtue of being rather small for a protein, having a molecular weight of only 6,000 in its simplest form. DNP treatment showed this molecule to consist of two peptide chains, for it contained two different N-terminal amino acids. The two chains were joined together by cystine molecules. By a chemical treatment that broke the bond between the two sulphur atoms in the cystine, Sanger split the insulin molecule into its two peptide chains, each intact. One of the chains had glycine as the N-terminal amino acid (call it the G-chain), and the other had phenylalanine as the N-terminal amino acid (the P-chain). The two could now be worked on separately.

Sanger and a co-worker, Hans Tuppy, first broke up the chains into individual amino acids and identified the twenty-one amino acids that made up the G-chain and the thirty that composed the P-chain. Next, to learn some of the sequences, they broke the chains, not into individual amino acids, but into fragments consisting of two or three. This could be done by partial hydrolysis, breaking only the weaker bonds in the chain, or by attacking the insulin with certain digestive substances which broke only certain links between amino acids and left the others intact.

By these devices Sanger and Tuppy broke each of the chains into a large number of different pieces. For instance, the P-chain yielded 48 different fragments, 22 of which were made up of two amino acids (dipeptides), 14 of three, and 12 of more than three.

These various small peptides, after being separated, could then be broken down into their individual amino acids by paper chromatography. Now the investigators were ready to determine the order of the amino acids in these fragments. Suppose they had a dipeptide consisting of valine and isoleucine. The question would be: Was the order Val-Ileu or Ileu-Val? In other words, was valine or isoleucine the N-terminal amino acid? (The amine group, and consequently the N-terminal unit, is conventionally considered to be at the left end of a chain.) Here the DNP label could provide the answer. If it was present on the valine, that would be the N-terminal amino acid, and the arrangement in the dipeptide would then be established to be Val-Ileu. If it was present on the isoleucine, it would be Ileu-Val.

The arrangement in a fragment consisting of three amino acids also could be worked out. Say its components were leucine, valine, and glutamic acid. The DNP test could first identify the N-terminal amino acid. If it was, say, leucine, the order had to be either Leu-Val-Glu or Leu-Glu-Val. Each of these combinations was then synthesized and deposited as a spot on a chromatogram to see which would occupy the same place on the paper as did the fragment being studied.

As for peptides of more than three amino acids, these could be broken down to smaller fragments for analysis.

After thus determining the structures of all the fragments into which the insulin molecule had been divided, the next step was to put the pieces together in the right order in the chain – in the fashion of a jigsaw puzzle. There were a number of clues to work with. For instance, the G-chain was known to contain only one unit of the amino acid alanine. In the mixture of peptides obtained from the breakdown of G-chains, alanine was found in two combinations: alanine-serine and cystine-alanine. This meant that in the intact G-chain the order must be CyS-Ala-Ser.

By means of such clues, Sanger and Tuppy gradually put the pieces together. It took a couple of years to identify all the fragments definitely and arrange them in a completely satisfactory sequence, but by 1952 they had worked out the exact arrangement of all the amino acids in the G-chain and the P-chain. They then went on to establish how the two chains were joined together. In 1953, their final triumph in deciphering the structure of insulin was announced. The complete structure of an important protein molecule had been worked out for the first time. For this achievement, Sanger was awarded the Nobel Prize in chemistry in 1958.

Biochemists immediately adopted Sanger's methods to determine the structure of other protein molecules. Ribonuclease, a protein molecule consisting of a single peptide chain with 124 amino acids, was conquered in 1959, and the protein unit of tobacco mosaic virus, with 158 amino acids, in 1960. In 1964, trypsin, a protein with 223 amino acids, was deciphered. By 1967, the technique was actually automated. The Swedish-Australian biochemist P. Edman devised a 'sequenator' which could work on 5 milligrams of pure protein, peeling off and identifying the amino acids one by one. Sixty amino acids of the myoglobin chain were identified in this fashion in four days.

Once the amino-acid order in a polypeptide chain was worked out, it became possible to attempt to put together amino acids in just that right order. Naturally, the beginning was a small one.

The first protein to be synthesized in the laboratory was 'oxytocin', a hormone with important functions in the body. Oxytocin is extremely small for a protein molecule: it consists only of eight amino acids. In 1953, the American biochemist Vincent du Vigneaud succeeded in synthesizing a peptide chain exactly like that thought to represent the oxytocin molecule. And, indeed, the synthetic peptide showed all the properties of the natural hormone. Du Vigneaud was awarded the Nobel Prize in chemistry in 1955.

More complicated protein molecules were synthesized as the years passed, but in order to synthesize a specific molecule with particular amino acids arranged in a particular order, the string had to be 'threaded', so to speak, one at a time. That was as difficult in the 1950s as it had been a half-century earlier in Fischer's time. Each time a particular amino acid was coupled to a chain, the new compound had to be separated from all the rest by tedious procedures, and then a new start had to be made to add one more particular amino acid. At each step a good part of the material was lost in side-reactions, and only small quantities of even simple chains were formed.

Beginning in 1959, however, a team under the leadership of the American biochemist Robert Bruce Merrifield, struck out in a new direction. An amino acid, the beginning of the desired chain, is bound to beads of polystyrene resin. These beads were insoluble in the liquid being used and could be separated from everything else by simple filtration. A new solution would be added containing the next amino acid, which would bind to the first. Again a filtration, then another. The steps in between additions were so simple and quick that they could be automated with almost nothing lost. In 1965, the molecule of insulin was synthesized in this fashion; in 1969, it was the turn of the still longer chain of ribonuclease with all its 124 amino acids. Then, in 1970, the Chinese-American biochemist Cho Hao Li synthesized the 188-amino-acid chain of human growth hormone.

With the protein molecule understood, so to speak, as a string of amino acids, it became desirable to take a still more sophisticated view. What was the exact manner in which that amino-acid chain bent and curved? What was the exact shape of the protein molecule?

Tackling this problem were the Austrian-English chemist Max Ferdinand Perutz and his English colleague John Cowdery Kendrew. Perutz took haemoglobin, the oxygen-carrying protein of blood, containing something like 12,000 atoms, as his province. Kendrew took on myoglobin, a muscle protein similar in function to haemoglobin but only about a quarter the size. As their tool, they used X-ray diffraction studies.

Perutz used the device of combining the protein molecules with a massive atom, such as that of gold or mercury, which was particularly efficient in diffracting X-rays. This gave him clues from which he could more accurately deduce the structure of the molecule without the massive atom. By 1959, myoglobin, and then haemoglobin, the year after, fell into place. It became possible to prepare three-dimensional models in which every single atom could be located in what seemed very likely to be the correct place. In both cases, the protein structure was clearly based upon the helix. As a result, Perutz and Kendrew shared the Nobel Prize in chemistry in 1962.

There is reason to think that the three-dimensional structures worked out by the Perutz–Kendrew techniques are after all determined by the nature of the string of amino acids. The amino-acid string has, so to speak, natural 'crease-points', and when they bend, certain interconnections inevitably take place and keep it properly folded. What these folds and interconnections are can be determined if all the inter-atomic distances are worked out and the angles at which the connecting bonds are placed are determined. This can be done, but it is a tedious job indeed. Here, too, computers have been called in to help, and these have not only made the calculation but thrown the results on a screen.

What with one thing or another, the list of protein molecules whose shapes are known in three-dimensional detail is growing rapidly. Insulin, which started the new forays into molecular biology, had its three-dimensional shape worked out by the English biochemist Dorothy Crowfoot Hodgkin in 1969.

Enzymes

There are useful consequences following from the complexity and almost infinite variety of protein molecules. Proteins have a multitude of different functions to perform in living organisms.

One major function is to provide the structural framework of the body. Just as cellulose serves as the framework of plants, so fibrous proteins act in the same capacity for the complex animals. Spiders spin gossamer threads and insect larvae spin cocoon threads of protein fibres. The scales of fish and reptiles are made up mainly of the protein keratin. Hair, feathers, horns, hoofs, claws, and fingernails – all merely modified scales – also contain keratin. Skin owes its strength and toughness to its high content of keratin. The internal supporting tissues – cartilage, ligaments, tendons, even the organic framework of bones – are made up largely of protein molecules, such as collagen and elastin. Muscle is made of a complex fibrous protein called actomyosin.

In all these cases, the protein fibres are more than a cellulose substitute. They are an improvement; they are stronger and more flexible. Cellulose will do to support a plant, which is not called on for any motion more complex than swaying with the wind. But protein fibres must be designed for the bending and flexing of the appendages of the body, for rapid motions and vibrations, and so on.

The fibres, however, are among the simplest of the proteins, in form as well as function. Most of the other proteins have more subtle and more complicated jobs to do.

To maintain life in all its aspects, numerous chemical reactions must proceed in the body. These must go on at high speed and in great variety, each reaction meshing with all the others, for it is not upon any one reaction, but upon all together, that life's smooth workings must depend. Moreover, all the reactions must proceed under the mildest of environments: without high temperatures, s ng chemicals, or great pressures. The reactions must be under strict yet flexible control, and they must be constantly adjusted to the changing characteristics of the environment and the changing needs of the body. The undue slowing down, or speeding up, of even one reaction out of the many thousands would more or less seriously disorganize the body.

All this is accomplished by protein molecules.

Towards the end of the eighteenth century, chemists, following the leadership of Lavoisier, began to study reactions in a quantitative way – in particular, to measure the rates at which chemical reactions proceeded. They quickly noted that reaction rates could be changed drastically by comparatively minor changes in the environment. For instance, when Kirchhoff found that starch could be converted to sugar in the presence of acid, he noticed that while the acid greatly speeded up this reaction, it was not itself consumed in the process. Other examples of this were soon discovered. The German chemist Johann Wolfgang Döbereiner found that finely divided platinum (called 'platinum black') encouraged the combination of hydrogen and oxygen to form water – a reaction which without this help could take place only at a high temperature. Döbereiner even designed a self-igniting lamp in which a jet of hydrogen, played upon a surface coated with platinum black, caught fire.

Because the 'hastened reactions' were usually in the direction of breaking down a complex substance to a simpler one, Berzelius named the phenomenon 'catalysis' (from Greek words essentially meaning 'break down'). Thus, platinum black came to be called a catalyst for the combination of hydrogen and oxygen, and acid a catalyst for the hydrolysis of starch to glucose.

Catalysis has proved of the greatest importance in industry. For instance, the best way of making sulphuric acid (the most important single inorganic chemical next to air, water, and, perhaps, salt) involves the burning of sulphur, first to sulphur dioxide (SO_2), then to sulphur trioxide (SO_3). The step from the dioxide to the trioxide would not proceed at more than a snail's pace without the help of a catalyst such as platinum black. Finely divided nickel (which has replaced platinum black in most cases, because it is cheaper) and such compounds as copper chromite, vanadium pentoxide, ferric oxide, and manganese dioxide also are important catalysts. In fact, a great deal of the success of an industrial chemical process depends on finding just the right catalyst for the reaction involved. It was the discovery of a new type of catalyst by Ziegler that revolutionized the production of polymers.

How is it possible for a substance, sometimes present only in very small concentrations, to bring about large quantities of reaction without itself being changed?

Well, one kind of catalyst does in fact take part in the reaction, but in a cyclic fashion, so that it is continually restored to its original form. An example is vanadium pentoxide (V_2O_5), which can catalyse the change of sulphur dioxide to sulphur trioxide. Vanadium pentoxide passes on one of its oxygen atoms to SO_2, forming SO_3 and changing itself to vanadyl oxide (V_2O_4). But the vanadyl oxide rapidly reacts with oxygen in the air and is restored to V_2O_5. The vanadium pentoxide thus acts as a middleman, handing an oxygen atom to sulphur dioxide, taking another from the air, handing that to sulphur dioxide, and so on. The process is so rapid that a small quantity of vanadium pentoxide will suffice to bring about the conversion of large quantities of sulphur dioxide, and in the end we will appear still to have the vanadium pentoxide unchanged.

In 1902, the German chemist George Lunge suggested that this sort of thing was the explanation of catalysis in general. In 1916, Irving Langmuir went a step farther and advanced an explanation for the catalytic action of substances, such as platinum, that are so non-reactive that they cannot be expected to engage in ordinary chemical reactions. Langmuir suggested that excess valence bonds at the surface of platinum metal would seize hydrogen and oxygen molecules. While held imprisoned in close proximity on the platinum surface, the hydrogen and oxygen molecules would be much more likely to combine to form water molecules than in their ordinary free condition as gaseous molecules. Once a water molecule was formed, it would be displaced from the platinum surface by hydrogen and oxygen molecules. Thus the process of seizure of hydrogen and oxygen, their combination into water, release of the water, seizure of more hydrogen and oxygen, and formation of more water could continue indefinitely.

This is called 'surface catalysis'. Naturally, the more finely divided the metal, the more surface a given mass will provide and the more effectively catalysis can proceed. Of course, if any extraneous substance attaches itself firmly to the surface bonds of the platinum, it will 'poison' the catalyst.

All surface catalysts are more or less selective, or 'specific'. Some easily absorb hydrogen molecules and will catalyse reactions involving hydrogen; others easily absorb water molecules and catalyse condensations or hydrolyses, and so on.

The ability of surfaces to add on layers of molecules ('adsorption') is widespread and can be put to uses other than catalysis. Silicon dioxide prepared in spongy form ('silica gel') will adsorb large quantities of water. Packed in with electronic equipment, the performance of which would suffer under conditions of high humidity, it acts as a 'desiccant', keeping humidity low.

Again, finely divided charcoal ('activated carbon') will adsorb organic molecules readily; the larger the organic molecule, the more readily. Activated carbon can be used to decolorize solutions, for it would adsorb the coloured impurities (usually of high molecular weight), leaving behind the desired substance (usually colourless and of comparatively low molecular weight).

Activated carbon is also used in gas masks, a use foreshadowed by an English physician, John Stenhouse, who first prepared a charcoal air filter in 1853. The oxygen and nitrogen of air pass through such a mass unaffected, but the relatively large molecules of poison gases are adsorbed.

The organic world, too, has its catalysts. Indeed, some of them have been known for thousands of years, though not by that name. They are as old as bread-making and wine-making.

Bread dough, left to itself and kept from contamination by outside influences, will not rise. Add a lump of 'leaven' (from a Latin word meaning 'rise'), and bubbles begin to appear, lifting and lightening the dough. The common English word for leaven is 'yeast', possibly descended from a Sanskrit word meaning 'to boil'.

Yeast also hastens the conversion of fruit juices and grain to alcohol. Here again, the conversion involves the formation of bubbles, so the process is called 'fermentation', from a Latin word meaning 'boil'. The yeast preparation is often referred to as 'ferment'.

It was not until the seventeenth century that the nature of leaven was discovered. In 1680, for the first time, van Leeuwenhoek saw yeast cells. For the purpose, he made use of an instrument that was to revolutionize biology – the 'microscope'. It was based on the bending and focusing of light by lenses. Instruments using combinations of lenses ('compound microscopes') were devised as early as 1590 by a Dutch spectacle-maker, Zacharias Janssen. The early microscopes were useful in principle, but the lenses were so imperfectly ground that the objects magnified were almost useless, fuzzy blobs. Van Leeuwenhoek ground tiny but perfect lenses that magnified quite sharply up to 200 times. He used single lenses ('simple microscope').

With time, the practice of using good lenses in combinations (for a compound microscope is, potentially at least, much stronger than a simple one) spread, and the world of the very little opened up further. A century and a half after Leeuwenhoek, a French physicist, Charles Cagniard de la Tour, using a good compound microscope, studied the tiny bits of yeast intently enough to catch them in the process of reproducing themselves. The little blobs were alive. Then, in the 1850s, yeast became a dramatic subject of study.

France's wine industry was in trouble. Ageing wine was going sour and becoming undrinkable, and millions of francs were being lost. The problem was placed before the young dean of the Faculty of Sciences at the University of Lille, in the heart of the vineyard area. The young dean was Louis Pasteur, who had already made his mark by being the first to separate optical isomers in the laboratory.

Pasteur studied the yeast cells in the wine under the microscope. It was obvious to him that the cells were of varying types. All the wine contained yeast that brought about fermentation, but those wines that went sour contained another type of yeast in addition. It seemed to Pasteur that the souring action did not get under way until the fermentation was completed. Since there was no need for yeast after the necessary fermentation, why not get rid of all the yeast at that point and avoid letting the wrong kind make trouble?

He therefore suggested to a horrified wine industry that the wine be heated gently after fermentation, in order to kill all the yeast in it. Ageing, he predicted, would then proceed without souring. The industry reluctantly tried his outrageous proposal, and found to its delight that souring ceased, while the flavour of the wine was not in the least damaged by the heating. The wine industry was saved. Furthermore, the process of gentle heating ('pasteurization') was later applied to milk, also, to kill any disease germs present.

Other organisms besides yeast hasten breakdown processes. In fact, a process analogous to fermentation takes place in the intestinal tract. The first man to study digestion scientifically was the French physicist René Antoine Ferchault de Réaumur. He used a hawk as his experimental subject, and in 1752 he made it swallow metal tubes containing meat; the tubes protected the meat from any mechanical grinding action, but they had openings, covered by gratings, so that chemical processes in the stomach could act on the meat. Réaumur found that when the hawk regurgitated these tubes, the meat was partly dissolved, and a yellowish fluid was present in the tubes.

In 1777, the Scottish physician Edward Stevens isolated fluid from the stomach ('gastric juice') and showed that the dissolving process could be made to take place outside the body, thus divorcing it from the direct influence of life.

Clearly the stomach juices contained something that hastened the breakdown of meat. In 1834, the German naturalist Theodor Schwann added mercuric chloride to the stomach juice and precipitated a white powder. After freeing the powder of the mercury compound, and dissolving what was left, he found he had a very concentrated digestive juice. He called the powder he had discovered 'pepsin', from the Greek word meaning 'digest'.

Meanwhile, two French chemists, Anselme Payen and Jean

François Persoz, had found in malt extract a substance that could bring about the conversion of starch to sugar more rapidly than could acid. They called this 'diastase', from a Greek word meaning 'to separate', because they had separated it from malt.

For a long time, chemists made a sharp distinction between living ferments such as yeast cells and non-living, or 'unorganized', ferments such as pepsin. In 1878, the German physiologist Wilhelm Kühne suggested that the latter be called 'enzymes', from Greek words meaning 'in yeast', because their activity was similar to that brought about by the catalysing substances in yeast. Kühne did not realize how important, indeed universal, that term 'enzyme' was to become.

In 1897, the German chemist Eduard Buchner ground yeast cells with sand to break up all the cells and succeeded in extracting a juice that he found could perform the same fermentative tasks that the original yeast cells could. Suddenly the distinction between the ferments inside and outside of cells vanished. It was one more breakdown of the vitalists' semi-mystical separation of life from non-life. The term 'enzyme' was now applied to all ferments.

For this discovery Buchner received the Nobel Prize in chemistry in 1907.

Now it was possible to define an enzyme simply as an organic catalyst. Chemists began to try to isolate enzymes and find out what sort of substances they were. The trouble was that the amount of enzyme in cells and natural juices was very small, and the extracts obtained were invariably mixtures in which it was hard to tell what was an enzyme and what was not.

Many biochemists suspected that enzymes were proteins, because enzyme properties could easily be destroyed, as proteins could be denatured, by gentle heating. But, in the 1920s, the German biochemist Richard Willstätter reported that certain purified enzyme solutions, from which he believed he had eliminated all protein, showed marked catalytic effects. He concluded from this that enzymes were not proteins but were relatively simple chemicals, which might, indeed, utilize a protein as a 'carrier molecule'. Most biochemists went along with Willstätter, who was a Nobel Prize winner and had great prestige.

However, the Cornell University biochemist James Batcheller Sumner produced strong evidence against this theory almost as soon as it was advanced. From jackbeans (the white seeds of a tropical American plant), Sumner isolated crystals that, in solution, showed the properties of an enzyme called 'urease'. This enzyme catalysed the breakdown of urea to carbon dioxide and ammonia. Sumner's crystals showed definite protein properties, and he could find no way to separate the protein from the enzyme activity. Anything that denatured the protein also destroyed the enzyme. All this seemed to show that what he had was an enzyme in pure and crystalline form and that enzyme was a protein.

Willstätter's greater fame for a time minimized Sumner's discovery. But, in 1930, the chemist John Howard Northrop and his co-workers at the Rockefeller Institute clinched Sumner's case. They crystallized a number of enzymes, including pepsin, and found all to be proteins. Northrop, furthermore, showed that these crystals were pure proteins and retained their catalytic activity even when dissolved and diluted to the point where the ordinary chemical tests, such as those used by Willstätter, could no longer detect the presence of protein.

Enzymes were thus established to be 'protein catalysts'. By now nearly a hundred enzymes have been crystallized, and all without exception are proteins.

For their work, Sumner and Northrop shared in the Nobel Prize in chemistry in 1946.

Enzymes are remarkable as catalysts in two respects – efficiency and specificity. There is an enzyme known as catalase, for instance, which catalyses the breakdown of hydrogen peroxide to water and oxygen. Now the breakdown of hydrogen peroxide in solution can also be catalysed by iron filings or manganese dioxide. However, weight for weight, catalase speeds up the rate of breakdown far more than any inorganic catalyst can. Each molecule of catalase can bring about the breakdown of 44,000 molecules of hydrogen peroxide per second at o°C. The result is that an enzyme need be present only in small concentration to perform its function.

For this same reason, it takes but small quantities of substances ('poisons'), capable of interfering with the workings of a key enzyme, to put an end to life. Heavy metals, when administered in such forms as mercuric chloride or barium nitrate, react with thiol groups, which are essential to the working of many enzymes. The action of those enzymes stops, and the organism is poisoned. Compounds such as potassium cyanide or hydrogen cyanide place their cyanide group (—CN) in combination with the iron atom of other key enzymes and bring death quickly and, it is to be hoped, painlessly, for hydrogen cyanide is the gas used for execution in the gas chambers of some of the Western United States.

Carbon monoxide is an exception among the common poisons. It does not act on enzymes primarily, but ties up the haemoglobin molecule (a protein but not an enzyme), which ordinarily carries oxygen from lungs to cells but cannot do so with carbon monoxide hanging on to it. Animals that do not use haemoglobin are not harmed by carbon monoxide.

Enzymes, with catalase a good example, are highly specific; catalase breaks down hydrogen peroxide and nothing else, whereas inorganic catalysts, such as iron filings and manganese dioxide, may break down hydrogen peroxide, but will also catalyse numerous other reactions.

What accounts for the remarkable specificity of enzymes? Lunge's and Langmuir's theories about the behaviour of a catalyst as a middle-man suggested an answer. Suppose we consider that an enzyme forms a temporary combination with the 'substrate' – the substance whose reaction it catalyses. The form, or configuration, of the particular enzyme may therefore play a highly important role. Plainly, each enzyme must present a very complicated surface, for it has a number of different side chains sticking out of the peptide backbone. Some of these side chains have a negative charge, some positive, some no charge. Some are bulky, some small. One can imagine that each enzyme may have a surface that just fits a particular substrate. In other words, it fits the substrate as a key fits a lock. Therefore, it will combine readily with that substance, but only clumsily or not at all with others. This would explain the high specificity of enzymes; each has a surface made to order, so to speak, for combining with a particular compound. That being the case, no wonder that proteins are built of so many different units and are constructed by living tissue in such great variety.

This view of enzyme action was borne out by the discovery that the presence of a substance similar in structure to a given substrate would slow down or inhibit the substrate's enzyme-catalysed reaction. The best-known case involves an enzyme called succinic acid dehydrogenase, which catalyses the removal of two hydrogen atoms from succinic acid. That reaction will not proceed if a substance called malonic acid, which is very similar to succinic acid, is present. The structures of succinic acid and malonic acid are:

succinic acid

malonic acid

The only difference between these two molecules is that succinic acid has one more CH_2 group at the left. Presumably the malonic acid, because of its structural similarity to succinic acid, can attach itself to the surface of the enzyme. Once it has pre-empted the spot on the surface to which the succinic acid would attach itself, it remains jammed there, so to speak, and the enzyme is out of action. The malonic acid 'poisons' the enzyme, so far as its normal function is concerned. This sort of thing is called 'competitive inhibition'.

The most positive evidence in favour of the enzyme–substrate-complex theory has come from spectrographic analysis. Presumably, if an enzyme combines with its substrate, there should be a change in the absorption spectrum: the combination's absorption of light should be different from that of the enzyme or the substrate alone. In 1936, the British biochemists David Keilin and Thaddeus Mann detected a change of colour in a solution of the enzyme peroxidase after its substrate, hydrogen peroxide, was added. The American biophysicist Britton Chance made a spectral analysis and found that there were two progressive changes in the absorption pattern, one following the other. He attributed the first change in pattern to the formation of the enzyme–substrate complex at a certain rate and the second to the decline of this combination as the reaction was completed. In 1964 the Japanese biochemist Kunio Yagi announced the isolation of an enzyme–substrate complex, made up of a loose union of the enzyme D-amino acid oxidase and its substrate alanine.

Now the question arises: Is the entire enzyme molecule necessary for catalysis, or would some part of it be sufficient? This is an important question from a practical as well as a theoretical standpoint. Enzymes are in wide use today; they have been put to work in the manufacture of drugs, citric acid, and many other chemicals. If the entire enzyme molecule is not essential and some small fragment of it would do the job, perhaps this active portion could be synthesized, so that the processes would not have to depend on the use of living cells, such as yeasts, moulds, and bacteria.

Some promising advances towards this goal have been made. For instance, Northrop found that when a few acetyl groups (CH_3CO) were added to the side chains of the amino acid tyrosine in the pepsin molecule, the enzyme lost some of its activity. There was no loss, however, when acetyl groups were added to the lysine side chains in pepsin. Tyrosine, therefore, must contribute to pepsin's activity while lysine obviously did not. This was the first indication that an enzyme might possess portions not essential to its activity.

Recently the 'active region' of another digestive enzyme was pinpointed with more precision. This enzyme is chymotrypsin. The pancreas first secretes it in an inactive form called 'chymotrypsinogen'. This inactive molecule is converted into the active one by the splitting of a single peptide link (accomplished by the digestive enzyme trypsin). That is to say, it looks as if the uncovering of a single amino acid endows chymotrypsin with its activity. Now it turns out that the attachment of a molecule known as DFP (diisopropylfluorophosphate) to chymotrypsin stops the enzyme's activity. Presumably, the DFP attaches itself to the key amino acid. Thanks to its tagging by DFP, that amino acid had been identified as serine. In fact, DFP has also been found to attach itself to serine in other digestive enzymes. In each case the serine is in the same position in a sequence of four amino acids: glycine–aspartic acid–serine–glycine.

It turns out that a peptide consisting of those four amino acids alone will not display catalytic activity. In some way, the rest of the enzyme molecule plays a role, too. We can think of the four-acid sequence – the active centre – as analogous to the cutting edge of a knife, which is useless without a handle.

Nor need the active centre, or cutting edge, necessarily exist all in one piece in the amino-acid chain. Consider the enzyme ribonuclease. Now that the exact order of its 124 amino acids is known, it has become possible to devise methods for deliberately altering this or that amino acid in the chain and noting the effect of the change on the enzyme's action. It was discovered that three amino acids, in particular, were necessary for action, but that they were widely separated. They were a histidine in position 12, a lysine in position 41, and another histidine in position 119.

This separation, of course, existed only in the chain viewed as a long string. In the working molecule, the chain was coiled into a specific three-dimensional configuration, held in place by four cystine molecules, stretching across the loops. In such a molecule, the three necessary amino acids are brought together into a close-knit unit.

The matter of an active centre was made even more specific in the case of lysozyme, an enzyme found in many places, including tears and nasal mucus. It brings about the dissolution of bacterial cells by catalysing the breakdown of key bonds in some of the substances that make up the bacterial cell wall. It is as though it causes the wall to crack and the cell contents to leak away.

Lysozyme was the first enzyme whose structure was completely analysed (in 1965) in three dimensions. Once this was done, it could be shown that the molecule of the bacterial cell wall that was subject to lysozyme's action fitted neatly along a cleft in the enzyme structure. The key bond was found to lie between an oxygen atom in the side-chain of glutamic acid (position 35) and another oxygen atom in the side-chain of aspartic acid (position 52). The two positions were brought together by the folding of the amino-acid chain with just enough separation that the molecule to be attacked could fit in between. The chemical reaction necessary for breaking the bond could easily take place under those circumstances – and it is in this fashion that lysozyme is specifically organized to do its work.

Then, too, it happens sometimes that the cutting edge of the enzyme molecule is not a group of amino acids at all, but an atom combination of an entirely different nature. A few such cases will be mentioned later in the book.

We cannot tamper with the cutting edge, but could we modify the handle without impairing the usefulness of the tool? The existence of different varieties of such a protein as insulin, for instance, encourages us to believe that we might. Insulin is a hormone, not an enzyme, but its function is highly specific. At a certain position in the G-chain of insulin there is a three-amino-acid sequence which differs in different animals: in cattle it is alanine–serine–valine; in swine, threonine–serine–isoleucine; in sheep, alanine–glycine–valine; in horses, threonine–glycine–isoleucine; and so on. Yet any of these insulins can be substituted for any other and still perform the same function.

What is more, a protein molecule can sometimes be cut down drastically without any serious effect on its activity (as the handle of a knife or an axe might be shortened without much loss in effectiveness). A case in point is the hormone called ACTH (adrenocorticotropic hormone). This is a peptide chain made up of thirty-nine amino acids, the order of which has now been fully determined. Up to fifteen of the amino acids have been removed from the C-terminal end without destroying the hormone's activity. On the other hand, the removal of one or two amino acids from the N-terminal end (the cutting edge, so to speak) kills activity at once.

The same sort of thing has been done to an enzyme called 'papain', from the fruit and sap of the papaya tree. Its enzymatic action is similar to that of pepsin. Removal of eighty of the pepsin molecule's 180 amino acids from the N-terminal end does not reduce its activity to any detectable extent.

So it is at least conceivable that enzymes may yet be simplified to the point where they will fall within the region of practical synthesis. Synthetic enzymes, in the form of fairly simple organic compounds, may then be made on a large scale for various purposes. This would be a form of 'chemical miniaturization'.

Metabolism

An organism, such as the human body, is a chemical plant of great diversity. It breathes in oxygen and drinks water. It takes in as food carbohydrates, fats, proteins, minerals, and other raw materials. It eliminates various indigestible materials plus bacteria and the products of the putrefaction they bring about. It also excretes carbon dioxide via the lungs, gives up water both by way of the lungs and the sweat glands, and excretes urine, which carries off a number of compounds in solution, the chief of these being urea. These chemical reactions determine the body's metabolism.

By examining the raw materials that enter the body and the waste products that leave it, we can tell a few things about what goes on within the body. For instance, since protein supplies most of the nitrogen entering the body, we know that urea (NH_2CONH_2) must be a product of the metabolism of proteins. But between protein and urea lies a long, devious, complicated road. Each enzyme of the body catalyses only a specific small reaction, rearranging perhaps no more than two or three atoms. Every major conversion in the body involves a multitude of steps and many enzymes. Even an apparently simple organism such as the tiny bacterium must make use of many thousands of separate enzymes and reactions.

All this may seem needlessly complex, but it is the very essence of life. The vast complex of reactions in tissues can be controlled delicately by increasing or decreasing the production of appropriate enzymes. The enzymes control body chemistry as the intricate movements of fingers on the strings control the playing of a violin, and without this intricacy the body could not perform its manifold functions.

To trace the course of the myriads of reactions that make up the body's metabolism is to follow the outline of life. The attempt to follow it in detail, to make sense of the intermeshing of countless reactions all taking place at once, may indeed seem a formidable and even hopeless undertaking. Formidable it is, but not hopeless.

The chemists' study of metabolism began modestly with an effort to find out how yeast cells converted sugar to ethyl alcohol. In 1905, two British chemists, Arthur Harden and W. J. Young, suggested that this process involved the formation of sugars bearing phosphate groups. They were the first to note that phosphorus played an important role in metabolism (and phosphorus has been looming larger and larger ever since). Harden and Young even found in living tissue a sugar–phosphate ester consisting of the sugar fructose with two phosphate groups (PO_3H_2) attached. This 'fructose diphosphate' (still sometimes known as 'Harden–Young ester') was the first 'metabolic intermediate' to be identified definitely, the first compound, that is, recognized to be formed momentarily, in the process of passing from the compounds as taken into the body to the compounds eliminated by it. Harden and Young had thus founded the study of 'intermediary metabolism', which concentrates on the nature of such intermediates and the reactions involving them. For this work and for further work on the enzymes involved in the conversion of sugar to alcohol by yeast (see p. 263), Harden shared the Nobel Prize in chemistry in 1929.

What began by involving only the yeast cell became of far broader importance when the German chemist Otto Fritz Meyerhof demonstrated in 1918 that animal cells, such as those of muscle, broke down sugar in much the same way as yeast did.

The chief difference was that in animal cells the breakdown did not proceed so far in this particular route of metabolism. Instead of converting the six-carbon glucose molecule all the way down to the two-carbon ethyl alcohol (CH_3CH_2OH), they broke it down only as far as the three-carbon lactic acid ($CH_3CHOHCOOH$).

Meyerhof's work made clear for the first time a general principle that has since become commonly accepted: that, with only minor differences, metabolism follows the same routes in all creatures, from the simplest to the most complex. For his studies on the lactic acid in muscle, Meyerhof shared the Nobel Prize in physiology and medicine in 1922 with the English physiologist Archibald Vivian Hill. The latter had tackled muscle from the standpoint of its heat production and had come to conclusions quite similar to those obtained from Meyerhof's chemical attack.

The details of the individual steps involved in the transition from sugar to lactic acid were evolved between 1937 and 1941 by Carl Ferdinand Cori and his wife Gerty Theresa Cori, working at Washington University in St Louis. They used tissue extracts and purified enzymes to bring about changes in various sugar–phosphate esters, then put all the changes together like a jigsaw puzzle. The scheme of step-by-step changes that they presented has stood with little modification to this day, and the Coris were awarded a share in the Nobel Prize in physiology and medicine in 1947.

In the path from sugar to lactic acid, a certain amount of energy is produced and is utilized by the cells. The yeast cell lives on it when it is fermenting sugar, and so, when necessary, does the muscle cell. It is important to remember that this energy is obtained without the use of oxygen from the air. Thus, a muscle is capable of working even when it must expend more energy than can be replaced by reactions involving the oxygen brought to it at a relatively slow rate by the blood. As the lactic acid accumulates, however, the muscle grows weary, and eventually it must rest until oxygen breaks up the lactic acid.

Next comes the question: In what form is the energy from the sugar-to-lactic-acid breakdown supplied to the cells, and how do they use it? The German-born American chemist Fritz Albert Lipmann found an answer in researches beginning in 1941. He showed that certain phosphate compounds formed in the course of carbohydrate metabolism store unusual amounts of energy in the bond that connects the phosphate group to the rest of the molecule. This 'high-energy phosphate bond' is transferred to energy carriers present in all cells. The best known of these carriers is 'adenosine triphosphate' (ATP). The ATP molecule and certain similar compounds represent the small currency of the body's energy. They store the energy in neat, conveniently sized, readily negotiable packets. When the phosphate bond is hydrolysed off, the energy is available to be converted into chemical energy for the building of proteins from amino acids, or into electrical energy for the transmission of a nerve impulse, or into kinetic energy via the contraction of muscle, and so on. Although the quantity of ATP in the body is small at any one time, there is always enough (while life persists), for as fast as the ATP molecules are used up, new ones are formed.

For his key discovery, Lipmann shared the Nobel Prize in physiology and medicine in 1953.

The mammalian body cannot convert lactic acid to ethyl alcohol (as yeast can); instead, by another route of metabolism, it bypasses ethyl alcohol and breaks down lactic acid all the way to carbon dioxide (CO_2) and water. In so doing, it consumes oxygen and produces a great deal more energy than is produced by the non-oxygen-requiring conversion of glucose to lactic acid.

The fact that consumption of oxygen is involved offers a convenient means of tracing a metabolic process – that is, finding out what intermediate products are created along the route. Let us say that at a given step in a sequence of reactions a certain substance (e.g. succinic acid), is suspected to be the intermediate substrate. We can mix this with living tissue (or in many cases with a single enzyme) and measure the rate at which the mixture consumes oxygen. If it shows a rapid uptake of oxygen, we can be confident that this particular substance can indeed further the process.

The German biochemist Otto Heinrich Warburg devised the

key instrument used to measure the rate of uptake of oxygen. Called the 'Warburg manometer', it consists of a small flask (where the substrate and the tissue or enzyme are mixed) connected to one end of a thin U-tube, the other end of which is open. A coloured fluid fills the lower part of the U. As the mixture of enzyme and substrate absorbs oxygen from the air in the flask, a slight vacuum is created there, and the coloured liquid in the U-tube rises on the side of the U connected to the flask. The rate at which the liquid rises can be used to calculate the rate of oxygen uptake.

Warburg's experiments on the uptake of oxygen by tissues won him the Nobel Prize in physiology and medicine in 1931.

Warburg and another German biochemist, Heinrich Wieland, identified the reactions that yield energy during the breakdown of lactic acid. In the course of the series of reactions, pairs of hydrogen atoms are removed from intermediate substances by means of enzymes called 'dehydrogenases'. These hydrogen atoms then combine with oxygen, with the catalytic help of enzymes called 'cytochromes'. In the late 1920s, Warburg and Wieland argued strenuously over which of these reactions was the important one, Warburg contending that it was the uptake of oxygen and Wieland that it was the removal of hydrogen. Eventually, David Keilin showed that both steps were essential.

The German biochemist Hans Adolf Krebs went on to work out the complete sequence of reactions and intermediate products from lactic acid to carbon dioxide and water. This is called the Krebs cycle, or the citric-acid cycle, citric acid being one of the key products formed along the way. For the achievement, completed in 1940, Krebs received a share in the Nobel Prize in physiology and medicine in 1953 (with Lipmann).

The Krebs cycle produces the lion's share of energy for those organisms that make use of molecular oxygen in respiration (which means all organisms except a few types of anaerobic bacteria that depend for energy on chemical reactions not involving oxygen). At different points in the Krebs cycle, a compound will lose two hydrogen atoms, which are eventually combined with oxygen to form water. This 'eventually' hides a good deal of detail. The two hydrogen atoms are passed from one variety of cytochrome molecule to another, until the final one, 'cytochrome oxidase', passes it on to molecular oxygen. Along the line of cytochromes, molecules of ATP are formed and the body is supplied with its chemical 'small change' of energy. All told, for every turn of the Krebs cycle, eighteen molecules of ATP are formed. Exactly where in the chain the ATP is formed, and exactly how, is still not certain. The entire process, because it involves oxygen and the piling-up of phosphate groups to form the ATP, is called 'oxidative phosphorylation', and this is a key reaction of living tissue. Any serious interference with it (as when one swallows potassium cyanide) brings death in minutes.

All the substances and all the enzymes that take part in oxidative phosphorylation are contained in tiny granules within the cytoplasm. These were first detected in 1898 by the German biologist C. Benda, who did not at that time, of course, understand their importance. He called them 'mitochondria' ('threads of cartilage', which he wrongly thought they were), and the name stuck.

The average mitochondrion is rugby-ball shaped, about 1/10,000 of an inch long and 1/25,000 of an inch thick. An average cell might contain anywhere from several hundred to a thousand mitochondria. Very large cells may contain a couple of hundred thousand, while anaerobic bacteria contain none. After the Second World War, electron-microscopic investigation showed the mitochondrion to have a complex structure of its own, for all its tiny size. The mitochondrion has a double membrane, the outer one smooth and the inner one elaborately wrinkled to present a large surface. Along the inner surface of the mitochrondrion are several thousand tiny structures called 'elementary particles'. It is these that seem to represent the actual sites of oxidative phosphorylation.

Meanwhile biochemists also made headway in solving the metabolism of fats. It was known that the fat molecules were carbon chains, that they could be hydrolysed to 'fatty acids' (most commonly sixteen or eighteen carbon atoms long), and that the molecules were broken down two carbons at a time. In

1947, Fritz Lipmann discovered a rather complex compound, which played a part in 'acetylation' – that is, transfer of a two-carbon fragment from one compound to another. He called the compound 'coenzyme A' (the A standing for acetylation). Three years later the German biochemist Fedor Lynen found that co-enzyme A was deeply involved in the breakdown of fats. Once it attached itself to a fatty acid, there followed a series of four steps which ended in lopping off the two carbons at the end of the chain to which the coenzyme A was attached. Then another coenzyme A molecule would attach itself to what was left of the fatty acid, chop off two more atoms, and so on. This is called the 'fatty-acid oxidation cycle'. This and other work won Lynen a share in the 1964 Nobel Prize in physiology and medicine.

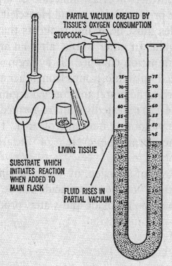

Warburg manometer.

The breakdown of proteins obviously must be, in general, more complicated than that of carbohydrates or fats, because some twenty different amino acids are involved. In some cases it turns out to be rather simple: one minor change in an amino acid may convert it into a compound that can enter the citric-acid cycle (as the two-carbon fragments from fatty acids can). But mainly amino acids are decomposed by complex routes.

We can now go back to the conversion of protein into urea – the question that we considered at the start. This conversion happens to be comparatively simple.

A group of atoms which is essentially the urea molecule forms part of a side-chain of the amino acid arginine. This group can be chopped off by an enzyme called 'arginase', and it leaves behind a kind of truncated amino acid, called 'ornithine'. In 1932, Krebs and a co-worker, K. Henseleit, while studying the formation of urea by rat-liver tissue, discovered that when they added arginine to the tissue, it produced a flood of urea – much more urea, in fact, than the splitting of every molecule of arginine they had added could have produced. Krebs and Henseleit decided that the arginine molecules must be acting as agents that produced urea over and over again. In other words, after an arginine molecule had its urea combination chopped off by arginase, the ornithine that was left picked up amine groups from other amino acids (plus carbon dioxide from the body) and formed arginine again. So the arginine molecule was repeatedly split, re-formed, split again, and so on, each time yielding a molecule of urea. This is called the 'urea cycle', the 'ornithine cycle', or the 'Krebs–Henseleit cycle'.

After the removal of nitrogen, by way of arginine, the remaining 'carbon skeletons' of the amino acids can be broken down by various routes to carbon dioxide and water, producing energy.

Tracers

The investigations of metabolism by all these devices still left biochemists in the position of being on the outside looking in, so to speak. They could work out general cycles, but to find out what was really going on in the living animal they needed some means of tracing, in fine detail, the course of events through the stages of metabolism – to follow the fate of particular molecules, as it were. Actually, techniques for doing this had been discovered early in the century, but the chemists were rather slow in making full use of them.

The first to pioneer along these lines was a German biochemist named Franz Knoop. In 1904, he conceived the idea of feeding labelled fat molecules to dogs to see what happened to the molecules. He labelled them by attaching a benzene ring at one end of the chain; he used the benzene ring because mammals possess no enzymes that can break it down. Knoop expected that what the benzene ring carried with it when it showed up in the urine might tell something about how the fat molecule broke down in the body – and he was right. The benzene ring invariably turned up with a two-carbon side-chain attached. From this he deduced that the body must split off the fat molecule's carbon atoms two at a time. (As we have seen, more than forty years later the work with coenzyme A confirmed his deduction.)

The carbon chains in ordinary fats all contain an even number of carbon atoms. What if you used a fat whose chain had an odd

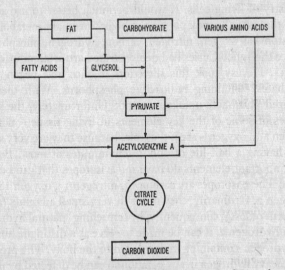

The over-all scheme of metabolism of carbohydrates, fats, and proteins.

number of carbon atoms? In that case, if the atoms were chopped off two at a time, you should end up with just one carbon atom attached to the benzene ring. Knoop fed this kind of fat molecule to dogs and did indeed end up with that result.

Knoop had employed the first 'tracer' in biochemistry. In 1913, the Hungarian chemist Georg von Hevesy and his co-worker, the German chemist Friedrich Adolf Paneth, hit upon another way to tag molecules: radioactive isotopes. They began with radioactive lead, and their first biochemical experiment was to measure how much lead, in the form of a lead-salt solution, a plant would take up. The amount was certainly too small to be measured by any available chemical method, but if radio-lead was used, it could easily be measured by its radioactivity. Hevesy and Paneth fed the radio-activity tagged lead-salt solution to plants, and at periodic intervals they would burn a plant and measure the radioactivity of its ash. In this way, they were able to determine the rate of absorption of lead by plant cells.

But the benzene ring and lead were very 'unphysiological' substances to use as tags. They might easily upset the normal chemistry of living cells. It would be much better to use as tags atoms that actually took part in the body's ordinary metabolism – such atoms as oxygen, nitrogen, carbon, hydrogen, phosphorus.

Once the Joliot-Curies had demonstrated artificial radioactivity in 1934, Hevesy took this direction at once and began using phosphates containing radioactive phosphorus. With these he measured phosphate uptake in plants. Unfortunately, the radio-isotopes of some of the key elements in living tissue – notably, nitrogen and oxygen – are not usable, because they are very short-lived, having a half-life of only a few minutes at most. But the most important elements do have *stable* isotopes that can be used as tags. These isotopes are carbon 13, nitrogen 15, oxygen 18, and hydrogen 2. Ordinarily, they occur in very small amounts (about 1 per cent or less); consequently, by 'enriching' natural hydrogen, say, in hydrogen 2, it can be made to serve as a distinguishing tag in a hydrogen-containing molecule fed to the body. The presence of the heavy hydrogen in any compound can be detected by means of the mass spectrograph, which separates it by virtue of its extra weight. Thus, the fate of the tagged hydrogen can be traced through the body.

Hydrogen, in fact, served as the first physiological tracer. It became available for this purpose when Harold Urey isolated hydrogen 2 (deuterium) in 1931. One of the first things brought to light by the use of deuterium as a tracer was that hydrogen atoms in the body were much less fixed to their compounds than had been thought. It turned out that they shuttled back and forth from one compound to another, exchanging places on the oxygen atoms of sugar molecules, water molecules, and so on. Since one ordinary hydrogen atom cannot be told from another, this shuttling had not been detected before the deuterium atoms disclosed it. What the discovery implied was that hydrogen atoms hopped about throughout the body, and, if deuterium atoms were attached to oxygen, they would spread through the body regardless of whether or not the compounds involved underwent overall chemical change. Consequently, the investigator must make sure that a deuterium atom found in a compound got there by some definite enzyme-catalysed reaction and not just by the shuttling, or exchange, process. Fortunately, hydrogen atoms attached to carbon do not exchange, so deuterium found along carbon chains has metabolic significance.

The roving habits of atoms were further emphasized in 1937 when the German-born American biologist Rudolf Schoenheimer and his associates began to use nitrogen 15. They fed rats on amino acids tagged with nitrogen 15, killed the rats after a set period, and analysed the tissues to see which compounds carried nitrogen 15. Here again, exchange was found to be important. After one tagged amino acid had entered the body, almost all the amino acids were shortly found to carry nitrogen 15. In 1942, Schoenheimer published a book entitled *The Dynamic State of Body Constituents*. That title describes the new look in biochemistry that the isotopic tracers brought about. A restless traffic in atoms goes on ceaselessly, quite aside from actual chemical changes.

Little by little the use of tracers filled in the details of the metabolic routes. It corroborated the general pattern of such things as sugar breakdown, the citric-acid cycle, and the urea cycle. It resulted in the addition of new intermediates, in the establishment of alternative routes of reaction, and so on.

When, thanks to the nuclear reactor, over a hundred different radioactive isotopes became available in quantity after the Second World War, tracer work went into high gear. Ordinary compounds could be bombarded by neutrons in a reactor and come out loaded with radioactive isotopes. Almost every biochemical laboratory in the United States (I might almost say in the world, for the United States soon made isotopes available to other countries for scientific use) started research programmes involving radioactive tracers.

The stable tracers were now joined by radioactive hydrogen (tritium), radio–phosphorus (phosphorus 32), radio–sulphur (sulphur 35), radio–potassium (potassium 42), radio–sodium, radio–iodine, radio–iron, radio–copper, and most important of all radio–carbon (carbon 14). Carbon 14 was discovered in 1940 by the American chemists Martin David Kamen and Samuel Ruben, and, to their surprise, it turned out to have a half-life of more than 5,000 years – unexpectedly long for a radio–isotope among the light elements.

Carbon 14 solved problems that had defied chemists for years and against which they had seemed to be able to make no headway at all. One of the riddles to which it gave the beginning of an answer was the production of the substance known as 'cholesterol'. Cholesterol's formula, worked out by many years of painstaking investigation by men such as Wieland (who received the 1927 Nobel Prize in chemistry for his work on compounds related to cholesterol), had been found to be:

The function of cholesterol in the body is not yet completely understood, but the substance is clearly of central importance. Cholesterol is found in large quantities in the fatty sheaths around nerves, in the adrenal glands, and in combination with certain proteins. An excess of it can cause gallstones and atherosclerosis. Most significant of all, cholesterol is the prototype of the whole family of 'steroids', the steroid nucleus being the four-ring combination you see in the formula. The steroids are a group of solid, fat-like substances, which include the sex hormones and the adrenocortical hormones. All of them undoubtedly are formed from cholesterol. But how is cholesterol itself synthesized in the body?

Until tracers came to their help, biochemists had not the foggiest notion. The first to tackle the question with a tracer were Rudolf Schoenheimer and his co-worker David Rittenberg. They gave rats heavy water to drink and found that its deuterium turned up in the cholesterol molecules. This in itself was not significant, because the deuterium could have got there merely by exchanges. But, in 1942 (after Schoenheimer tragically had committed suicide), Rittenberg and another co-worker, the German-American biochemist, Konrad Emil Bloch, discovered a more definite clue. They fed rats acetate ion (a simple two-carbon group, CH_3COO-) with the deuterium tracer attached to the carbon atom in the CH_3 group. The deuterium again showed up in cholesterol molecules, and this time it could not have arrived there by exchange; it must have been incorporated in the molecule as part of the CH_3 group.

Two-carbon groups (of which the acetate ion is one version) seem to represent a general crossroads of metabolism. Such groups, then, might very well serve as the pool of material for building cholesterol. But just how did they form the molecule?

In 1950, when carbon 14 had become available, Bloch repeated the experiment, this time labelling the two carbons of the acetate ion, each with a different tag. He marked the carbon of the CH_3 group with the stable tracer carbon 13, and he labelled the carbon of the $COO-$ group with radioactive carbon 14. Then, after feeding the compound to a rat, he analysed its cholesterol to see where the two tagged carbons would appear in the molecule. The analysis was a task that called for delicate chemical artistry, and Bloch and a number of other experimenters worked at it for years, identifying the source of one after another of the cholesterol carbon atoms. The pattern that developed eventually suggested that the acetate groups probably first formed a substance called 'squalene', a rather scarce thirty-carbon compound in the body to which no one had ever dreamed of paying serious attention before. Now it appeared to be a way station on the road to cholesterol, and biochemists have begun to study it with intense interest. For this work, Bloch shared the 1964 Nobel Prize in physiology and medicine with Lynen.

In much the same way as they tackled the synthesis of cholesterol, biochemists have gone after the construction of the porphyrin ring of haem, a key structure in haemoglobin and in many enzymes. David Shemin of Columbia University fed ducks the amino acid glycine, labelled in various ways. Glycine (NH_2CH_2COOH) has two carbon atoms. When he tagged the CH_2 carbon with carbon 14, that carbon showed up in the porphyrin extracted from the ducks' blood. When he labelled the COOH carbon, the radioactive tracer did not appear in the porphyrin. In short, the CH_2 group entered into the synthesis of porphyrin but the COOH group did not.

Shemin, working with Rittenberg, found that the incorporation of glycine's atoms into porphyrin could take place just as well in red blood cells in the test tube as it could in living animals. This simplified matters, gave more clear-cut results, and avoided sacrificing or inconveniencing the animals.

He then labelled glycine's nitrogen with nitrogen 15 and its CH_2 carbon with carbon 14, then mixed the glycine with duck blood. Later, he carefully took apart the porphyrin produced and found that all four nitrogen atoms in the porphyrin molecule came from the glycine. So did an adjacent carbon atom in each of the four small pyrrole rings (see the formula on p. 32), and also the four carbon atoms that serve as bridges between the pyrrole rings. This left twelve other carbon atoms in the porphyrin ring itself and fourteen in the various side-chains. These were shown to arise from acetate ion, some from the CH_3 carbon and some from the COO— carbon.

From the distribution of the tracer atoms it was possible to deduce the manner in which the acetate and glycine entered into the porphyrin. First they formed a one-pyrrole ring; then two such rings combined, and finally two two-ring combinations joined to form the four-ring porphyrin structure.

In 1952, a compound called 'porphobilinogen' was isolated in pure form, as a result of an independent line of research by the English chemist R. G. Westall. This compound occurs in the urine of persons with defects in porphyrin metabolism, so it was suspected of having something to do with porphyrins. Its structure turned out to be just about identical with the one-pyrrole-ring structure that Shemin and his co-workers had postulated as one of the early steps in porphyrin synthesis. Porphobilinogen was a key stage.

It was next shown that 'delta-aminolevulinic acid', a substance with a structure like that of a porphobilinogen molecule split in half, could supply all the atoms necessary for incorporation into the porphyrin ring by the blood cells. The most plausible conclusion is that the cells first form delta-aminolevulinic acid from glycine and acetate (eliminating the COOH group of glycine as carbon dioxide in the process), that two molecules of delta-aminolevulinic acid then combine to form porphobilinogen (a one-pyrrole ring), and that the latter in turn combines first into a two-pyrrole ring and finally into the four-pyrrole ring of porphyrin.

Photosynthesis

Of all the triumphs of tracer research, perhaps the greatest has been the tracing of the complex series of steps that builds green plants – on which all life on this planet depends.

The animal kingdom could not exist if animals could feed only on one another, any more than a community of people could grow rich solely by taking in one another's washing or a man could lift himself by yanking upwards on his belt buckle. A lion that eats a zebra or a man who eats a steak is consuming precious substance that was obtained at great pains and with considerable attrition from the plant world. The second law of thermodynamics tells us that at each stage of the cycle something is lost. No animal stores all of the carbohydrate, fat, and protein contained in the food it eats, nor can it make use of all the energy available in the food. Inevitably a large part, indeed most, of the energy is wasted in un-usable heat. At each level of eating, then, some chemical energy is frittered away. Thus, if all animals were strictly carnivorous, the whole animal kingdom would die off in a very few generations. In fact, it would never have come into being in the first place.

The fortunate fact is that most animals are herbivorous. They feed on the grass of the field, on the leaves of trees, on seeds, nuts, and fruit, or on the seaweed and microscopic green plant cells that fill the upper layers of the oceans. Only a minority of animals can be supported in the luxury of being carnivorous.

As for the plants themselves, they would be in no better plight were they not supplied with an external source of energy. They build carbohydrates, fats, and proteins from simple molecules, such as carbon dioxide and water. This synthesis calls for an in-put of energy, and the plants get it from the most copious possible source: sunlight. Green plants convert the energy of sunlight into the chemical energy of complex compounds, and that chemical energy supports all life forms (except for certain bacteria). This was first clearly pointed out in 1845 by the German physicist Julius Robert von Mayer, who was one of those who pioneered the law of conservation of energy and who was therefore parti-cularly aware of the problem of energy balance. The process by which green plants make use of sunlight is called 'photosyn-thesis', from Greek words meaning 'put together by light'.

The first attempt at a scientific investigation of plant growth was made early in the seventeenth century by the Flemish chemist Jan Baptista Van Helmont. He grew a small willow tree in a tub containing a weighed amount of soil, and he found, to everyone's surprise, that although the tree grew large, the soil weighed just as much as before. It had been taken for granted that plants derived their substance from the soil. (Actually plants do take some minerals and ions from the soil, but not in any easily weighable amount.) If they did not get it there, where did they get it from? Van Helmont decided that plants must manufacture their substance from water, with which he had supplied the soil liberally. He was only partly right.

A century later the English physiologist Stephen Hales showed that plants built their substance in great part from a material more ethereal than water, namely, air. Half a century later, the Dutch physician Jan Ingen-Housz identified the nourishing ingredient in air as carbon dioxide. He also demonstrated that a plant did not absorb carbon dioxide in the dark; it needed light (the 'photo' of photosynthesis). Meanwhile Priestley, the discoverer of oxygen, had learned that green plants gave off oxygen. And, in 1804, the Swiss chemist Nicholas Théodore de Saussure proved that water was incorporated in plant tissue, as Van Helmont had suggested.

The next important contribution came in the 1850s, when the French mining engineer Jean Baptiste Boussingault grew plants in soil completely free of organic matter. He showed, in this way, that plants could obtain their carbon from atmospheric carbon dioxide only. On the other hand, plants would not grow in soil free of nitrogen compounds, and this showed they derived their nitrogen from the soil and that atmospheric nitrogen was not utilized (except, as it turned out, by certain bacteria). From Boussingault's time, it became apparent that the service of soil as direct nourishment for the plant was confined to certain inorganic salts, such as nitrates and phosphates. It was these ingredients that organic fertilizers (such as manure) added to soil. Chemists began to advocate the addition of chemical fertilizers, which served the purpose excellently and which eliminated noisome odours as well as decreasing the dangers of infection and disease, much of which could be traced to the farm's manure pile.

Thus, the skeleton of the process of photosynthesis was established. In sunlight, a plant took up carbon dioxide and combined it with water to form its tissues, giving off 'left-over' oxygen in the process. Hence, it became plain that green plants not only provided food but also renewed the earth's oxygen supply. Were it not for this, within a matter of centuries the oxygen would fall to a low level and the atmosphere would be loaded with enough carbon dioxide to asphyxiate animal life.

The scale on which the earth's green plants manufacture organic matter and release oxygen is enormous. The Russian-American biochemist Eugene I. Rabinowitch, a leading investigator of photosynthesis, estimates that each year the green plants of the earth combine a total of 150,000 million tons of carbon (from carbon dioxide) with 25,000 million tons of hydrogen (from water) and liberate 400,000 million tons of oxygen. Of this gigantic performance, the plants of the forest and fields on land account for only 10 per cent; for 90 per cent we have to thank the one-celled plant and seaweed of the oceans.

We still have only the skeleton of the process. What about the details? Well, in 1817, Pierre Joseph Pelletier and Joseph Bienaimé Caventou of France, who were later to be the discoverers of quinine, strychnine, caffeine, and several other specialized plant products, isolated the most important plant product of all – the one that gives the green colour to green plants. They called the compound 'chlorophyll', from Greek words meaning 'green leaf'. Then, in 1865, the German botanist Julius von Sachs showed that chlorophyll was not distributed generally through plant cells (though leaves appear uniformly green), but was localized in small sub-cellular bodies, later called 'chloroplasts'.

It became clear that photosynthesis took place within the chloroplasts and that chlorophyll was essential to the process. Chlorophyll was not enough, however. Chlorophyll by itself, however carefully extracted, could not catalyse the photosynthetic reaction in a test tube.

Chloroplasts generally are considerably larger than mitochondria. Some one-celled plants possess only one large chloroplast per cell. Most plant cells, however, contain many smaller chloroplasts, each from two to three times as long and as thick as the typical mitochondrion.

The structure of the chloroplast seems to be even more complex than that of the mitochondrion. The interior of the chloroplast is made up of many thin membranes stretching across from wall to wall. These are the 'lamellae'. In most types of chloroplasts, these lamellae thicken and darken in places to produce 'grana', and it is within the grana that the chlorophyll molecules are found.

If the lamellae within the grana are studied under the electron microscope, they in turn seem to be made up of tiny units, just barely visible, that look like the neatly laid tiles of a bathroom floor. Each of these objects may be a photosynthesizing unit containing 250 to 300 chlorophyll molecules.

The chloroplasts are more difficult than mitochondria to isolate intact. It was not until 1954 that the Polish-American biochemist Daniel I. Arnon, working with disrupted spinach-leaf cells, could obtain chloroplasts completely intact and was able to carry through the complete photosynthetic reaction.

The chloroplast contains not only chlorophyll, but a full complement of enzymes and associated substances, all properly and intricately arranged. It even contains cytochromes by which the energy of sunlight, trapped by chlorophyll, can be converted into ATP through oxidative phosphorylation.

Meanwhile, though, what about the structure of chlorophyll, the most characteristic substance of the chloroplasts? For decades, chemists had tackled this key substance with every tool at their command, but it yielded only slowly. Finally, in 1906, Richard Willstätter of Germany (who was later to rediscover chromatography and to insist, incorrectly, that enzymes were not proteins) identified a central component of the chlorophyll molecule. It was the metal magnesium. (Willstätter received the Nobel Prize in chemistry in 1915 for this discovery and other work on plant pigments.) Willstätter and Hans Fischer went on to work on the structure of the molecule – a task that took a full generation to complete. By the 1930s, it had been determined that chlorophyll had a porphyrin ring structure basically like that of haem (a molecule which Fischer had deciphered). Where haem had an iron atom at the centre of the porphyrin ring, chlorophyll had a magnesium atom.

If there were any doubt on this point, it was removed by R. B. Woodward. That master synthesist, who had put together quinine in 1945, strychnine in 1947, and cholesterol in 1951, now capped his previous efforts by putting together a molecule in 1960 that matched the formula worked out by Willstätter and Fischer, and, behold, it had all the properties of chlorophyll isolated from green leaves. Woodward received the 1965 Nobel Prize for chemistry as a result.

Exactly what reaction in the plant did chlorophyll catalyse? All that was known, up to the 1930s, was that carbon dioxide and water went in and oxygen came out. Investigation was made more difficult by the fact that isolated chlorophyll could not be made to bring about photosynthesis. Only intact plant cells or, at best, intact chloroplasts, would do, which meant that the system under study was very complex.

As a first guess, biochemists assumed that the plant cells synthesized glucose ($C_6H_{12}O_6$) from the carbon dioxide and water and then went on to build from this the various plant substances, adding nitrogen, sulphur, phosphorus, and other inorganic elements from the soil.

On paper, it seemed as if glucose might be formed by a series of steps which first combined the carbon atom of carbon dioxide with water (releasing the oxygen atoms of CO_2), and then polymerized the combination, CH_2O (formaldehyde), into glucose. Six molecules of formaldehyde would make one molecule of glucose.

This synthesis of glucose from formaldehyde could indeed be performed in the laboratory, in a tedious sort of way. Presumably, the plant might possess enzymes that speeded the reactions. To be sure, formaldehyde is a very poisonous compound, but the chemists assumed that the formaldehyde was turned into glucose so quickly that at no time did the plant contain more than a very small amount of it. This formaldehyde theory, first proposed in 1870 by Baeyer (the synthesizer of indigo), lasted for two generations, simply because there was nothing better to take its place.

A fresh attack on the problem began in 1938, when Ruben and Kamen undertook to probe the chemistry of the green leaf with tracers. By the use of oxygen 18, the uncommon stable isotope of oxygen, they made one clear-cut finding. It turned out that when the water given a plant was labelled with oxygen 18, the oxygen released by the plant carried this tag, but the oxygen did not carry the tag when only the carbon dioxide supplied to the plant was labelled. In short, the experiment showed that the oxygen given off by plants came from the water molecule and not from the carbon dioxide molecule, as had been mistakenly assumed in the formaldehyde theory.

Ruben and his associates tried to follow the fate of the carbon atoms in the plant by labelling the carbon dioxide with the radio-active isotope carbon 11 (the only radio-carbon known at the time). But this attempt failed. For one thing, carbon 11 has a half-life of only 20·5 minutes. For another, they had no available method at the time for separating individual compounds in the plant cell quickly and thoroughly enough.

But, in the early 1940s, the necessary tools came to hand. Ruben and Kamen discovered carbon 14, the long-lived radio-isotope, which made it possible to trace carbon through a series of leisurely reactions. And the development of paper chromato-graphy provided a means of separating complex mixtures easily and cleanly. (In fact, radioactive isotopes allowed a neat refine-ment of paper chromatography: the radioactive spots on the paper, representing the presence of the tracer, would produce dark spots on a photographic film laid under it, so that the chromatogram would take its own picture – a technique called 'autoradiography'.)

After the Second World War, another group, headed by the American biochemist Melvin Calvin, picked up the ball. They exposed microscopic one-celled plants ('chlorella') to carbon di-oxide containing carbon 14 for short periods, in order to allow the photosynthesis to progress only through its earliest stages. Then they mashed the plant cells, separated their substances on a chromatogram, and made an autoradiograph.

They found that even when the cells had been exposed to the tagged carbon dioxide for only a minute and a half, the radio-active carbon atoms turned up in as many as fifteen different substances in the cell. By cutting down the exposure time, they reduced the number of substances in which radiocarbon was incorporated, and eventually they decided that the first, or almost the first, compound in which the cell incorporated the carbon-dioxide carbon was 'glyceryl phosphate'. (At no time did they detect any formaldehyde, so the venerable formaldehyde theory passed quietly out of the picture.)

Glyceryl phosphate is a three-carbon compound. Evidently it must be formed by a roundabout route, for no one-carbon or two-carbon precursor could be found. Two other phosphate-containing compounds were located that took up tagged carbon within a very short time. Both were varieties of sugars: 'ribulose diphosphate' (a five-carbon compound) and 'sedoheptulose phosphate' (a seven-carbon compound). The investigators identified enzymes that catalysed reactions involving such sugars, studied those reactions, and worked out the travels of the carbon-dioxide molecule. The scheme that best fits all their data is the following.

First, carbon dioxide is added to the five-carbon ribulose diphosphate, making a six-carbon compound. This quickly splits in two, creating the three-carbon glyceryl phosphate. A series of reactions involving sedoheptulose phosphate and other compounds then puts two glyceryl phosphates together to form the six-carbon glucose phosphate. Meanwhile ribulose diphosphate is regenerated and is ready to take on another carbon-dioxide molecule. You can imagine six such cycles turning. At each turn, each cycle supplies one carbon atom (from the carbon dioxide), and out of these a molecule of glucose phosphate is built. Another turn of the six cycles produces another molecule of glucose phosphate, and so on.

This is the reverse of the citric-acid cycle, from an energy standpoint. Whereas the citric-acid cycle converts the fragments of carbohydrate breakdown to carbon dioxide, the ribulose-diphosphate cycle builds up carbohydrates from carbon dioxide. The citric-acid cycle delivers energy to the organism; the ribulose-diphosphate cycle, conversely, has to consume energy.

Here the earlier results of Ruben and Kamen fit in. The energy of sunlight is used, thanks to the catalytic action of chlorophyll,

The Proteins

to split a molecule of water into hydrogen and oxygen, a process called 'photolysis' (from Greek words meaning 'loosening by light'). This is the way that the radiant energy of sunlight is converted into chemical energy, for the hydrogen and oxygen molecules contain more chemical energy than did the water molecule from which they came.

In other circumstances it takes a great deal of energy to break up water molecules into hydrogen – for instance, heating the water to something like 2,000 degrees or sending a strong electric current through it. But chlorophyll does the trick easily at ordinary temperatures. All it needs is the relatively weak energy of visible light. The plant uses the light-energy that it absorbs with an efficiency of at least 30 per cent; some investigators believe its efficiency may approach 100 per cent under ideal conditions. If man could harness energy as efficiently as the plants do, he would have much less to worry about with regard to his supplies of food and energy.

After the water molecules have been split, half of the hydrogen atoms find their way into the ribulose-diphosphate cycle, and half of the oxygen atoms are liberated into the air. The rest of the hydrogens and oxygens recombine into water. In doing so, they release the excess of energy that was given to them when sunlight split the water molecules, and this energy is transferred to high-energy phosphate compounds such as ATP. The energy stored in these compounds is then used to power the ribulose-diphosphate cycle. For his work in deciphering the reactions involved in photosynthesis, Calvin received the Nobel Prize in chemistry in 1961.

To be sure, there are some forms of life that gain energy without chlorophyll. About 1880, 'chemosynthetic bacteria' were discovered; bacteria that trapped carbon dioxide in the dark and did not liberate oxygen. Some oxidized sulphur compounds to gain energy; some oxidized iron compounds; and some indulged in still other chemical vagaries.

Then, too, some bacteria have chlorophyll-like compounds ('bacteriochlorophyll'), which enable them to convert carbon dioxide to organic compounds at the expense of light-energy –

even, in some cases, in the near infra-red, where ordinary chlorophyll will not work. However, only chlorophyll itself can bring about the splitting of water and the conservation of the large energy store so gained; bacteriochlorophyll must make do with less energetic devices.

All methods of fundamental energy gain, other than that which uses sunlight by way of chlorophyll, are essentially dead-end, and no creature more complicated than a bacterium has successfully made use of them. For the rest of life (and even for most bacteria), chlorophyll and photosynthesis, directly or indirectly, are the basis of life.

3 The Cell

Chromosomes

It is an odd paradox that man until recent times knew very little about his own body. In fact, it was only some 300 years ago that he learned about the circulation of the blood, and only within the last fifty years or so has he discovered the functions of many of the organs.

Prehistoric man, from cutting up animals for cooking and from embalming his own dead in preparation for after-life, was aware of the existence of the large organs, such as the brain, liver, heart, lungs, stomach, intestines, and kidneys. This awareness was intensified through the frequent use of the appearance of the internal organs of a ritually sacrificed animal (particularly the appearance of its liver) in foretelling the future or estimating the extent of divine favour or disfavour. Egyptian papyri dealing validly with surgical technique and presupposing some familiarity with body structure can be dated earlier than 2000 B.C.

The ancient Greeks went so far as to dissect animals and an occasional human corpse with the deliberate purpose of learning something about 'anatomy' (from Greek words meaning 'to cut up'). Some delicate work was done. Alcmaeon of Croton, about 500 B.C., first described the optic nerve and the Eustachian tube. Two centuries later, in Alexandria, Egypt (then the world centre of science), a school of Greek anatomy started brilliantly with Herophilus and his pupil Erasistratus. They investigated the parts of the brain, distinguishing the cerebrum and cerebellum, and studied the nerves and blood vessels as well.

Ancient anatomy reached its peak with Galen, a Greek physician who practised in Rome in the latter half of the second century. Galen worked up theories of bodily functions that were accepted as gospel for 1,500 years afterwards. But his notions about the human body were full of curious errors. This is understandable, for the ancients obtained most of their information from dissecting animals. Inhibitions of one kind or another made men uneasy about dissecting the human body.

In their denunciations of the pagan Greeks, early Christian writers accused them of having practised heartless vivisections on human beings. But this comes under the heading of polemical literature; not only is it doubtful that the Greeks did human vivisections, but obviously they did not even dissect enough dead bodies to learn much about the human anatomy. In any case, the Church's disapproval of dissection virtually put a stop to anatomical studies throughout the Middle Ages. As this period of history approached its end, anatomy began to revive in Italy. In 1316, an Italian anatomist, Mondino de Luzzi, wrote the first book to be devoted entirely to anatomy, and he is therefore known as the 'Restorer of Anatomy'.

The interest in naturalistic art during the Renaissance also fostered anatomical research. In the fifteenth century, Leonardo da Vinci performed some dissections by means of which he revealed new facts of anatomy, picturing them with the power of artistic genius. He showed the double curve of the spine and the sinuses that hollow the bones of the face and forehead. He used his studies to derive theories of physiology more advanced than Galen's. But Leonardo, though he was a genius in science as well as in art, had little influence on scientific thought in his time. Either from neurotic disinclination or from sober caution, he did not publish any of his scientific work, but kept it hidden in coded notebooks. It was left for later generations to discover his scientific achievements when his notebooks were finally published.

The French physician Jean Fernel was the first modern to take up dissection as an important part of a physician's duties. He published a book on the subject in 1542. However, his work was almost completely overshadowed by a much greater work published in the following year. This was the famous *De Humani Corporis Fabrica* (*Concerning the Structure of the Human Body*) of Andreas Vesalius, a Belgian who did most of his work in Italy. On the theory that the proper study of mankind was man, Vesalius dissected the appropriate subject and corrected many of Galen's errors. The drawings of the human anatomy in his book (which are reputed to have been made by Jan Stevenzoon van Calcar, a pupil of the artist Titian) are so beautiful and accurate that they are still republished today and will always stand as classics. Vesalius can be called the father of modern anatomy. His *Fabrica* was as revolutionary in its way as Copernicus's *De Revolutionibus Orbium Coelestiumi*, published in the very same year.

Just as the revolution initiated by Copernicus was brought to fruition by Galileo, so the one initiated by Vesalius came to a head in the crucial discoveries of William Harvey. Harvey was an English physician and experimentalist, of the same generation as Galileo and William Gilbert, the experimenter with magnetism. His particular interest was that vital body juice – the blood. What did it do in the body, anyway?

It was known that there were two sets of blood vessels: the veins and the arteries. (Praxagoras of Cos, a Greek physician of the third century B.C., had given the latter the name 'artery' from Greek words meaning 'I carry air', because these vessels were found to be empty in dead bodies. Galen had later shown that in life they carried blood.) It was also known that the heart-beat drove the blood in some sort of motion, for when an artery was cut, the blood gushed out in pulses that synchronized with the heart-beat.

Galen had proposed that the blood see-sawed to and fro in the blood vessels, travelling first in one direction through the body and then in the other. This theory required him to explain why the back-and-forth movement of the blood was not blocked by the wall between the two halves of the heart; Galen answered simply that the wall was riddled with invisibly small holes that let the blood through.

Harvey took a closer look at the heart. He found that each half was divided into two chambers, separated by a one-way valve that allowed blood to flow from the upper chamber ('auricle') to the lower ('ventricle'), but not vice versa. In other words, blood entering one of the auricles could be pumped into its corresponding ventricle and from there into blood vessels issuing from it, but there could be no flow in the opposite direction.

Harvey then performed some simple but beautifully clear-cut experiments to determine the direction of flow in the blood vessels. He would tie off an artery or a vein in a living animal to see on which side of this blockage the pressure within the blood vessel would build up. He found that when he stopped the flow in an artery, the vessel always bulged on the side between the heart and the block. This meant that the blood in arteries must flow in the direction away from the heart. When he tied a vein, the bulge was always on the other side of the block; therefore, the blood flow in veins must be towards the heart. Further evidence in favour of this one-way flow in veins rests in the fact that the larger veins contain valves that prevent blood from moving away from the heart. This had been discovered by Harvey's teacher, the Italian anatomist Hieronymus Fabrizzi (better known by his Latinized name, Fabricius). Fabricius, however, under the load of Galenic tradition, refused to draw the inevitable conclusion and left the glory to his English student.

Harvey went on to apply quantitative measurements to the blood flow (the first time anyone had applied mathematics to a biological problem). His measurements showed that the heart pumped out blood at such a rate that in twenty minutes its output equalled the total amount of blood contained in the body. It did not seem reasonable to suppose that the body could manufacture new blood, or consume the old, at any such rate. The logical conclusion, therefore, was that the blood must be recycled through the body. Since it flowed away from the heart in the arteries and towards the heart in the veins, Harvey decided that the blood was pumped by the heart into the arteries, then passed from them into the veins, then flowed back to the heart, then was pumped into the arteries again, and so on. In other words, it circulated continuously in one direction through the heart-and-blood-vessel system.

Earlier anatomists, including Leonardo da Vinci, had hinted at such an idea, but Harvey was the first to state and investigate the theory in detail. He set forth his reasoning and experiments in a small, badly printed book entitled *De Motus Cordis* (*Concerning the Motion of the Heart*), which was published in 1628 and has stood ever since as one of the great classics of science.

The main question left unanswered by Harvey's work was: How did the blood pass from the arteries into the veins? Harvey said there must be connecting vessels of some sort, though they were too small to be seen. This was reminiscent of Galen's theory about small holes in the heart wall, but whereas Galen's holes in the heart were never found and do not exist, Harvey's connecting vessels were confirmed as soon as a microscope became available. In 1661, just four years after Harvey's death, an Italian physician named Marcello Malpighi examined the lung tissues of a frog with a primitive microscope, and, sure enough, there were tiny blood vessels connecting the arteries with the veins. Malpighi named them 'capillaries', from a Latin word meaning 'hair-like'.

The use of the microscope made it possible to see other minute structures as well. The Dutch naturalist Jan Swammerdam discovered the red blood corpuscles, while the Dutch anatomist Regnier de Graaf discovered tiny 'ovarian follicles' in animal ovaries. Small creatures, such as insects, could be studied in detail.

Work in such fine detail encouraged the careful comparison of structures in one species with structures in others. The English botanist Nehemiah Grew was the first 'comparative anatomist' of note. In 1675, he published his studies comparing the trunk structure of various trees, and in 1681 studies comparing the stomachs of various animals.

The coming of the microscope introduced biologists, in fact, to a more basic level of organization of living things: a level at which all ordinary structures could be reduced to a common denominator. In 1665, the English scientist Robert Hooke, using a

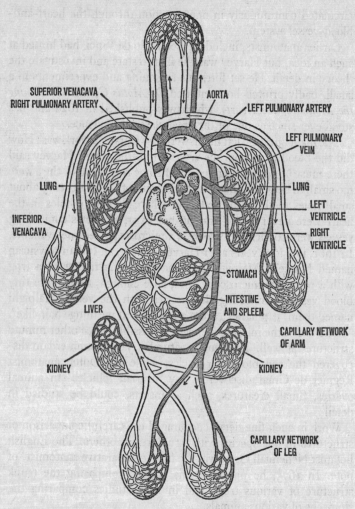

The circulatory system.

compound microscope of his own design, discovered that cork, the bark of a tree, was built of extremely tiny compartments, like a super-fine sponge. He called these holes 'cells', likening them to small rooms, such as the cells in a monastery. Other microscopists then found similar 'cells', but full of fluid, in living tissue.

Over the next century and a half it gradually dawned on biologists that all living matter was made up of cells and that each cell was an independent unit of life. Some forms of life – certain micro-organisms – consisted of only a single cell; the larger organisms were composed of many cooperating cells. One of the earliest to propose this view was the French physiologist René Joachim Henri Dutrochet. His report, published in 1824, went unnoticed, however, and the cell theory gained prominence only after Matthias Jakob Schleiden and Theodor Schwann of Germany independently formulated it in 1838 and 1839.

The colloidal fluid filling certain cells was named 'protoplasm' ('first form') by the Czech physiologist Jan Evangelista Purkinje in 1839, and the German botanist Hugo von Mohl extended the term to signify the contents of all cells. The German anatomist Max Johann Sigismund Schultze emphasized the importance of protoplasm as the 'physical basis of life' and demonstrated the essential similarity of protoplasm in all cells, both plant and animal, and in both very simple and very complex creatures.

The cell theory is to biology about what the atomic theory is to chemistry and physics. Its importance in the dynamics of life was established when, around 1860, the German pathologist Rudolf Virchow asserted, in a succinct Latin phrase, that all cells arose from cells. He showed that the cells in diseased tissue were produced by the division of originally normal cells.

By that time it had become clear that every living organism, even the largest, began life as a single cell. One of the earliest microscopists, Johann Ham, an assistant of Leeuwenhoek, had discovered in seminal fluid tiny bodies that were later named 'spermatozoa' (from Greek words meaning 'animal seed'). Much later, in 1827, the German physiologist Karl Ernst von Baer had identified the ovum, or egg cell, of mammals. Biologists came to realize that the union of an egg and a spermatozoon formed a fertilized ovum from which the animal eventually developed by repeated divisions and redivisions.

The big question was: How did cells divide? The answer lay

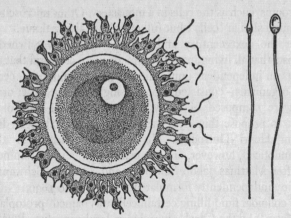

Human egg and sperm cells.

in a small globule of comparatively dense material within the cell, making up about a tenth its volume, first reported by Robert Brown (the discoverer of Brownian motion) in 1831 and named the 'nucleus'. (To distinguish it from the nucleus of the atom, I shall refer to it from now on as the 'cell nucleus'.)

If a one-celled organism was divided into two parts, one of which contained the intact cell nucleus, the part containing the cell nucleus was able to grow and divide, but the other part could not. (Later it was also learned that the red blood cells of mammals, lacking nuclei, are short-lived and have no capacity for either growth or division. For that reason, they are not considered true cells and are usually called 'corpuscles'.)

Unfortunately, further study of the cell nucleus and the mechanism of division was thwarted for a long time by the fact that the cell was more or less transparent, so that its substructures could not be seen. Then the situation was improved by the discovery that certain dyes would stain parts of the cell and not others. A dye called 'haematoxylin' (obtained from logwood) stained the cell nucleus black and brought it out prominently against the background of the cell. After Perkin and other chemists began to produce synthetic dyes, biologists found themselves with a variety of dyes from which to choose.

In 1879, the German biologist Walther Flemming found that with certain red dyes he could stain a particular material in the cell nucleus which was distributed through it as small granules. He called this material 'chromatin' (from the Greek word for 'colour'). By examining this material, Flemming was able to follow some of the changes in the process of cell division. To be sure, the stain killed the cell, but in a slice of tissue he would catch various cells at different stages of cell division. They served as still pictures, which he put together to form a kind of 'moving picture' of the progress of cell division.

In 1882, Flemming published an important book in which he described the process in detail. At the start of cell division, the chromatin material gathered itself together in the form of threads. The thin membrane enclosing the cell nucleus seemed to dissolve, and at the same time a tiny object just outside it divided in two. Flemming called this object the 'aster', from a Greek word for 'star', because radiating threads gave it a star-like appearance. After dividing, the two parts of the aster travelled to opposite sides of the cell. Its trailing threads apparently entangled the threads of chromatin, which had meanwhile lined up in the centre of the cell, and the aster pulled half the chromatin threads to one side of the cell, half to the other. As a result, the cell pinched in at the middle and split into two cells. A cell nucleus developed in each, and the chromatin material that the nuclear membrane enclosed broke up into granules again.

Flemming called the process of cell division 'mitosis', from the Greek word for 'thread', because of the prominent part played in it by the chromatin threads. In 1888, the German anatomist Wilhelm von Waldeyer gave the chromatin thread the name 'chromosome' (from the Greek for 'coloured body'), and that name has stuck. It should be mentioned, though, that chromosomes, despite their name, are colourless in their unstained natural state, and of course are then quite difficult to make out against the very similar background. (Nevertheless, even so, they had dimly been seen in flower cells as early as 1848 by the German amateur botanist Wilhelm Friedrich Benedict Hofmeister.)

Continued observation of stained cells showed that the cells of

Asimov's Guide to Science

each species of plant or animal had a fixed and characteristic number of chromosomes. Before a cell divides in two during mitosis, the number of chromosomes is doubled, so that each of the two daughter cells after the division has the same number as the original mother cell.

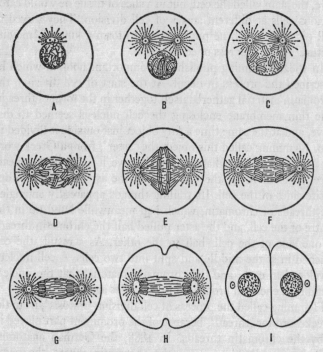

Division of a cell by mitosis.

The Belgian embryologist Eduard van Bededen discovered in 1885 that the chromosomes did *not* double in number when egg and sperm cells were being formed. Consequently each egg and each sperm cell had only half the number of chromosomes that ordinary cells of the organism possessed. (The cell division that produces sperm and egg cells therefore is called 'meiosis', from a

Greek word meaning 'to make less'.) When an egg and a sperm cell combined, however, the combination (the fertilized ovum) had a complete set of chromosomes, half contributed by the mother through the egg cell and half by the father through the sperm cell. This complete set was passed on by ordinary mitosis to all the cells that made up the body of the organism developing from the fertilized egg.

Even though the use of dyes makes the chromosomes visible, they do not make it easy to see one individual chromosome among the rest. Generally, they look like a tangle of stubby spaghetti. Thus, it was long thought that each human cell contained twenty-four pairs of chromosomes. It was not until 1956 that a more painstaking count of these cells (certainly earnestly studied) showed twenty-three pairs to be the correct count.

Fortunately, this problem no longer exists. A technique has been devised whereby treatment with a low-concentration salt-solution, in the proper manner, swells the cells and disperses the chromosomes. They can then be photographed, and that photograph can be cut into sections, each containing a separate chromosome. If these chromosomes are matched into pairs and then arranged in the order of decreasing length, the result is a 'karyotype', a picture of the chromosome-content of the cell, consecutively numbered.

The karyotype offers a subtle tool in medical diagnosis, for separation of the chromosomes is not always perfect. In the process of cell division, a chromosome may be damaged or even broken. Sometimes the separation may not be even, so that one of the daughter cells gets an extra chromosome while the other is missing one. Such abnormalities are sure to damage the working of the cell, often to such an extent that the cell cannot function. (This is what keeps the process of mitosis so seemingly accurate – not that it really is as accurate as it seems, but that the mistakes are buried.)

Such imperfections are particularly dire when they take place in the process of meiosis, for then egg cells or sperm cells are produced with imperfections in the chromosome-complement. If an organism can develop at all from such an imperfect start (and usually it cannot), every cell in its body has the imperfection: the result is a serious congenital disease.

The most frequent disease of this type involves severe mental retardation. It is called 'Down's syndrome' (because it was first described in 1866 by the English physician John Langdon Haydon Down), and it occurs once in every thousand births. It is more commonly known as 'mongolism', because one of the symptoms is a slant to the eyelids that is reminiscent of the epicanthic fold of the peoples of eastern Asia. Since the syndrome has no more to do with the Asians, however, than with others, this is a poor name.

It was not until 1959 that the cause of Down's syndrome was discovered. In that year, three French geneticists – Jérôme Jean Lejeune, M. Gautier, and P. Turpin – counted the chromosomes in cells from three cases and found that each had 47 chromosomes instead of 46. It turned out that the error was in the possession of *three* members of chromosome-pair 21. Then, in 1967, the mirror-image example of the disease was located. A mentally retarded three-year-old girl was found to have a single chromosome-21. She was the first discovered case of a living human being with a missing chromosome.

Cases of this sort involving other chromosomes seem less common but are turning up. Patients with a particular type of leukaemia show a tiny extra chromosome-fragment in their cells. This is called the 'Philadelphia chromosome' because it was first located in a patient hospitalized in that city. Broken chromosomes, in general, turn up with greater-than-normal frequency in certain not-very-common diseases.

Genes

In the 1860s an Austrian monk named Gregor Johann Mendel, who was too occupied with the affairs of his monastery to pay attention to the biologist's excitement about cell division, was quietly carrying through some experiments in his garden which were destined eventually to make sense out of chromosomes. Abbé Mendel was an amateur botanist, and he became particularly interested in the results of cross-breeding pea plants of varying characteristics. His great stroke of intuition was to study one clearly defined characteristic at a time.

He would cross plants with different seed colours (green or yellow), or smooth-seeded peas with wrinkle-seeded ones, or long-stemmed plants with short-stemmed ones, and then would follow the results in the offspring of the succeeding generations. Mendel kept a careful statistical record of his results, and his conclusions can be summarized essentially as follows:

1. Each characteristic was governed by 'factors' that (in the cases that Mendel studied) could exist in one of two forms. One version of the factor for seed colour, for instance, would cause the seeds to be green; the other form would make them yellow. (For convenience, let us use the present-day terms. The factors are now called 'genes', a term put forward in 1909 by the Danish biologist Wilhelm Ludwig Johannsen from a Greek word meaning 'to give birth to', and the different forms of a gene controlling a given characteristic are called 'alleles'. Thus the seed-colour gene possessed two alleles, one for green seeds, the other for yellow seeds.)

2. Every plant had a pair of genes for each characteristic, one contributed by each parent. The plant transmitted one of its pair to a germ cell, so that when the germ cell of two plants united by pollination, the offspring had two genes for the characteristic once more. The two genes might be either identical or alleles.

3. When the two parent plants contributed alleles of a particular gene to the offspring, one allele might overwhelm the effect of the other. For instance, if a plant producing yellow seeds was crossed with one producing green seeds, all the members of the next generation would produce yellow seeds. The yellow allele of the seed-colour gene was 'dominant', the green allele 'recessive'.

4. Nevertheless, the recessive allele was not destroyed. The green allele, in the case just cited, was still present, even though it produced no detectable effect. If two plants containing mixed

genes (i.e., each with one yellow and one green allele) were crossed, some of the offspring might have two green alleles in the fertilized ovum; in that case those particular offspring would produce green seeds, and the offspring of such parents in turn would also produce green seeds. Mendel pointed out that there were four possible ways of combining alleles from a pair of hybrid parents, each possessing one yellow and one green allele. A yellow allele from the first parent might combine with a yellow allele from the second; a yellow allele from the first might combine with a green allele from the second; a green allele from the first might combine with a yellow allele from the second; and a green allele from the first might combine with a green allele from the second. Of the four combinations, only the last would result in a plant that would produce green seeds. Assuming that all four combinations were equally probable, one fourth of the plants of the new generation should produce green seeds. Mendel found that this was indeed so.

5. Mendel also found that characteristics of different kinds – for instance, seed colour and flower colour – were inherited independently of each other. That is, red flowers were as apt to go with yellow seeds as with green seeds. The same was true of white flowers.

Mendel performed these experiments in the early 1860s, wrote them up carefully, and sent a copy of his paper to Karl Wilhelm von Nägeli, a Swiss botanist of great reputation. Von Nägeli's reaction was negative. Von Nägeli had, apparently, a predilection for all-encompassing theories (his own theoretical work was semi-mystical and turgid in expression), and he saw little merit in the mere counting of pea plants as a way to truth. Besides, Mendel was an unknown amateur.

It seems that Mendel allowed himself to be discouraged by von Nägeli's comments, for he turned to his monastery duties, grew fat (too fat to bend over in the garden), and abandoned his researches. He did, however, publish his paper in 1866 in a provincial Austrian journal, where it attracted no further attention for a generation.

But other scientists were slowly moving towards the same conclusions to which (unknown to them) Mendel had already come. One of the routes by which they arrived at an interest in genetics was the study of 'mutations', that is, of freak animals, or monsters, which had always been regarded as bad omens. (The word 'monster' came from a Latin word meaning 'warning'.) In 1791, a Massachusetts farmer named Seth Wright took a more practical view of a sport that turned up in his flock of sheep. A lamb was born with abnormally short legs, and it occurred to the shrewd Yankee that short-legged sheep could not escape over the low stone walls around his farm. He therefore deliberately bred a line of short-legged sheep from his not unfortunate accident.

This practical demonstration stimulated others to look for useful mutations. By the end of the nineteenth century the American horticulturist Luther Burbank was making a successful career of breeding hundreds of new varieties of plants which were improvements over the old in one respect or another, not only by mutations, but by judicious crossing and grafting.

Meanwhile botanists tried to find an explanation of mutation. And in what is perhaps the most startling coincidence in the history of science, no fewer than three men, independently, and in the very same year, came to precisely the same conclusions that Mendel had reached a generation earlier. They were Hugo De Vries of Holland, Karl Erich Correns of Germany, and Erich von Tschermak of Austria. None of them knew of each other's or Mendel's work. All three were ready to publish in 1900. All three, in a final check of previous publications in the field, came across Mendel's paper, to their own vast surprise. All three did publish in 1900, each citing Mendel's paper, giving Mendel full credit for the discovery, and advancing his own work only as confirmation.

A number of biologists immediately saw a connection between Mendel's genes and the chromosomes that could be seen under the microscope. The first to draw a parallel was an American cytologist named Walter S. Sutton, in 1904. He pointed out that chromosomes, like genes, came in pairs, one of which was inherited from the father and one from the mother. The only trouble with this analogy was that the number of chromosomes in the cells of any organism was far smaller than the number of inherited characteristics. Man, for instance, has only twenty-three pairs of chromosomes and yet certainly possesses thousands of inheritable characteristics. Biologists therefore had to conclude that chromosomes were not genes. Each must be a collection of genes.

In short order, biologists discovered an excellent tool for studying specific genes. It was not a physical instrument but a new kind of laboratory animal. In 1906, the Columbia University zoologist Thomas Hunt Morgan, who was at first sceptical of Mendel's theories, conceived the idea of using fruit flies (*Drosophila melanogaster*) for research in genetics. (The term 'genetics' was coined in 1902 by the British biologist William Bateson.)

Fruit flies had considerable advantages over pea plants (or any ordinary laboratory animal) for studying the inheritance of genes. They bred quickly and prolifically, could easily be raised by the hundreds on very little food, had scores of inheritable characteristics which could be observed readily, and had a comparatively simple chromosomal set-up – only four pairs of chromosomes per cell.

With the fruit fly, Morgan and his co-workers discovered an important fact about the mechanism of inheritance of sex. They found that the female fruit fly has four perfectly matched pairs of chromosomes so that all the egg cells, receiving one of each pair, are identical so far as chromosome make-up is concerned. However, in the male one of each of the four pairs consists of a normal chromosome, called the 'X chromosome', and a stunted one, which was named the 'Y chromosome'. Therefore when sperm cells are formed, half have an X chromosome and half a Y chromosome. When a sperm cell with the X chromosome fertilizes an egg cell, the fertilized egg, with four matched pairs, naturally becomes a female. On the other hand, a sperm cell with a Y chromosome produces a male. Since both alternatives are equally probable, the number of males and females in the typical species of living things is roughly equal. (In some creatures, notably various birds, it is the female that has a Y chromosome.)

This chromosomal difference explains why some disorders or

mutations show up only in the male. If a defective gene occurs on one of a pair of X chromosomes, the other member of the pair is still likely to be normal and can salvage the situation. But in the male, a defect on the X chromosome paired with the Y chromosome generally cannot be compensated for, because the latter carries very few genes. Therefore the defect shows up.

The most notorious example of such a 'sex-linked disease' is 'haemophilia', a condition in which blood clots only with difficulty, if at all. Individuals with haemophilia run the constant risk of bleeding to death from slight causes or of suffering agonies from internal bleeding. A woman who carries a gene that will produce haemophilia on one of her X chromosomes is very likely to have a normal gene at the same position in the other X chromosome. She will therefore not show the disease. She will, however, be a 'carrier'. Of the egg cells she forms, half will have the normal X chromosome and half the haemophiliac X chromosome. If the egg with the abnormal X chromosome is fertilized by sperm with an X chromosome from a normal male, the resulting child will be a girl who will not be haemophiliac but who will again be a carrier; if it is fertilized by sperm with a Y chromosome from a normal male, the haemophiliac gene in the ovum will not be counteracted by anything in the Y chromosome, and the result is a boy with haemophilia. By the laws of chance, half the sons of haemophilia-carriers will be haemophiliacs; half the daughters will be, in their turn, carriers.

The most eminent haemophilia-carrier in history was Queen Victoria. Only one of her four sons (the oldest, Leopold) was haemophiliac. Edward VII – from whom later British monarchs descended – escaped, so there is no haemophilia now in the British royal family. However, two of Victoria's daughters were carriers. One had a daughter (also a carrier) who married Tsar Nicholas II of Russia. As a result, their only son was a haemophiliac; this helped alter the history of Russia and the world, for it was through his influence on the haemophiliac that the monk Gregory Rasputin gained power in Russia and helped bring on the discontent that eventually led to revolution. The other daughter of Victoria had a daughter (also a carrier) who married into the royal house of Spain, producing haemophilia there. Because of its presence among the Spanish Bourbons and the Russian Romanoffs, haemophilia was sometimes called the 'royal disease', but it has no particular connection with royalty, except for Victoria's misfortune.

A lesser sex-linked disorder is colour-blindness, which is far more common among men than among women. Actually, the absence of one X chromosome may produce sufficient weakness among men generally as to account for the fact that where women are protected against death from childbirth infections they tend to live some three to seven years longer, on the average, than men. That twenty-third complete pair makes women the sounder biological organism, in a way.

The X and Y chromosomes (or 'sex chromosomes') are arbitrarily placed at the end of the karyotype, even though the X chromosome is among the longest. Apparently chromosome abnormalities, such as those involved in Down's syndrome, are more common among the sex chromosomes than among the others. This may not be because the sex chromosomes are most likely to be involved in abnormal mitoses, but perhaps because sex-chromosome abnormalities are less likely to be fatal, so that more young manage to be born with them.

The type of sex-chromosome abnormality that has drawn the most attention is one in which a male ends up with an extra Y chromosome in his cells, so that he is XYY, so to speak. It turns out that XYY males are difficult to handle. They are tall, strong, and bright, but are characterized by a tendency to rage and violence. Richard Speck, who killed eight nurses in Chicago in 1966, is supposed to have been an XYY. A murderer was acquitted in Australia in October 1968 on the grounds that he was an XYY and therefore not responsible for his action. Nearly 4 per cent of the male inmates in a certain Scottish prison have turned out to be XYY, and there are some estimates that XYY combinations may occur in as many as 1 man in every 3,000.

There seems to be some reason for considering it desirable to run a chromosome check on everyone and certainly on every newborn child. As is the case of other procedures, simple in theory but tedious in practice, attempts are being made to computerize such a process.

Research on fruit flies showed that traits were not necessarily inherited independently, as Mendel had thought. It happened that the seven characteristics of pea plants that he had studied were governed by genes on separate chromosomes. Morgan found that where two genes governing two different characteristics

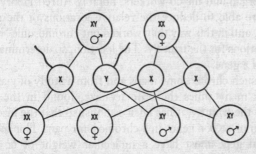

Combinations of X and Y chromosomes.

were located on the same chromosome, those characteristics were generally inherited together (just as a passenger in the front seat of a car and one in the back seat travel together).

This genetic linkage is not, however, unchangeable. Just as a passenger can change cars, so a piece of one chromosome occasionally switches to another, swapping places with a piece from the other. Such 'crossing over' may occur during the division of a cell. As a result, linked traits are separated and reshuffled in a new linkage. For instance, there is a variety of fruit fly with scarlet eyes and curly wings. When it is mated with a white-eyed, miniature-winged fruit fly, the offspring will generally be either red-eyed and curly-winged or white-eyed and miniature-winged. But the mating may sometimes produce a white-eyed, curly-winged fly or a red-eyed, miniature-winged one as a result of crossing over. The new form will persist in succeeding generations unless another crossing over takes place.

Now picture a chromosome with a gene for red eyes at one end and a gene for curly wings at the other end. Let us say that in the middle of the chromosome's length there are two adjacent genes governing two other characteristics. Obviously, the probability of a break occurring at that particular point, separating these two genes, is smaller than the probability of a break coming at one of the many points along the length of the chromosome that would separate the genes at the opposite ends. By noting the frequency of separation of given pairs of linked characteristics by crossing over, Morgan and his co-workers, notably Alfred Henry Sturtevant, were able to deduce the relative locations of the genes in question, and in this way they worked out chromosome 'maps' of gene locations for the fruit fly. The location, so determined, is the 'locus' of a gene.

From such chromosome maps and from a study of giant chromosomes many times the ordinary size, found in the salivary glands of the fruit fly, it has been established that the insect has a minimum of 10,000 genes in a chromosome pair. This means the individual gene must have a molecular weight of 60 million. Working from this, man's somewhat larger chromosomes may contain from 20,000 to 90,000 genes per chromosome pair, or up to 2 million altogether.

For his work on the genetics of fruit flies, Morgan received the Nobel Prize in medicine and physiology in 1933.

Increasing knowledge of genes raises hopes that the genetic endowment of individual humans might some day be analysed and modified: either preventing seriously anomalous conditions from developing, or correcting them if they slip by. Such 'genetic engineering' would require human chromosome-maps, clearly a tremendously larger job than in the case of the fruit fly. The task was made somewhat simpler in a startling way in 1967, when Howard Green of New York University formed hybrid cells containing both mouse and human chromosomes. Relatively few human chromosomes persisted after several cell divisions, and the effects due to their activity were more easily pinpointed.

Another step in the direction of gene-knowledge and gene-manipulation came in 1969, when the American biochemist Jonathan Beckwith and his co-workers isolated an individual gene for the first time in history. It was from an intestinal bacterium, and it controlled an aspect of sugar metabolism.

Every once in a while, with a frequency which can be calculated, a sudden change occurs in a gene. The mutation shows itself by some new and unexpected physical characteristic, such as the short legs of Farmer Wright's lamb. Mutations in nature are comparatively rare. In 1926, the geneticist Hermann Joseph Muller, who had been a member of Morgan's research team, discovered a way to increase the rate of mutations artificially in fruit flies so that the inheritance of such changes could be studied more easily. He found that X-rays would do the trick; presumably, they damaged the genes. The study of mutations made possible by Muller's discovery won him the Nobel Prize in medicine and physiology in 1946.

As it happens, Muller's researches have given rise to some rather disquieting thoughts concerning the future of the human species. While mutations are an important driving force in evolution, occasionally producing an improvement that enables a species to cope better with its environment, the beneficial mutation is very much the exception. Most mutations – at least 99 per cent of them – are detrimental, some even lethal. Eventually, even those that are only slightly harmful die out, because their bearers do not get along as well and leave fewer descendants than healthy individuals do. But in the meantime a mutation may cause illness and suffering for many generations. Furthermore, new mutations keep cropping up continually, and every species carries a constant load of defective genes.

The great number of different gene varieties – including large quantities of seriously harmful ones – in normal populations was clearly shown by the work of the Russian-American geneticist Theodosius Dobzhansky in the 1930s and 1940s. It is this diversity that makes evolution march on as it does, but it is the number of deleterious genes (the 'genetic load') that gives rise to some fears for the future.

Two modern developments seem to be adding steadily to this load. First, the advances in medicine and social care tend to compensate for the handicaps of people with detrimental mutations, at least so far as the ability to reproduce is concerned. Glasses are available to individuals with defective vision; insulin keeps alive sufferers from diabetes (a hereditary disease), and so on. Thus they pass on their defective genes to future generations. The alternatives – allowing defective individuals to die young or sterilizing or imprisoning them – are, of course, unthinkable, except where the handicap is sufficiently great to make the individual less than human, as in idiocy or homicidal paranoia. Undoubtedly, the human species can still bear its load of negatively mutated genes, despite its humanitarian impulses.

Crossing over in chromosomes.

But there is less excuse for the second modern hazard – namely, adding to the load by unnecessary exposure to radiation. Genetic research shows incontrovertibly that for the population as a whole even a slight increase in general exposure to radiation means a corresponding slight increase in the mutation rate. And since 1895 mankind has been exposed to types and intensities of radiation of which previous generations knew nothing. Solar radiation, the natural radioactivity of the soil, and cosmic rays have always been with us. Now, however, we use X-rays in medicine and dentistry with abandon; we concentrate radioactive material; we form artificially radioactive isotopes of terrifying radiant potency; we even explode nuclear bombs. All of this increases the background radiation.

No one, of course, suggests that research in nuclear physics be abandoned, or that X-rays never be used by the docto and dentist. There is, however, a strong recommendation that the danger be recognized and that exposure to radiation be minimized; that, for instance, X-rays be used with discrimination and care and that the sexual organs be routinely shielded during all such use. Another suggested precaution is that each individual keep a record of his total accumulated exposure to X-rays, so that he will have some idea of whether he is in danger of exceeding a reasonable limit.

Of course, the geneticists could not be sure that the principles established by experiments on plants and insects necessarily applied to man. After all, man was neither a pea plant nor a fruit fly. But direct studies of certain characteristics in man showed that human genetics did follow the same rules. The best-known example is the inheritance of blood types.

Blood transfusion is a very old practice, and early physicians occasionally even tried to transfuse animal blood into persons weakened by loss of blood. But transfusions even of human blood often turned out badly, so that laws were sometimes passed forbidding transfusion. In the 1890s, the Austrian pathologist Karl Landsteiner finally discovered that human blood came in different types, some of which were incompatible with each other. He found that sometimes when blood from one person was mixed with a sample of serum (the blood fluid remaining after the red cells and a clotting factor are removed) from another person, the red cells of the first person's whole blood would clump together. Obviously such a mixture would be very bad if it occurred in transfusion, and it might even kill the patient if the clumped cells blocked the blood circulation in key vessels. Landsteiner also found, however, that some bloods could be mixed without causing any deleterious clumping.

By 1902, Landsteiner was able to announce that there were four types of human blood, which he called A, B, AB, and O. Any given individual had blood of just one of these types. Of course, a particular type could be transferred without danger from one person to another having the same type. In addition, O blood could safely be transfused to a person possessing any of the other three types, and either A blood or B blood could be given to an AB patient. But red-cell clumping ('agglutination') would result when AB blood was transfused to an A or B individual, A and B were mixed, or when an O individual received a transfusion of any blood other than O. (Nowadays, because of possible serum reactions, in good practice patients are given only blood of their own type.)

In 1930, Landsteiner (who by then had become a United States citizen) received the Nobel Prize in medicine and physiology.

Geneticists have established that these blood types (and all the others since discovered, including the Rh variations) are inherited in a strictly Mendelian manner. It seems that there are three gene alleles, responsible respectively for A, B, and O blood. If both parents have O-type blood, all the children of that union will have O-type blood. If one parent is O-type and the other A-type, all the children may show A-type blood, for the A allele is dominant over the O. The B allele likewise is dominant over the O allele. The B allele and A allele, however, show no dominance with respect to each other, and an individual possessing both alleles has AB-type blood.

The Mendelian rules work out so strictly that blood groups can be (and are) used to test paternity. If an O-type mother has a B-type child, the child's father must be B-type, for that B allele must have come from somewhere. If the woman's husband happens to be A or O, it is clear that she has been unfaithful (or there has been a baby mix-up at the hospital). If an O-type woman with a B-type child accuses an A or O man of being the parent, she is either mistaken or lying. On the other hand, while blood type can sometimes prove a negative, it can never prove a positive. If the woman's husband, or the man accused, is indeed a B-type, the case remains unproved. Any B-type man, or any AB-type man, could have been the father.

The applicability of the Mendelian rules of inheritance to human beings has also been borne out by the existence of sex-linked traits. Colour blindness and haemophilia (a hereditary failure of the blood to clot) are found almost exclusively in males, and they are inherited in precisely the manner that sex-linked characteristics are inherited in the fruit fly.

Naturally, the thought will arise that, by forbidding people with such afflictions to have children, the disorder can be wiped out. By directing proper mating, the human breed might even be improved, as breeds of cattle have been. This is by no means a new idea. The ancient Spartans believed this and tried to put it into practice 2,500 years ago. In modern times, the notion was revived by an English scientist, Francis Galton (a cousin of Charles Darwin). In 1883, he coined the word 'eugenics' to describe his scheme. (The derivation of the word is from the Greek and means 'good birth'.)

Galton was not aware, in his time, of the findings of Mendel. He did not understand that characteristics might seem to be absent, yet be carried as recessives. He did not understand that groups of characteristics would be inherited intact and that it might be difficult to get rid of an undesirable one without also getting rid of a desirable one. Nor was he aware that mutations would reintroduce undesirable characteristics in every generation.

Human genetics is an enormously complicated subject that is not likely to be completely or neatly worked out in the foreseeable future. Because man breeds neither as frequently nor as prolifically as the fruit fly; because his matings cannot be subjected to laboratory control for experimental purposes; because he has many more chromosomes and many more inherited characteristics than the fruit fly; because the human characteristics in which we are most interested, such as creative genius, intelligence, and moral strength, are extremely complex, involving the interplay of numerous genes plus environmental influences – for all these reasons, geneticists cannot deal with human genetics with the same confidence with which they study fruit-fly genetics.

Eugenics remains a dream, therefore, made hazy and insubstantial by lack of knowledge. Those who are today most articulate in favour of elaborate eugenic programmes tend to be racists or eccentrics.

Just how does a gene bring the physical characteristic for which it is responsible into being? What is the mechanism whereby it gives rise to yellow seeds in pea plants, or curled wings in fruit flies, or blue eyes in human beings?

Biologists are now certain that genes exert their effects by way of enzymes. One of the clearest cases in point involves the colour of eyes, hair, and skin. The colour (blue or brown, yellow or black, pink or brown, or shades in between) is determined by the amount of pigment, called 'melanin' (from the Greek word for 'black'), that is present in the eye's iris, the hair, or the skin. Now melanin is formed from an amino acid, tyrosine, by way of a number of steps, most of which have now been worked out. A number of enzymes are involved, and the amount of melanin formed will depend upon the quantity of these enzymes. For instance, one of the enzymes, which catalyses the first two steps, is tyrosinase. Presumably some particular gene controls the production of tyrosinase by the cells. In that way, it will control the colouring of the skin, hair, and eyes. And since the gene is transmitted from generation to generation, children will naturally resemble their parents in colouring. If a mutation happens to produce a defective gene that cannot form tyrosinase, there will be no melanin, and the individual will be an 'albino'. The absence of a single enzyme (and hence the deficiency of a single gene) will thus suffice to bring about a major change in personal characteristics.

Granted that an organism's characteristics are controlled by its enzyme make-up, which in turn is controlled by genes, the next question is: How do the genes work? Unfortunately, even the fruit fly is much too complex an organism to trace out the matter in detail. But, in 1941, the American biologists George Wells Beadle and Edward Lawrie Tatum began such a study with a simple organism which they found admirably suited to this purpose. It is the common pink bread mould (scientific name, *Neurospora crassa*).

Neurospora is not very demanding in its diet. It will grow very well on sugar plus inorganic compounds that supply nitrogen, sulphur, and various minerals. Aside from sugar, the only organic substance that has to be supplied to it is a vitamin called 'biotin'.

At a certain stage in its life cycle, the mould produces eight spores, all identical in genetic constitution. Each spore contains seven chromosomes; as in the sex cell of a higher organism, its chromosomes come singly, not in pairs. Consequently, if one of its chromosomes is changed, the effect can be observed, because there is no normal partner present to mask the effect. Beadle and Tatum therefore were able to create mutations in *Neurospora* by exposing the mould to X-rays and then could follow the specific effects in the behaviour of the spores.

If, after the mould had received a dose of radiation, the spores still thrived on the usual medium of nutrients, clearly no mutation had taken place, at least so far as the organism's nutritional requirements for growth were concerned. If the spores would not grow on the usual medium, the experimenters proceeded to determine whether they were alive or dead, by feeding them a complete medium containing all the vitamins, amino acids, and other items they might possibly need. If the spores grew on this, the conclusion was that the X-rays had produced a mutation that had changed *Neurospora*'s nutritional requirements. Apparently it now needed at least one new item in its diet. To find out what that was, the experimenters tried the spores on one diet after another, each time with some items of the complete medium missing. They might omit all the amino acids, or all the various vitamins, or all but one or two amino acids or one or two vitamins. In this way they narrowed down the requirements until they identified just what it was that the spore now needed in its diet before the mutation.

It turned out sometimes that the mutated spore required the amino acid arginine. The normal, 'wild strain' had been able to manufacture its own arginine from sugar and ammonium salts. Now, thanks to the genetic change, it could no longer synthesize arginine, and unless this amino acid was supplied in its diet, it could not make protein and therefore could not grow.

The clearest way to account for such a situation was to suppose that the X-rays had disrupted a gene responsible for the formation of an enzyme necessary for manufacturing arginine. For lack of the normal gene, *Neurospora* could no longer make the enzyme. No enzyme, no arginine.

Beadle and his co-workers went on to use this sort of information to study the relation of genes to the chemistry of metabolism.

There was a way to show, for instance, that more than one gene was involved in the making of arginine. For simplicity's sake, let us say there are two – gene A and gene B – responsible for the formation of two different enzymes, both of which are necessary for the synthesis of arginine. Then a mutation of either gene A or gene B will rob *Neurospora* of the ability to make the amino acid. Suppose we irradiate two batches of *Neurospora* and produce an arginine-less strain in each one. If we are lucky, one mutant may have a defective A gene and a normal B gene, the other a normal A and defective B. To see if that has happened, let us cross the two mutants at the sexual stage of their life cycle. If the two strains do indeed differ in this way, the recombination of chromosomes may produce some spores whose A and B genes are both normal. In other words, from two mutants that are incapable of making arginine, we will get some offspring that *can* make it. Sure enough, exactly that sort of thing happened when the experiments were performed.

It was possible to explore the metabolism of *Neurospora* in finer detail than this. For instance, here were three different mutant strains incapable of making arginine on an ordinary medium. One would grow only if it was supplied with arginine itself. The second would grow if it received either arginine or a very similar compound called citrulline. The third could grow on arginine or citrulline or still another similar compound called ornithine.

What conclusion would you draw from all this? Well, we can guess that these three substances are steps in a sequence of which arginine is the final product. Each requires an enzyme. First ornithine is formed from some simpler compound with the help of an enzyme, then another enzyme converts ornithine to citrulline, and finally a third enzyme converts citrulline to arginine. (Actually, chemical analysis shows that each of the three is slightly more complex than the one before.) Now a *Neurospora* mutant that lacks the enzyme for making ornithine but possesses the other enzymes can get along if it is supplied with ornithine, for from it the spore can make citrulline and then the essential arginine. Of course, it can also grow on citrulline, from which it can make arginine, and on arginine itself. By the same token, we can reason that the second mutant strain lacks the enzyme needed to convert ornithine to citrulline. This strain therefore must be provided with citrulline, from which it can make arginine, or with arginine itself. Finally, we can conclude that the mutant that will grow only on arginine has lost the enzyme (and gene) responsible for converting citrulline to arginine.

By analysing the behaviour of the various mutant strains they were able to isolate, Beadle and his co-workers founded the science of 'chemical genetics'. They worked out the course of synthesis of many important compounds by organisms. Beadle proposed what has become known as the 'one-gene-one-enzyme theory' – that is, that every gene governs the formation of a single enzyme – a suggestion that is now generally accepted by geneticists. For their pioneering work, Beadle and Tatum shared in the Nobel Prize in medicine and physiology in 1958.

Beadle's discoveries put biochemists on the *qui vive* for evidence of gene-controlled changes in proteins, particularly in human mutants, of course. A case turned up, unexpectedly, in connection with the disease called 'sickle-cell anaemia'.

This disease had first been reported in 1910 by a Chicago physician named James Bryan Herrick. Examining a sample of blood from a Negro teenage patient under the microscope, he found that the red cells, normally round, had odd, bent shapes, many of them resembling the crescent shape of a sickle. Other physicians began to notice the same peculiar phenomenon, almost always in Negro patients. Eventually investigators decided that sickle-cell anaemia was a hereditary disease. It followed the Mendelian laws of inheritance. Apparently there is a sickle-cell gene that, when inherited in double dose from both parents, produces these distorted red cells. Such cells are unable to carry oxygen properly and are exceptionally short-lived, so there is a shortage of red cells in the blood. Those who inherit the double dose tend to die of the disease in childhood. On the other hand, when a person has only one sickle-cell gene, from one of his parents, the disease does not appear. Sickling of his red cells shows up only when the person is deprived of oxygen to an unusual degree, as at high altitudes. Such people are considered to have the 'sickle-cell trait', but not the disease.

It was found that about 9 per cent of the Negroes in America have the trait, and 0·25 per cent have the disease. In some localities in Central Africa as much as a quarter of the Negro population shows the trait. Apparently the sickle-cell gene arose as a mutation in Africa and has been inherited ever since by individuals of African descent. If the disease is fatal, why has the defective gene not died out? Studies in Africa during the 1950s turned up the answer. It seems that people with the sickle-cell trait tend to have greater immunity to malaria than do normal individuals. The sickle cells are somehow inhospitable to the malarial parasite. It is estimated that in areas infested with malaria, children with the trait have a 25 per cent better chance of surviving to child-bearing age than those without the trait have. Hence, possessing a single dose of the sickle-cell gene (but not the anaemia-causing double dose) confers an advantage. The two opposing tendencies – promotion of the defective gene by the protective effect of the single dose and elimination of the gene by its fatal effect in double dose – tend to produce an equilibrium which maintains the gene at a certain level in the population.

In regions where malaria is not an acute problem, the gene does tend to die out. In America, the incidence of sickle-cell genes among Negroes may have started as high as 25 per cent. Even allowing for a reduction to an estimated 15 per cent by admixture with non-Negro individuals, the present incidence of only 9 per cent shows that the gene is dwindling away. In all probability it will continue to do so. If Africa is freed of malaria, it will presumably dwindle there, too.

The biochemical significance of the sickle-cell gene suddenly came into prominence in 1949 when Linus Pauling and his co-workers at the California Institute of Technology (where Beadle also was working) showed that the gene affected the haemoglobin of the red blood cells. Persons with a double dose of the sickle-cell gene were unable to make normal haemoglobin. Pauling proved this by means of the technique called 'electrophoresis', a method that uses an electric current to separate proteins by virtue of differences in the net electric charge on the various protein molecules. (The electrophoretic technique was developed by the Swedish chemist Arne Wilhelm Kaurin Tiselius, who received the Nobel Prize in chemistry in 1948 for this valuable contribution.) Pauling, by electrophoretic analysis, found that patients with sickle-cell anaemia had an abnormal haemoglobin (named 'haemoglobin S'), which could be separated from normal haemoglobin. The normal kind was given the name haemoglobin A (for 'adult') to distinguish it from a haemoglobin in foetuses, called haemoglobin F.

Since 1949, biochemists have discovered a number of other abnormal haemoglobins besides the sickle-cell one, and they are lettered from haemoglobin C to haemoglobin M. Apparently, the gene responsible for the manufacture of haemoglobin has been mutated into many defective alleles, each giving rise to a haemoglobin that is inferior for carrying out the functions of the molecule in ordinary circumstances but perhaps helpful in some unusual condition. Thus, just as haemoglobin S in a single dose improves resistance to malaria, so haemoglobin C in a single dose improves the ability of the body to get along on marginal quantities of iron.

Since the various abnormal haemoglobins differ in electric charge, they must differ somehow in the arrangement of amino acids in the peptide chain, for the amino-acid make-up is responsible for the charge pattern of the molecule. The differences must be very small, because the abnormal haemoglobins all function as haemoglobin after a fashion. The hope of locating the difference in a huge molecule of some 600 amino acids was correspondingly small. Nevertheless, the German-American biochemist Vernon Martin Ingram and co-workers tackled the problem of the chemistry of the abnormal haemoglobins.

They first broke down haemoglobin A, haemoglobin S, and haemoglobin C into peptides of various sizes by digesting them with a protein-splitting enzyme. Then they separated the fragments of each haemoglobin by 'paper electrophoresis' – that is, using the electric current to convey the molecules along a moistened piece of filter paper instead of through a solution. (We can think of this as a kind of electrified paper chromatography.) When the investigators had done this with each of the three haemoglobins, they found that the only difference among them was that a single peptide turned up in a different place in each case.

They proceeded to break down and analyse this peptide. Eventually they learned that it was composed of nine amino acids and that the arrangement of these nine was exactly the same in all three haemoglobins except at one position. The respective arrangements were:

Haemoglobin A: His-Val-Leu-Leu-Thr-Pro-Glu-Glu-Lys
Haemoglobin S: His-Val-Leu-Leu-Thr-Pro-Val-Glu-Lys
Haemoglobin C: His-Val-Leu-Leu-Thr-Pro-Lys-Glu-Lys

As far as could be told, the only difference among the three haemoglobins lay in that single amino acid in the seventh position in the peptide: it was glutamic acid in haemoglobin A, valine in haemoglobin S, and lysine in haemoglobin C. Since glutamic acid gives rise to a negative charge, lysine to a positive charge, and valine to no charge at all, it is not surprising that the three proteins behave differently in electrophoresis. Their charge pattern is different.

But why should so slight a change in the molecule result in so drastic a change in the red cell? Well, the normal red cell is one third haemoglobin A. The haemoglobin A molecules are packed so tightly in the cell that they just barely have room for free movement. In short, they are on the point of precipitating out of solution. Part of the influence that determines whether a protein is to precipitate out or not is the nature of its charge. If all the proteins have the same net charge, they repel one another and keep from precipitating. The greater the charge (i.e., the repulsion), the less likely the proteins are to precipitate. In haemoglobin S the intermolecular repulsion may be slightly less than in haemoglobin A, and haemoglobin S is correspondingly less soluble and more likely to precipitate. When a sickle-cell is paired with a normal gene, the latter may form enough haemoglobin A to keep the haemoglobin S in solution, though it is a near squeak.

But when both of the genes are sickle-cell mutants, they will produce only haemoglobin S. This molecule cannot remain in solution. It precipitates out into crystals, and they distort and weaken the red cell.

This theory would explain why the change of just one amino acid in each half of a molecule made up of nearly 600 is sufficient to produce a serious disease and the near-certainty of an early death.

Albinism and sickle-cell anaemia are not the only human defects that have been traced to the absence of a single enzyme or the mutation of a single gene. There is phenylketonuria, a hereditary defect of metabolism, which often causes mental retardation. It results from the lack of an enzyme needed to convert the amino acid phenylalanine to tyrosine. There is galactosaemia, a disorder causing eye cataracts and damage to the brain and liver, which has been traced to the absence of an enzyme required to convert a galactose phosphate to a glucose phosphate. There is a defect, involving the lack of one or another of the enzymes that control the breakdown of glycogen (a kind of starch) and its conversion to glucose, which results in abnormal accumulations of glycogen in the liver and elsewhere and usually leads to early death. These are examples of 'inborn errors of metabolism', a congenital lack of the capacity to form some more or less vital enzyme found in normal human beings. This concept was first introduced to medicine by the English physician Archibald Edward Garrod in 1908, but it lay disregarded for a generation until, in the mid 1930s, the English geneticist John Burdon Sanderson Haldane brought the matter to the attention of scientists once more.

Such disorders are generally governed by a recessive allele of the gene that produces the enzyme involved. When only one of a pair of genes is defective, the normal one can carry on, and the individual is usually capable of leading a normal life (as in the case of a possessor of the sickle-cell trait). Trouble generally comes only when two parents happen to have the same unfortunate gene and have the further bad luck of combining those two in a fertilized egg. Their child then is the victim of a kind of Russian roulette. Probably all of us carry our load of abnormal, defective, even dangerous genes, usually masked by normal genes. You can understand why the human geneticists are so concerned about radiation or anything else that may increase the mutation-rate and add to the load.

Nucleic Acids

The really remarkable thing about heredity is not these spectacular, comparatively rare aberrations, but the fact that by and large inheritance runs so strictly true to form. Generation after generation, millennium after millennium, the genes go on reproducing themselves in exactly the same form and generating exactly the same enzymes, with only an occasional accidental variation of the blueprint. They rarely fail by so much as the introduction of a single wrong amino acid in a large protein molecule. How do they manage to make true copies of themselves over and over again with such astounding faithfulness?

A gene is built of two major components. Perhaps half of it is protein, but the other part is not. To this non-protein portion we must now direct our attention.

In 1869, a Swiss biochemist named Friedrich Miescher, while breaking down the protein of cells with pepsin, discovered that the pepsin did not break up the cell nucleus. The nucleus shrank some, but it remained intact. By chemical analysis, Miescher then found that the cell nucleus consisted largely of a phosphorus-containing substance that did not at all resemble protein in its properties. He called the substance 'nuclein'. It was renamed 'nucleic acid' twenty years later when the substance was found to be strongly acid.

Miescher devoted himself to a study of this new material and eventually discovered that sperm cells (which have very little material outside the cell nucleus) were particularly rich in nucleic acid. Meanwhile, the German chemist Felix Hoppe-Seyler, in whose laboratories Miescher had made his first discovery, and who had personally confirmed the young man's work before allowing it to be published, isolated nucleic acid from yeast cells. This seemed different in properties from Miescher's material, so Miescher's variety was named 'thymus nucleic acid' (because it could be obtained with particular ease from the thymus gland of animals), and Hoppe-Seyler's, naturally, was called 'yeast nucleic acid'. Since thymus nucleic acid was at first derived only from animal cells and yeast nucleic acid only from plant cells, it was thought for a while that this might represent a general chemical distinction between animals and plants.

The German biochemist Albrecht Kossel, another pupil of Hoppe-Seyler, was the first to make a systematic investigation of the structure of the nucleic-acid molecule. By careful hydrolysis, he isolated from it a series of nitrogen-containing compounds, which he named 'adenine', 'guanine', 'cytosine', and 'thymine'. Their formulas are now known to be:

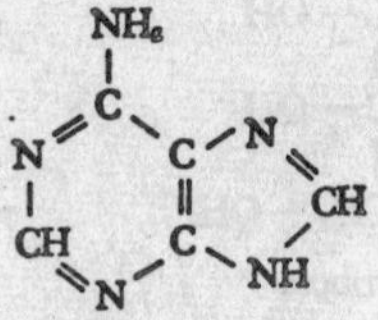

adenine

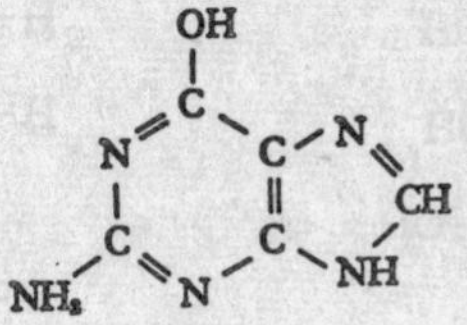

guanine

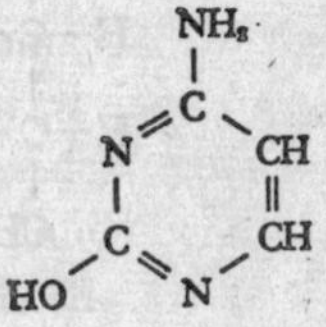

cytosine

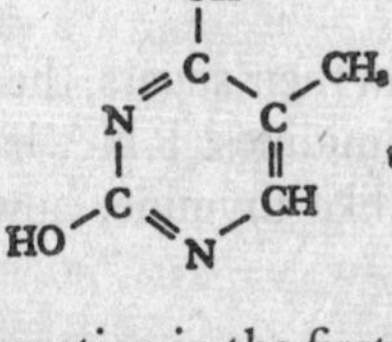

thymine

The double-ring formation in the first two compounds is called the 'purine ring' and the single ring in the other two is the 'pyrimidine ring'. Therefore adenine and guanine are referred to as purines, and cytosine and thymine are pyrimidines.

For these researches, which started a very fruitful train of discoveries, Kossel received the Nobel Prize in medicine and physiology in 1910.

In 1911, the Russian-born American biochemist Phoebus Aaron Theodore Levene, a pupil of Kossel, carried the investigation a stage further. Kossel had discovered in 1891 that nucleic acids contained carbohydrate, but now Levene showed that the nucleic acids contained five-carbon sugar molecules. (This was, at the time, an unusual finding. The best-known sugars, such as glucose, contain six carbons.) Levene followed this by showing that the two varieties of nucleic acid differed in the nature of the five-carbon sugar. Yeast nucleic acid contained 'ribose', while thymus nucleic acid contained a sugar very much like ribose except for the absence of one oxygen atom, so this sugar was called 'deoxyribose'. Their formulas are:

$$
\begin{array}{ccc}
O\!\!\!\diagdown_{CH} & & O\!\!\!\diagdown_{CH} \\
| & & | \\
H-C-OH & & H-C-H \\
| & & | \\
H-C-OH & & H-C-OH \\
| & & | \\
H-C-OH & & H-C-OH \\
| & & | \\
CH_2-OH & & CH_2-OH \\
\text{ribose} & & \text{deoxyribose}
\end{array}
$$

In consequence, the two varieties of nucleic acid came to be called 'ribonucleic acid' (RNA) and 'deoxyribonucleic acid' (DNA).

Besides the difference in their sugars, the two nucleic acids also differ in one of the pyrimidines. RNA has 'uracil' in place of thymine. Uracil is very like thymine, however, as you can see from the formula:

$$
\begin{array}{c}
OH \\
| \\
C \\
N \diagup \diagdown CH \\
| \qquad \| \\
C \diagdown \diagup CH \\
HO \qquad N \\
\text{uracil}
\end{array}
$$

By 1934, Levene was able to show that the nucleic acids could be broken down to fragments which contained a purine or a pyrimidine, either the ribose or the deoxyribose sugar, and a phosphate group. This combination is called a 'nucleotide'. Levene proposed that the nucleic-acid molecule was built up of nucleotides as a protein is built up of amino acids. His quantitative studies suggested to him that the molecule consisted of just four nucleotide units, one containing adenine, one guanine, one cytosine, and one either thymine (in DNA) or uracil (in RNA). It turned out, however, that what Levene had isolated were not nucleic-acid molecules but pieces of them, and by the middle 1950s biochemists found that the molecular weights of nucleic acids ran as high as six million. Nucleic acids are thus certainly equal and very likely superior to proteins in molecular size.

The exact manner in which nucleotides are built up and interconnected was confirmed by the British biochemist Alexander Robertus Todd, who built up a variety of nucleotides out of simpler fragments and carefully bound nucleotides together under conditions that allowed only one variety of bonding. He received the Nobel Prize in chemistry in 1957 for this work.

As a result, the general structure of the nucleic acid could be seen to be somewhat like the general structure of protein. The protein molecule is made up of a polypeptide backbone out of which jut the side chains of the individual amino acids. In nucleic acids, the sugar portion of one nucleotide is bonded to the sugar portion of the next by means of a phosphate group attached to both. There is thus a 'sugar-phosphate backbone' running the length of the molecule. From this, there extended purines and pyrimidines, one to each nucleotide.

By the use of cell-staining techniques, investigators began to pin down the location of nucleic acids in the cell. The German chemist Robert Feulgen, employing a red dye that stained DNA but not RNA, found that DNA was located in the cell nucleus, specifically in the chromosomes. He detected it not only in animal cells but also in plant cells. In addition, by staining RNA he showed that this nucleic acid, too, occurred in both plant and animal cells. In short, the nucleic acids were universal materials existing in all living cells.

The Swedish biochemist Torbjörn Caspersson studied the subject further by removing one of the nucleic acids (by means of an enzyme which reduced it to soluble fragments that could be washed out of the cell) and concentrating on the other. He would photograph the cell in ultra-violet light; since a nucleic acid absorbs ultra-violet much more strongly than do other cell materials, the location of the DNA or the RNA – whichever he had left in the cell – showed up clearly. DNA showed up only in the chromosomes. RNA made its appearance mainly in certain particles in the cytoplasm. Some RNA also showed up in the 'nucleolus', a structure within the nucleus. (In 1948, the Rockefeller Institute biochemist Alfred Ezra Mirsky showed that small quantities of RNA are present even in the chromosomes, while Ruth Sager showed that DNA can occur in the cytoplasm, notably in the chloroplasts of plants.)

Caspersson's pictures disclosed that the DNA lay in localized bands in the chromosomes. Was it possible that DNA molecules were none other than the genes, which up to this time had had a rather vague and formless existence?

Throughout the 1940s, biochemists pursued this lead with growing excitement. They found it particularly significant that the amount of DNA in the cells of an organism was always rigidly constant, except that the sperm and egg cells had only half this amount, which would be expected, since they had only half the chromosome supply of normal cells. The amount of RNA and of the protein in chromosomes might vary all over the lot, but the quantity of DNA remained fixed. This certainly seemed to indicate a close connection between DNA and genes.

There were, of course, a number of things that argued against the idea. For instance, what about the protein in chromosomes? Proteins of various kinds were associated with the nucleic acid, forming a combination called 'nucleoprotein'. Considering the complexity of proteins, and their great and specific importance in other capacities in the body, should not the protein be the important part of the molecule? It would seem that the nucleic acid might well be no more than an adjunct – at most a working portion of the molecule, like the haem in haemoglobin.

But the proteins (known as protamine and histone) most commonly found in isolated nucleoprotein turned out to be rather simple, as proteins went. Meanwhile DNA was steadily being found to be more and more complex. The tail was beginning to wag the dog.

At this point some remarkable evidence, which seemed to show that the tail *was* the dog, was disclosed. It involved the pneumococcus, the well-known pneumonia microbe. Bacteriologists had long studied two different strains of pneumococci grown in the laboratory – one with a smooth coat made of a complex carbohydrate, the other lacking this coat and therefore rough in appearance. Apparently the rough strain lacked some enzyme needed to make the carbohydrate capsule. But an English bacteriologist named Fred Griffith had discovered that, if killed bacteria of the smooth variety were mixed with live ones of the rough strain and then injected into a mouse, the tissues of the infected mouse would eventually contain live pneumococci of the smooth variety! How could this happen? The dead pneumococci had certainly not been brought to life. Something must have transformed the rough pneumococci so that they were now capable of making the smooth coat. What was that something? Evidently it was a factor of some kind contributed by the dead bacteria of the smooth strain.

In 1944, three American biochemists, Oswald Theodore Avery, Colin Munro Macleod, and Maclyn McCarty, identified the transforming principle. It was DNA. When they isolated pure DNA from the smooth strain and gave it to rough pneumococci, that alone sufficed to transform the rough strain to a smooth.

Investigators went on to isolate other transforming principles, involving other bacteria and other properties, and in every case the principle turned out to be a variety of DNA. The only plausible conclusion was that DNA could act like a gene. In fact, various lines of research, particularly with viruses (see Chapter 4), showed that the protein associated with DNA is almost superfluous from a genetic point of view: DNA can produce genetic effects all by itself, either in the chromosome or – in the case of the non-chromosomal inheritance – in cytoplasmic bodies such as the chloroplasts.

If DNA is the key to heredity, it must have a complex structure, because it has to carry an elaborate pattern, or code of instructions (the 'genetic code'), for the synthesis of specific enzymes. Assuming that it is made up of the four kinds of nucleotide, they cannot be strung in a regular arrangement, such as 1, 2, 3, 4, 1, 2, 3, 4, 1, 2, 3, 4. . . . Such a molecule would be far too simple to carry a blueprint for enzymes. In fact, the American biochemist Erwin Chargaff and his co-workers found definite evidence in 1948 that the composition of nucleic acids was more complicated than had been thought. Their analysis showed that the various purines and pyrimidines were not present in equal amounts, and that the proportions varied in different nucleic acids.

Everything seemed to show that the four purines and pyrimidines were distributed along the DNA backbone as randomly as the amino-acid side-chains were distributed along the peptide backbone. Yet some regularities did seem to exist. In any given DNA molecule, the total number of purines seemed to be equal, always, to the total number of pyrimidines. In addition, the number of adenines (one purine) was always equal to the number of thymines (one pyrimidine), while the number of guanines (the other purine) was always equal to the number of cytosines (the other pyrimidine).

We could symbolize adenine as A, guanine as G, thymine as T, and cytosine as C. The purines would then be A+G and the pyrimidines T+C. The findings concerning any given DNA molecule could then be summarized as:

$$A = T$$
$$G = C$$
$$A+G = T+C$$

More general regularities also emerged. As far back as 1938, Astbury had pointed out that nucleic acids scattered X-rays in diffraction patterns, a good sign of the existence of structural regularities in the molecule. The New Zealand-born British bio-chemist Maurice Hugh Frederick Wilkins calculated that these regularities repeated themselves at intervals considerably greater than the distance from nucleotide to nucleotide. One logical con-clusion was that the nucleic-acid molecule took the form of a helix, with the coils of the helix forming the repetitive unit noted by the X-rays. This thought seemed the more attractive because Linus Pauling was at that time demonstrating the helical struc-ture of certain protein molecules.

In 1953, the English physicist Francis Harry Compton Crick and his co-worker, the American biochemist (and one-time Quiz Kid) James Dewey Watson, put all the information together and came up with a revolutionary model of the nucleic-acid molecule – a model that represented it not merely as a helix but (and this was the key point) as a double helix – two sugar-phosphate back-bones winding like a double-railed spiral staircase up the same vertical axis. From each sugar-phosphate chain, purines and pep-tides extended inwards towards each other, meeting as though to form the steps of this double-railed spiral staircase.

Just how might the purines and pyrimidines be arrayed along these parallel chains? To make a good uniform fit, a double-ring purine on one side should always face a single-ring pyrimidine on the other, to make a three-ring width altogether. Two pyrimi-dines could not stretch far enough to cover the space; while two purines would be too crowded. Furthermore, an adenine from one chain would always face a thymine on the other, and a gua-nine on one chain would always face a cytosine on the other. In this way, one could explain the finding that $A = T$, $G = C$, and $A+T = G+C$.

This 'Watson–Crick model' of nucleic-acid structure has proved to be extraordinarily fruitful and Wilkins, Crick, and Watson shared the 1962 Nobel Prize in medicine and physiology as a result.

The Watson–Crick model makes it possible, for instance, to explain just how a chromosome may duplicate itself in the process of cell division. Consider the chromosome as a string of DNA molecules. The molecules can first divide by a separation of the two helices making up the double helix; the two chains unwind themselves from each other, so to speak. This can be done because opposing purines and pyrimidines are held by hydrogen bonds, weak enough to be easily broken. Each chain is a half-molecule that can bring about the synthesis of its own missing complement. Where it has a thymine, it attaches an adenine; where it has a cytosine, it attaches a guanine; and so on. All the raw materials for making the units, and the necessary enzymes, are on hand in the cell. The half-molecule simply plays the role of a 'template', or mould, for putting the units together in the proper order. The units eventually will fall into the appropriate places and stay there because that is the most stable arrangement.

To summarize, then, each half-molecule guides the formation of its own complement, held to itself by hydrogen bonds. In this way, it rebuilds the complete, double-helix DNA molecule, and the two half-molecules into which the original molecule divided thus form two molecules where only one existed before. Such a process, carried out by all the DNAs down the length of a chromosome, will create two chromosomes which are exactly alike and perfect copies of the original mother chromosome. Occasionally something may go wrong: the impact of a sub-atomic particle or of an energetic radiation, or the intervention of certain chemicals, may introduce an imperfection somewhere or other in the new chromosome. The result is a mutation.

Evidence in favour of this mechanism of replication has been piling up. Tracer studies, employing heavy nitrogen to label chromosomes and following the fate of the labelled material during cell division, have tended to bear out the theory. In addition, some of the important enzymes involved in replication have been identified.

In 1955, the Spanish-American biochemist Severo Ochoa isolated from a bacterium (*Aztobacter vinelandii*) an enzyme which proved capable of catalysing the formation of RNA from nucleotides. In 1956, a former pupil of Ochoa's, Arthur Kornberg, isolated another enzyme (from the bacterium *Escherichia coli*), which could catalyse the formation of DNA from nucleotides. Ochoa proceeded to synthesize RNA-like molecules from nucleotides, and Kornberg did the same for DNA. (The two men shared the Nobel Prize in medicine and physiology in 1959.) Kornberg also showed that his enzyme, given a bit of natural DNA to serve as a template, could catalyse the formation of a molecule which seemed to be identical with natural DNA. In 1965, Sol Spiegelman of the University of Illinois used RNA from a living virus (the simplest class of living things) and produced additional molecules of that sort. Since these additional molecules showed the essential properties of the virus, this was the closest approach yet to producing test-tube life. In 1967, Kornberg and others did the same, using DNA from a living virus as template.

The amount of DNA which is associated with the simplest manifestations of life is small – a single molecule in a virus – and can be made smaller. In 1967, Spiegelman allowed the nucleic acid of a virus to replicate and selected samples after shorter and shorter intervals for further replication. In this way, he selected molecules that completed the job unusually quickly – because they were smaller than average. In the end he had reduced the virus size to one-sixth normal and multiplied replication-speed fifteenfold.

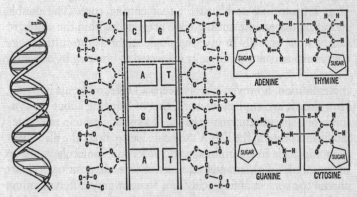

Model of the nucleic-acid molecule. The drawing at the left shows the double helix; in the centre a portion of it is shown in detail (omitting the hydrogen atoms); at the right is a detail of the nucleotide combinations.

Although it is DNA that replicates in cells, many of the simpler (sub-cellular) viruses contain RNA only. RNA molecules in double strands replicate in such viruses. The RNA in cells is single-stranded and does not replicate.

Nevertheless, a single-stranded structure and replication are not mutually exclusive. The American biophysicist Robert Louis Sinsheimer discovered a strain of virus that contained DNA made up of a single strand. That DNA molecule had to replicate itself; but how could that be done with but a single strand? The answer was not difficult. The single strand brought about the production of its own complement and the complement then brought about the production of the 'complement to the complement', that is, a replica of the original strand.

It is clear that the single-strand arrangement is less efficient than the double-strand arrangement (which is probably why the former exists only in certain very simple viruses and the latter in all other living creatures). For one thing, a single strand must replicate itself in two successive steps, whereas the double strand does so in a single step. Second, it now seems that only one strand of the DNA molecule is the important working structure; the cutting edge of the molecule, so to speak. Its complement may be thought of as a protecting scabbard for that cutting edge. The double strand represents the cutting edge protected within the scabbard except when actually in use; the single strand is the cutting edge always exposed and continually subjected to blunting by accident.

Replication, however, merely keeps a DNA molecule in being. How does it accomplish its work, that of bringing about the synthesis of a specific enzyme – that is, of a specific protein molecule? To form a protein, the DNA molecule has to direct the placement of amino acids in a certain specific order in a molecule made up of hundreds of thousands of units. For each position it must choose the correct amino acid from some twenty different amino acids. If there were twenty corresponding units in the DNA molecule, it would be easy. But DNA is made up of only four different building blocks – the four nucleotides. Thinking about this, the astronomer George Gamow suggested in 1954 that the nucleotides, in various combinations, might be used as what we now call a 'genetic code' (just as the dot and dash of the Morse code can be combined in various ways to represent the letters of the alphabet, numerals, and so on).

If you take the four different nucleotides (A, G, C, T), two at a time, there are 4×4, or 16, possible combinations (AA, AG, AC, AT, GA, GG, GC, GT, CA, CG, CC, CT, TA, TG, TC, and TT). This is still not enough. If you take them three at a time, there are $4 \times 4 \times 4$, or 64 different combinations, more than enough. (The reader may amuse himself trying to list the different combinations and see if he can find a sixty-fifth.)

It seemed as though each different 'nucleotide triplet' or 'codon' represented a particular amino acid. In view of the great number of different codons possible, it could well be that two or even three different codons represented one particular amino acid. In this case, the genetic code would be what cryptographers call 'degenerate'.

This left two chief questions: Which codon (or codons) corresponded to which amino acid? And how did the codon information (which was securely locked in the nucleus where, alone, the DNA was to be found) reach the sites of enzyme formation that were in the cytoplasm?

To take the second problem first, suspicion soon fell upon RNA as the substance serving as go-between. The French biochemists François Jacob and Jacques Lucien Monod were the first to suggest this. The structure of such RNA would have to be very like DNA with such differences as existed not affecting the genetic code. RNA had ribose in place of deoxyribose (one extra oxygen atom per nucleotide) and uracil in place of thymine (one missing methyl group, CH_3, per nucleotide). Furthermore, RNA was present chiefly in the cytoplasm, but also, to a small extent, in the chromosomes themselves.

It was not hard to see, and then demonstrate, what was happening. Every once in a while, when the two coiled strands of the DNA molecule unwound, one of those strands (always the same one, the cutting edge) replicates its structure, not on nucleotides that form a DNA molecule, but on nucleotides that form an

RNA molecule. In this case, the adenine of the DNA strand does not attach thymine nucleotides to itself, but uracil nucleotides instead. The resulting RNA molecule, carrying the genetic code imprinted on its nucleotide pattern, can then leave the nucleus and enter the cytoplasm.

Since it carries the DNA 'message', it has been named 'messenger-RNA', or more simply, 'mRNA'.

The Rumanian-American biochemist George Emil Palade, thanks to careful work with the electron microscope, demonstrated, in 1956, the site of enzyme manufacture in the cytoplasm to be tiny particles, about two millionths of a centimetre in diameter. They were rich in RNA and were therefore named 'ribosomes'. There are as many as 15,000 ribosomes in a bacterial cell, perhaps ten times as many in a mammalian cell. They are the smallest of the sub-cellular particles or 'organelles'. It was soon determined that the messenger-RNA – carrying the genetic code on its structure – made its way to the ribosomes and layered itself on to one or more of them, and that the ribosomes were the site of protein synthesis.

The next step was taken when the American biochemist Mahlon Bush Hoagland, who had also been active in working out the notion of mRNA, showed that in the cytoplasm were a variety of small RNA molecules, which might be called 'soluble-RNA' or 'sRNA', because their small size enabled them to dissolve freely in the cytoplasmic fluid.

At one end of each sRNA molecule was a particular triplet of nucleotides that just fitted a complementary triplet somewhere on the mRNA chain. That is, if the sRNA triplet were AGC, it would fit tightly to a UCG triplet on the mRNA and only there. At the other end of the sRNA molecule was a spot where it would combine with one particular amino acid and none other. On each sRNA molecule, the triplet at one end meant a particular amino acid on the other. Therefore a complementary triplet on the mRNA meant that only a certain sRNA molecule carrying a certain amino-acid molecule would affix itself there. A large number of sRNA molecules would affix themselves one after the other, right down the line, to the triplets making up the mRNA structure (triplets that had been moulded right on the DNA molecule of a particular gene). All the amino acids properly lined up could then easily be hooked together to form an enzyme molecule.

Because the information from the messenger-RNA is, in this way, transferred to the protein molecule of the enzyme, sRNA has come to be called 'transfer-RNA', and it is this name which is now well established.

In 1964, the molecule of alanine-transfer-RNA (the transfer-RNA that attaches itself to the amino acid alanine) was completely analysed by a team headed by the American biochemist Robert W. Holley. This was done by the Sanger method of breaking down the molecule into small fragments by appropriate enzymes, then analysing the fragments and deducing how they must fit together. The alanine-transfer-RNA, the first naturally occurring nucleic acid to be completely analysed, was found to be made up of a chain of seventy-seven nucleotides. These include not only the four nucleotides generally found in RNA (A, G, C, and U) but also several of seven others closely allied to them.

It had been supposed at first that the single chain of a transfer-RNA would be bent like a hairpin at the middle and the two ends would twine about each other in a double helix. The structure of alanine-transfer-RNA did not lend itself to this. Instead it seemed to consist of three loops, so that it looked rather like a lopsided three-leaf clover. In subsequent years, other transfer-RNA molecules were analysed in detail, and all seemed to have the same three-leaf-clover structure. For his work, Holley received a share of the 1968 Nobel Prize for medicine and physiology.

In this way, the structure of a gene controls the synthesis of a specific enzyme. Much, of course, remains to be worked out, for genes do not simply organize the production of enzymes at top speed at all times. The gene may be working efficiently now, slowly at another time, and not at all at still another time. Some cells manufacture protein at great rates, with an ultimate capacity of combining some 15 million amino acids per chromosome per minute, some only slowly, some scarcely at all, yet all the cells in a given organism have the same genic organization. Then, too,

each type of cell in the body is highly specialized, with characteristic functions and chemical behaviour of its own. An individual
cell may synthesize a given protein rapidly at one time, slowly at
another. And again all have the same genic organization all the
time.

It is clear that cells have methods for blocking and unblocking
the DNA molecules of the chromosomes. Through the pattern of
blocking and unblocking, different cells with identical gene-
patterns can produce different combinations of proteins, while a
particular cell with an unchanging gene-pattern can produce
different combinations from time to time.

In 1961, Jacob and Monod suggested that each gene has its own
repressor, coded by a 'regulator gene'. This repressor – depending on its geometry, which can be altered by delicate changes in
circumstances within the cell – will block or release the gene. In
1967, such a repressor was isolated and found to be a small protein. Jacob and Monod, together with a co-worker, André
Michael Lwoff, received the 1965 Nobel Prize for medicine and
physiology as a result.

Nor is the flow of information entirely one-way, from gene to
enzyme. There is 'feedback' as well. Thus, there is a gene that
brings about the formation of an enzyme that catalyses a reaction
that converts the amino acid, threonine, to another amino acid,
isoleucine. Isoleucine, by its presence, somehow serves to activate
the repressor, which begins to shut down the very gene that produces the particular enzyme that led to that presence. In other
words, as isoleucine concentration goes up, less is formed; if the
concentration declines, the gene is unblocked, and more isoleucine
is formed. The chemical machinery within the cell – genes, re-
pressors, enzymes, end-products – is enormously complex and
intricately interrelated. The complete unravelling of the pattern
is not likely to take place rapidly.

But meanwhile, what of the other question: Which codon goes
along with which amino acid? The beginning of an answer came
in 1961, thanks to the work of the American biochemists Marshall
Warren Nirenberg and J. Heinrich Matthaei. They began by
making use of a synthetic nucleic acid, built up according to

Ochoa's system from uracil nucleotides only. This 'polyuridylic acid' was made up of a long chain of ... UUUUUUUU ... and could only possess one codon, UUU.

Nirenberg and Matthaei added this polyuridylic acid to a

The genetic code. In the left-hand column are the initials of the four RNA bases (uracil, cytosine, adenine, guanine) representing the first 'letter' of the codon triplet; the second letter is represented by the initials across the top, while the third but less important letter appears in the final column. For example, tyrosine (Tyr) is coded for by either UAU or UAC. Amino acids coded by each codon are shown abbreviated as follows: Phe – phenylalanine; Leu – leucine; Ileu – isoleucine; Met – methionine; Val – valine; Ser – serine; Pro – proline; Thr – threonine; Ala – alanine; Tyr – tyrosine; His – histidine; Glun – glutamine; Aspn – asparagine; Lys – lysine; Asp – aspartic acid; Glu – glutamic acid; Cys – cysteine; Tryp – tryptophan; Arg – arginine; Gly – glycine.

First position	Second position				Third position
	U	C	A	G	
U	Phe	Ser	Tyr	Cys	U
	Phe	Ser	Tyr	Cys	C
	Leu	Ser	(normal 'full stop')	'full stop'	A
	Leu	Ser	(less common 'full stop')	Tryp	G
C	Leu	Pro	His	Arg	U
	Leu	Pro	His	Arg	C
	Leu	Pro	Glun	Arg	A
	Leu	Pro	Glun	Arg	G
A	Ileu	Thr	Aspn	Ser	U
	Ileu	Thr	Aspn	Ser	C
	Ileu?	Thr	Lys	Arg	A
	Met ('capital letter')	Thr	Lys	Arg	G
G	Val	Ala	Asp	Gly	U
	Val	Ala	Asp	Gly	C
	Val	Ala	Glu	Gly	A
	Val ('capital letter')	Ala	Glu	Gly	G

system that contained various amino acids, enzymes, ribosomes, and all the other components necessary to synthesize proteins. Out of the mixture tumbled a protein made up of the amino acid phenylalanine. This meant that UUU was equivalent to phenylalanine. The first item in the 'codon dictionary' was worked out.

The next step was to prepare a nucleotide made out of a preponderance of uridine nucleotides with a small quantity of adenine nucleotides added. This meant that along with the UUU codon, an occasional UUA, or AUU, or UAU codon might appear. Ochoa and Nirenberg showed that, in such a case, the protein formed was mainly phenylalanine, but also contained an occasional leucine, isoleucine, and tyrosine, three other amino acids.

Slowly, by methods such as these, the dictionary was extended. It was found that the code is indeed degenerate and that GAU and GAC might each stand for aspartic acid, for instance, and the GUU, GAU, GUC, GUA, and GUG, all stand for glycine. In addition, there was some punctuation. The codon AUG not only stood for the amino acid methionine, but apparently signified the beginning of a chain. It was a 'capital letter', so to speak. Then, too, UAA and UAG signalled the end of a chain: they were full-stops.

By 1967, the dictionary was complete. Nirenberg and his collaborator, the Indian-American chemist Har Gobind Khorana, were awarded shares (along with Holley) in the 1968 Nobel Prize for medicine and physiology.

The possible implications of a true understanding of the genetic code and of the manner in which it is modified from tissue to tissue and cell to cell are staggering. Probably nothing more exciting has happened in the life sciences in a century.

The Origin of Life

Once we get down to the nucleic-acid molecules, we are as close to the basis of life as we can get. Here, surely, is the prime substance of life itself. Without DNA, living organisms could not reproduce, and life as we know it could not have started. All the substances of living matter – enzymes and all the others, whose production is catalysed by enzymes – depend in the last analysis on DNA. How, then, did DNA, and life, start?

This is a question that science has always hesitated to ask, because the origin of life has been bound up with religious beliefs even more strongly than has the origin of the earth and the universe. It is still dealt with only hesitantly and apologetically. But in recent years a book entitled *The Origin of Life*, by the Russian biochemist Aleksandr Ivanovich Oparin, has brought the subject very much to the fore. The book was published in the Soviet Union in 1924 and in English translation in 1936. In it the problem of life's origin for the first time was dealt with in detail from a completely materialistic point of view. Since the Soviet Union is not inhibited by the religious scruples to which the Western nations feel bound, this, perhaps, is not surprising.

Most of man's early cultures developed myths telling of the creation of the first human beings (and sometimes of other forms of life as well) by gods or demons. However, the formation of life itself was rarely thought of as being entirely a divine prerogative. At least the lower forms of life might arise spontaneously from non-living material without supernatural intervention. Insects and worms might, for instance, arise from decaying meat, frogs from mud, mice from rotting wheat. The idea was based on actual observation, for decaying meat, to take the most obvious example, did indeed suddenly give rise to maggots. It was only natural to assume that the maggots were formed from the meat.

Aristotle believed in the existence of 'spontaneous generation'. So did the great theologians of the Middle Ages, such as Thomas Aquinas. So did William Harvey and Isaac Newton. After all, the evidence of one's own eyes was hard to refute.

The first to put this belief to the test of experimentation was the Italian physician Francesco Redi. In 1668, he decided to check on whether maggots really formed out of decaying meat. He put pieces of meat in a series of jars and then covered some of them with fine gauze and left others uncovered. Maggots developed only in the meat in the uncovered jars, to which flies had had free access. Redi concluded that the maggots had arisen from microscopically small eggs laid on the meat by the flies. Without flies and their eggs, he insisted, meat could never produce maggots, however long it decayed and putrefied.

Experimenters who followed Redi confirmed this, and the belief that visible organisms arose from dead matter died. But when microbes were discovered, shortly after Redi's time, many scientists decided that these forms of life at least must come from dead matter. Even in gauze-covered jars, meat would soon begin to swarm with micro-organisms. For two centuries after Redi's experiments, belief in the possibility of the spontaneous generation of micro-organisms remained very much alive.

It was another Italian, the naturalist Lazzaro Spallanzani, who first cast serious doubt on this notion. In 1765, he set out two sets of vessels containing a broth. One he left open to the air. The other, which he had boiled to kill any organisms already present, he sealed up to keep out any organisms that might be floating in the air. The broth in the first vessels soon teemed with micro-organisms, but the boiled and sealed-up broth remained sterile. This proved to Spallanzani's satisfaction that even microscopic life could not arise from inanimate matter. He even isolated a single bacterium and witnessed its division into two bacteria.

The proponents of spontaneous generation were not convinced. They maintained that boiling destroyed some 'vital principle' and that this was why no microscopic life developed in Spallanzani's boiled, sealed flasks. It remained for Pasteur to settle the question, in 1862, seemingly once and for all. He devised a flask with a long swan neck in the shape of a horizontal S. With the opening unstoppered, air could percolate into the flask, but dust particles and micro-organisms could not, for the curved neck would serve as a trap, like the drain trap under a sink. Pasteur put some broth in the flask, attached the S-shaped neck, boiled the broth until it steamed (to kill any micro-organisms in the neck as well as in the broth), and waited for developments. The broth remained sterile. There was no vital principle in air. Pasteur's demonstration apparently laid the theory of spontaneous generation to rest permanently.

All this left a germ of embarrassment for scientists. How had life arisen, after all, if not through divine creation or through spontaneous generation?

Towards the end of the nineteenth century some theorists went to the other extreme and made life eternal. The most popular theory was advanced by Svante Arrhenius (the chemist who had developed the concept of ionization). In 1907, he published a

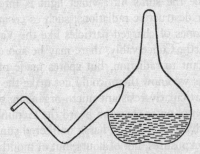

Pasteur's flask for the experiment on spontaneous generation.

book entitled *Worlds in the Making*, picturing a universe in which life had always existed and migrated across space, continually colonizing new planets. It travelled in the form of spores that escaped from the atmosphere of a planet by random movement and then were driven through space by the pressure of light from the sun.

Such light pressure is by no means to be sneered at as a possible driving force. The existence of radiation pressure had been predicted in the first place by Maxwell, on theoretical grounds, and, in 1899, it had been demonstrated experimentally by the Russian physicist Peter Nicolaevich Lebedev.

Arrhenius's views held, then, that spores travelled on and on through interstellar space, driven by light radiation this way and that, until they died or fell on some planet, where they would spring into active life and compete with life forms already present, or inoculate the planet with life if it was uninhabited but habitable.

At first blush, this theory looks attractive. Bacterial spores, protected by a thick coat, are very resistant to cold and de-hydration and might conceivably last a long time in the vacuum of space. Also, they are of just the proper size to be more affected by the outward pressure of a sun's radiation than by the inward pull of its gravity. But Arrhenius's suggestion fell before the on-slaught of ultra-violet light. In 1910, experimenters showed that ultra-violet light quickly killed bacterial spores, and in inter-planetary space the sun's ultra-violet light is intense – not to speak of other destructive radiations, such as cosmic rays, solar X-rays, and zones of charged particles like the Van Allen belts around the earth. Conceivably, there may be spores somewhere that are resistant to radiation, but spores made of protein and nucleic acid, as we know them, could not make the grade. To be sure, some particularly resistant micro-organisms were exposed to the radiation of outer space on board the Gemini 9 capsule in 1966 and survived six hours of harsh unfiltered sunlight. But we are talking of exposures, not of hours, but of months and years.

Such a highly hydrogenated atmosphere we might call 'At-mosphere I'. Through photodissociation, this would slowly turn into an atmosphere of carbon dioxide and nitrogen (see Vol. 1, p. 214) or 'Atmosphere II'. After that an ozone layer would form in the upper atmosphere, and spontaneous change would halt. Could it be, then, that life could form in one or the other of the early atmospheres?

Consider Atmosphere II, for instance. Carbon dioxide is soluble in water, and while earth was bathed in Atmosphere II, the ocean would be a vast collection of soda water. The ultra-violet radiation from the sun at sea level would be much more intense than it is today while Atmosphere II was in its last stages of formation and before the ozone layer was completely in place. What's more, the earth's soil would have had a larger supply of radioactive atoms then than now. Under such conditions, could organic matter have sprung into existence?

H. C. Urey felt life started in Atmosphere I. In 1952, Stanley Lloyd Miller, then a graduate student in Urey's laboratories, circulated water, plus ammonia, methane and hydrogen, past an electric discharge (to simulate the ultra-violet radiation of the sun). At the end of a week he analysed his solution by paper chromatography and found that, in addition to the simple substances without nitrogen atoms, he also had glycine and alanine, the two simplest of the amino acids, plus some indication of one or two more complicated ones.

Miller's experiment was significant in several ways. In the first place, these compounds had formed quickly and in surprisingly large quantities. One sixth of the methane with which he had started had gone into the formation of more complex organic compounds, yet the experiment had been in operation for only a week.

Then, too, the kind of organic molecules formed in Miller's experiments were just those present in living tissue. The path taken by the simple molecules, as they grew more complex, seemed pointed directly towards life. This pointing-towards-life continued consistently in later, more elaborate experiments. At no time were molecules formed in significant quantity that seemed to point in an unfamiliar non-life direction.

Thus, P. H. Abelson followed Miller's work by trying a variety of similar experiments with starting materials made up of different gases in different combinations. It turned out that as long as he began with molecules that included atoms of carbon, hydrogen, oxygen, and nitrogen, amino acids of the kind normally found in proteins were formed. Nor were electric discharges the only source of energy that would work. In 1959, two German scientists, Wilhelm Groth and H. von Weyssenhoff, designed an experiment in which ultra-violet light could be used instead, and they also got amino acids.

If there was any doubt that the direction-towards-life was the line of least resistance, there was the fact that in the late 1960s, more and more complicated molecules, representing the first stages of that direction, were found in gas clouds of outer space (see Vol. 1, Chapter 2). It may be, then, that at the time the earth was formed out of clouds of dust and gas, the first stages of building up complex molecules had already taken place.

The earth, at its first formation, may have had a supply of amino acids. Evidence in favour of that came in 1970. The Ceylon-born biochemist Cyril Ponnamperuma studied a meteorite that had fallen in Australia on 28 September 1969. Careful analyses showed the presence of small traces of five amino acids: glycine, alanine, glutamic acid, valine, and proline. There was no optical activity in these amino acids, so they were formed not by life processes (and hence their presence was not the result of earthly contamination) but by the non-living chemical processes of the type that took place in Miller's flask.

Could chemists in the laboratory progress beyond the amino-acid stage? One way of doing so would be to start with larger samples of raw materials and subject them to energy for longer periods. This would produce increasing numbers of more and more complicated products, but the mixtures of these products would become increasingly complex and would be increasingly difficult to analyse.

Instead, chemists began with later stages. The products formed in earlier experiments would be used as new raw materials. Thus, one of Miller's products was hydrogen cyanide. The Spanish-American biochemist Juan Oro added hydrogen cyanide to the starting mixture in 1961. He obtained a richer mixture of amino acids and even a few short peptides. He also formed purines, in particular adenine, a vital component of nucleic acids. In 1962, Oro used formaldehyde as one of his raw materials and produced ribose and deoxyribose, also components of nucleic acids.

In 1963, Ponnamperuma also performed experiments similar to those of Miller, using electron beams as a source of energy, and found that adenine was formed. Together with Ruth Mariner and Carl Sagan, he went on to add adenine to a ribose solution and under ultra-violet light, 'adenosine', a molecule formed of adenine and ribose linked together, was formed. If phosphate was also present, it, too, was hooked on to form the adenine nucleotide. Indeed, three phosphate groups could be added to form adenosine triphosphate (ATP), which, as was explained in Chapter 2 (see p. 102), is essential to the energy-handling mechanisms of living tissue. In 1965, he formed a 'dinucleotide', two nucleotides bound together. Additional products can be built up if substances such as cyanamide ($CNNH_2$) and ethane

(CH_3CH_3) – substances which may well have been present in the primordial era – are added to the mixtures employed by various experimenters in this field. There is no question, then, but that normal chemical and physical changes in the primordial ocean and atmosphere could have acted in such a way as to build up proteins and nucleic acids.

Any compound that formed in the lifeless ocean would tend to endure and accumulate. There were no organisms, either large or small, to consume them or cause them to decay. Moreover, in the primeval atmosphere there was no free oxygen to oxidize and break down the molecules. The only important factors tending to break down complex molecules would have been the very ultraviolet and radioactive energies that built them up. But ocean currents might have carried much of the material to a safe haven at mid-levels in the sea, away from the ultra-violet-irradiated surface and the radioactive bottom. Indeed, Ponnamperuma and his co-workers have estimated that fully 1 per cent of the primordial ocean may have been made up of these built-up organic compounds. If so, this would represent a mass of over a thousand billion tons. This is certainly an ample quantity for natural forces to play with, and in such a huge mass even substances of most unlikely complexity are bound to be built up in not too long a period (particularly considering a couple of thousand million years are available for the purpose).

There is no logical barrier, then, to supposing that out of the simple compounds in the primordial ocean and atmosphere there appeared, with time, higher and higher concentrations of the more complicated amino acids, as well as simple sugars; that amino acids combined to form peptides; that purines, pyrimidines, sugar, and phosphate combined to form nucleotides; and that, gradually, over the ages, proteins and nucleic acids were created. Then, eventually, must have come the key step – the formation, through chance combinations, of a nucleic acid molecule, capable of inducing replication. That moment marked the beginning of life.

Thus a period of 'chemical evolution' preceded the evolution of life itself.

A single living molecule, it seems, might well have been sufficient to get life under way and give rise to the whole world of widely varying living things, as a single fertilized cell can give rise to an enormously complex organism. In the organic 'soup' that constituted the ocean at that time, the first living molecule could have replicated thousands of millions of molecules like itself in short order. Occasional mutations would create slightly changed forms of the molecule, and those that were in some way more efficient than the others would multiply at the expense of their neighbours and replace the old forms. If one group was more efficient in warm water and another group in cold water, two varieties would arise, each restricted to the environment it fitted best. In this fashion, the course of 'organic evolution' would be set in motion.

Even if several living molecules came into existence independently at the beginning, it is very likely that the most efficient one would have outbred the others, so that all life today may very well be descended from a single original molecule. In spite of the great present diversity of living things, all have the same basic ground plan. Their cells all carry out metabolism in pretty much the same way. Furthermore, it seems particularly significant that the proteins of all living things are composed of L-amino acids rather than amino acids of the D type. It may be that the original nucleoprotein from which all life is descended happened to be built from L-amino acids by chance, and since D could not be associated with L in any stable chain, what began as chance persisted by replication into grand universality. (This is not to imply that D-amino acids are totally absent in nature. They occur in the cell walls of some bacteria and in some antibiotic compounds. These, however, are quite exceptional cases.)

Of course, the step from a living molecule to the kind of life we know today is still an enormous one. Except for the viruses, all life is organized into cells, and a cell, however small it may seem by human standards, is enormously complex in its chemical structure and interrelationships. How did that start?

The question of the origin of cells was illuminated by the researches of the American biochemist Sidney W. Fox. It seemed to him that the early earth must have been quite hot and that the energy of heat alone could be sufficient to form complex compounds out of simple ones. To test this, Fox in 1958 heated a mixture of amino acids and found they formed long chains that resembled those in protein molecules. These 'proteinoids' were digested by enzymes that digested ordinary proteins and could be used as food by bacteria.

Most startling of all, when Fox dissolved the proteinoids in hot water and let the solution cool, he found they would cling together in little 'microspheres' about the size of small bacteria. These microspheres were not alive by the usual standards, but they behaved as cells do, in some respects at least (they are surrounded by a kind of membrane, for instance). By adding certain chemicals to the solution, Fox could make the microspheres swell or shrink, much as ordinary cells do. They can produce buds, which sometimes seem to grow larger and then break off. Microspheres can separate, divide in two, or cling together in chains.

Perhaps in primordial times, such tiny not-quite-living aggregates of materials formed in several varieties. Some were particularly rich in DNA and were very good at replicating, though only moderately successful at storing energy. Other aggregates could handle energy well but replicated only limpingly. Eventually, collections of such aggregates might have cooperated, each supplying the deficiencies of the other, to form the modern cell, which was much more efficient than any of its parts alone. The modern cell still has the nucleus – rich in DNA but unable of itself to handle oxygen – and numerous mitochondria – which handle oxygen with remarkable efficiency but cannot reproduce in the absence of nuclei. (That mitochondria may once have been independent entities is indicated by the fact that they still possess small quantities of DNA.)

Throughout the existence of Atmospheres I and II, primitive life-forms could only exist at the cost of breaking down complex chemical substances into simpler ones and storing the energy evolved. The complex substances were rebuilt by the action of the ultra-violet radiation of the sun. Once Atmosphere II was completely formed and the ozone layer was in place, the danger of starvation set in, for the ultra-violet supply was cut off.

By then, though, some mitochondria-like aggregate was formed which contained chlorophyll – the ancestor of the modern chloroplast. In 1966, the Canadian biochemists G. W. Hodson and B. L. Baker began with pyrrole and paraformaldehyde (both of which can be formed from still simpler substances in Miller-type experiments) and demonstrated the formation of porphyrin rings, the basic structure of chlorophyll, after merely three hours' gentle heating.

Even the inefficient use of visible light by the first primitive chlorophyl-containing aggregates must have been much preferable to the procedure of non-chlorophyll systems at the time when the ozone layer was forming. Visible light could easily penetrate the ozone layer, and the lower energy of visible light (compared with ultra-violet) was enough to activate the chlorophyll system.

The first chlorophyll-using organisms may have been no more complicated than individual chloroplasts today. There are, in fact, two thousand species of a group of one-celled photosynthesizing organisms called 'blue-green algae' (they are not all blue-green, but the first ones studied were). These are very simple cells, rather bacteria-like in structure, except that they contain chlorophyll and bacteria do not. Blue-green algae may be the simplest descendants of the original chloroplast, while bacteria may be the descendants of chloroplasts that lost their chlorophyll and took to parasitism or to foraging on dead tissue and its components.

As chloroplasts multiplied in the ancient seas, carbon dioxide was gradually consumed and molecular oxygen took its place. The present Atmosphere III was formed. Plant cells grew steadily more efficient, each one containing numerous chloroplasts. At the same time, elaborate cells without chlorophyll could not exist on the previous basis, for the plant cells cleared the oceans of the food supply and no more was formed except within those cells. However, cells without chlorophyll but with elaborate mitochondrial equipment that could handle complex molecules with great efficiency and store the energy of their breakdown could live by ingesting the plant cells and stripping the molecules the latter had painstakingly built up. Thus originated the animal cell of today. Eventually, organisms grew complex enough to begin to leave the fossil record (plant and animal) that we have today.

Meanwhile, the earth environment had changed fundamentally, from the standpoint of creation of new life. Life could no longer originate and develop from purely chemical evolution. For one thing, the forms of energy that had brought it into being in the first place – ultra-violet and radioactive energy – were effectively gone. For another, the well-established forms of life would quickly consume any organic molecules that arose spontaneously. For both these reasons there is virtually no chance of any new and independent breakthrough from non-life into life (barring some future intervention by man, if he learns to turn the trick). Spontaneous generation today is so highly improbable that it can be regarded as essentially impossible.

Life in Other Worlds

If we accept the view that life arose simply from the workings of physical and chemical laws, it follows that in all likelihood life is not confined to the earth. What are the possibilities of life elsewhere in the universe?

Beginning close to home, we can eliminate the moon pretty definitely, because of its lack of detectable air or water. There has been some speculation that very primitive forms of life may exist here and there in deep crannies, where traces of water and air might conceivably be present. The British astronomer V. A. Firsoff has argued in his book, *Strange World of the Moon*, that water may lie below the moon's surface and that gases absorbed on the dust-covered surface may give rise to a very shallow 'atmosphere' that might support life of a sort. Urey maintained that meteoric bombardment of a primordial earth may have splashed water to the moon, giving it temporary lakes and streams. Lunar Orbiter photographs of the moon have indeed shown markings that seem for all the world like tracks of ancient rivers.

However, the rocks brought back from the moon by successive expeditions to our satellite beginning in 1969 have proved absolutely without water or organic material. To be sure, they are surface rocks only, and the story may be different once samples from several feet below are obtained. Nevertheless, what evidence we have so far seems to indicate there may be no life of any sort, however simple, on the moon.

Venus might seem a somewhat better candidate for life, judging by its mass alone (for it is quite close to the earth in size) and by the indisputable fact that it possesses an atmosphere even denser than ours. In 1959, the bright star Regulus was eclipsed by Venus. From studies of the manner in which the light of Regulus penetrated Venus's atmosphere, new knowledge concerning its depth and density was obtained. The American astronomer Donald Howard Menzel was able to announce that the diameter of the solid body of Venus was 3,783·5 miles, seventy miles less than the best previous estimate.

The atmosphere of Venus seems to be mainly carbon dioxide. Venus has no detectable free oxygen, but this lack is not a fatal barrier to life. As I explained above, the earth very likely supported life for ages before it possessed free oxygen in its atmosphere (although, to be sure, that life was undoubtedly very primitive).

Until recently no sign of vapour was found in Venus's air, and that *was* a fatal objection to the possibility of life there. But astronomers were uncomfortable about the failure to find water on Venus, because it left them with no completely satisfactory explanation of the planet's cloud cover. Then, in 1959, Charles B. Moore of the Arthur D. Little Company laboratory went up in a balloon and at an altitude of fifteen miles in the stratosphere, above most of the earth's interfering atmosphere, took pictures of Venus with a telescope. Its spectrum in the infra-red showed that Venus's upper atmosphere contained as much water vapour as does the earth's.

However, the possibilities of life on Venus took another and, apparently, final nose dive, because of its surface temperature. Radio waves coming from Venus suggested that its surface temperature might be far above the boiling point of water. There was some hope, however, that the radio waves indicating this temperature might arise somewhere in Venus's upper atmosphere and that the planet's actual surface might have a bearable temperature.

This hope, however, was dashed in December 1962, when the American Venus probe, Mariner II, passed close by Venus's position in space and scanned its surface for radio-wave radiation. The results, when analysed, clearly showed that Venus's surface was far too hot for it to retain an ocean. All its water supply was in its clouds. Venus's temperature – which may be as high as 500°C, judging by a Soviet probe that actually transmitted surface temperatures in December 1970 – seems effectively to eliminate the possibility of life as we know it. Mercury, which is closer to the sun than Venus, smaller, as hot, and lacking an atmosphere, is also eliminated.

Mars is without doubt a more hopeful possibility. It has a thin atmosphere of carbon dioxide and nitrogen, and it seemed to have enough water to show thin ice caps (probably an inch or so thick at most), which form and melt with the seasons. The Martian temperatures are low, but not too low for life and probably no worse at any time than Antarctica. The temperature may even reach an occasional balmy 80°F at Mars's equator during the height of the summer day.

The possibility of life on Mars has excited the world for nearly a century. In 1877, the Italian astronomer Giovanni Virginio Schiaparelli detected fine, straight lines on the surface of the planet. He named them 'canali'. That is the Italian word for 'channels', but it was mistranslated into English as 'canals', whereupon people jumped to the conclusion that the lines were artificial waterways, perhaps built to bring water from the ice caps to other parts of the planet for irrigation. The American astronomer Percival Lowell vigorously championed this interpretation of the markings, and the Sunday supplements (and science-fiction stories) went to town on the 'evidence' of intelligent life on Mars.

Actually, there was considerable doubt that the markings exist at all. Many astronomers have never seen them, despite earnest attempts; others have seen them only in flashes. Many believe that they may be optical illusions, arising from strained efforts to see something just at the limit of vision. In any case, no astronomer believes that Mars could support any form of advanced life.

Yet there remained the chance that Mars might support simple forms of life. There are, for instance, green patches on Mars's surface that change with the seasons, expanding in the hemisphere that is experiencing summer and contracting in the hemisphere that is experiencing winter. Could this be a sign of a cover of a simple form of plant life? Laboratory experiments have shown that some lichens and micro-organisms, though adapted to the earth's environment rather than Mars's, could live and grow at temperatures and in an atmosphere that are believed to simulate the Martian environment.

Such hopes began to wither as the result of the results of the Mariner IV Martian probe, launched on 28 November 1964. In 1965, it passed within 6,000 miles of Mars and sent back photographs of its surface. This revealed no canals but showed, for the first time, that the Martian surface was heavily cratered, rather like the moon's. It was deduced from this that not only was the Martian atmosphere thin and desiccated now, but it might always have been.

Additional information of the Martian surface gained from an even more sophisticated 'fly-by' in 1969 made the situation seem still worse. The atmosphere was even thinner than had been believed, and the temperature lower. The temperature at the Martian south pole seemed to be no higher than 150°K, and the 'ice caps', on which so much hope for a water supply had been based, may well be solid carbon dioxide. While there can be no certainty until an actual Martian landing has been made, it seems rather likely now that life as we know it is not present on Mars.

As for planetary bodies beyond Mars (or satellites or planetoids), it would seem that conditions are harsher still. To be sure, Carl Sagan has suggested that the atmosphere of Jupiter would produce so strong a greenhouse effect as to give rise to moderate temperatures that might support life of a sort. This seemed less far-fetched when it turned out that Jupiter was radiating three times as much energy as it receives from the sun, so that some other energy source (planetary contraction, perhaps) may be involved and temperatures higher than expected may exist. This, however, remains strongly speculative, and lacking a planetary probe in Jupiter's neighbourhood, nothing more can be said.

We can conclude, then, that as far as the solar system is concerned, the earth and only the earth seems to be an abode of life. But the solar system isn't all there is. What are the possibilities of life elsewhere in the universe?

The total number of stars in the known universe is estimated to be at least 1,000,000,000,000,000,000,000 (a thousand trillion). Our own Galaxy contains well in excess of 100,000,000,000 stars. If all the stars developed by the same sort of process as the one that is believed to have created our own solar system (i.e., the condensing of a large cloud of dust and gas), then it is likely that no star is solitary, but each is part of a local system containing more than one body. We know that there are many double stars, revolving around a common centre, and it is estimated that at least one out of three stars belongs to a system containing two or more stars.

What we really want, though, is a multiple system in which a number of members are too small to be self-luminous and are planets rather than stars. Though we have no means (so far) of detecting directly any planet beyond our own solar system, even for the nearest stars, we can gather indirect evidence. This has been done at the Sproul Observatory of Swarthmore College under the guidance of the Dutch-American astronomer Peter Van de Kamp.

In 1943, small irregularities of one of the stars of the double-star system, 61 Cygni, showed that a third component, too small to be self-luminous, must exist. This third component, 61 Cygni C, had to be about eight times the mass of Jupiter and therefore (assuming the same density) about twice the diameter. In 1960, a planet of similar size was located circling about the small star Lalande 21185 (located, at least, in the sense that its existence was the most logical way of accounting for irregularities in the star's motion). In 1963, a close study of Barnard's star indicated the presence of a planet there, too, one that was only one and a half times the mass of Jupiter.

Barnard's star is second closest to ourselves, Lalande 21185 third closest, and 61 Cygni twelfth closest. That three planetary systems should exist in close proximity to ourselves would be extremely unlikely, unless planetary systems were very common generally. Naturally, at the vast distances of the stars, only the largest planets could be detected and even then with difficulty. Where super-Jovian planets exist, it seems quite reasonable (and even inevitable) to suppose that smaller planets also exist.

But even assuming that all or most stars have planetary systems and that many of the planets are earth-like in size, we must know what criteria such planets must fulfil to be habitable. The American space scientist Stephen H. Dole has made a particular study of this problem in his book *Habitable Planets for Man*, published in 1964, and has reached certain conclusions, admittedly speculative, but reasonable.

He points out, in the first place, that a star must be of a certain size in order to possess a habitable planet. The larger the star, the shorter-lived it is, and, if it is larger than a certain size, it will not live long enough to allow a planet to go through the long stage of chemical evolution prior to the development of complex life forms. A star that is too small cannot warm a planet sufficiently, unless that planet is so close that it will suffer damaging tidal effects. Dole concludes that only stars of special classes F2 to K1 are suitable for the nurturing of planets that are comfortably habitable for mankind: planets that he can colonize (if travel between the stars ever becomes practicable) without undue effort. There are, Dole estimates, 17,000 million such stars in our Galaxy.

Such a star might be capable of possessing a habitable planet and yet might not possess one. Dole estimates the probabilities that a star of suitable size might have a planet of the right mass and at the right distance, with an appropriate period of rotation and an appropriately regular orbit, and, by making what seem to him to be reasonable estimates, he concludes that there are likely to be 600 million habitable planets in our Galaxy alone, each of them already containing some form of life.

If these habitable planets are spread more or less evenly throughout the Galaxy, Dole estimates that there is one habitable planet per 80,000 cubic light-years. This means that the nearest habitable planet to ourselves may be some twenty-seven light-years away and that, within one hundred light-years of ourselves, there may be a total of fifty habitable planets.

Dole lists fourteen stars within twenty-two light-years of ourselves that might possess habitable planets and weighs the probabilities that this might be true in each case. He concludes that the greatest likelihood of habitable planets is to be found in the stars closest to us, the two sun-like stars of the Alpha Centauri system, Alpha Centauri A and Alpha Centauri B. These two companion stars, taken together, have, Dole estimates, 1 chance in 10 of possessing habitable planets. The total probability for all 14 neighbouring stars is about 2 chances in 5.

If life is the consequence of the chemical reactions described in the previous section, its development should prove inevitable on any earth-like planet. Of course, a planet may possess life and yet not possess intelligent life. We have no way of making even an intelligent guess as to the likelihood of the development of intelligence on a planet, and Dole, for instance, is careful to make none. After all, our own earth, the only habitable planet we really know we can study, existed for at least 2,000 million years with a load of life, but not intelligent life.

It is possible that the porpoises and some of their relatives are intelligent, but, as sea creatures, they lack limbs and could not develop the use of fire; consequently, their intelligence, if it exists, could not be bent in the direction of a developed technology. If land life alone is considered, then it is only for about a million years or so that the earth has been able to boast a living creature with intelligence greater than that of an ape.

Still, this means that the earth has possessed intelligent life for 1/2,000 of the time it has possessed life of any kind (as a rough guess). If we can say that of all life-bearing planets, 1 out of 2,000 bears intelligent life, then that would still mean that out of the 640 million habitable planets Dole speaks of, there may be 320,000 intelligences. We may well be far from alone in the universe.

Until recently this sort of possibility was considered seriously only in science-fiction stories. Those of my readers who happen to be aware that I have written a few science-fiction stories in my time and who may put down my remarks here to over-enthusiasm may be assured that today many astronomers accept the high probability of intelligent life on many planets.

In fact, United States scientists took the possibilities seriously enough to set up an enterprise, under the leadership of Frank D. Drake, called Project Ozma (deriving its name from one of the *Oz* books for children) to listen for possible radio signals from other worlds. The idea is to look for some pattern in radio waves coming in from space. If they detect signals in an ordered pattern, as opposed to the random formless broadcasts from radio stars or excited matter in space, it may be assumed that such signals will represent messages from some extraterrestrial intelligence. Of course, even if such messages were received, communication with the distant intelligence would still be a problem. The messages would have been many years on the way, and a reply also would take many years to reach the distant broadcasters, since the nearest potentially habitable planet is $4\frac{1}{3}$ light-years away.

The sections of the heavens listened to at one time or another in the course of Project Ozma included the directions in which lie Epsilon Eridani, Tau Ceti, Omicron-2 Eridani, Epsilon Indi, Alpha Centauri, 70 Ophiuchi, and 61 Cygni. After two months of negative results, however, the project was suspended.

The notion, however, is by no means dead. Because the universe is so vast, evidence of far-away life cannot yet be detected, and may never be detected. But precisely because the universe is so vast, such far-away life *must* exist, even intelligent life, perhaps in many millions of varieties.

And perhaps if we never find them, they may yet, in time to come, find us.

4 The Micro-organisms

Bacteria

Before the seventeenth century, the smallest known living creatures were tiny insects. It was taken for granted, of course, that no smaller organisms existed. Living beings might be made invisible by a supernatural agency (all cultures believed that in one way or another), but no one supposed that there were creatures in nature too small to be seen.

Had man suspected such a thing, he might have come much sooner to the deliberate use of magnifying devices. Even the Greeks and Romans knew that glass objects of certain shapes would focus sunlight on a point and would magnify objects seen through it. A hollow glass sphere filled with water would do so, for instance. Ptolemy discussed the optics of burning glasses, and Arabic writers such as Alhazen, about A.D. 1000, extended his observations.

It was Robert Grosseteste, an English bishop, philosopher, and keen amateur scientist, who, early in the thirteenth century, first suggested a peacetime use for this weapon. He pointed out that lenses (so named because they were shaped like lentils) might be useful in magnifying objects too small to see conveniently. His pupil, Roger Bacon, acted on this suggestion and devised spectacles to improve poor vision.

At first only convex lenses, to correct far-sightedness, were made. Concave lenses, to correct near-sightedness, were not developed until about 1400. The invention of printing brought more and more demand for spectacles, and by the sixteenth century spectacle-making was a skilled profession. It became a particular speciality in the Netherlands.

(Bifocals, serving for both far and near vision, were invented by Benjamin Franklin in 1760. In 1827, the British astronomer George Biddell Airy designed the first lenses to correct astigmatism, from which he suffered himself. And, around 1888, a French physician introduced the idea of contact lenses, which may some day make ordinary spectacles more or less obsolete.)

To get back to the Dutch spectacle-makers. In 1608, so the story goes, an apprentice to a spectacle-maker named Hans Lippershey, whiling away an idle hour, amused himself by looking at objects through two lenses held one behind the other. He was amazed to find that when he held them a certain distance apart, far-off objects appeared close at hand. The apprentice promptly told his master about it, and Lippershey proceeded to build the first 'telescope', placing the two lenses in a tube to hold them at the proper spacing. Prince Maurice of Nassau, commander of the Dutch forces in rebellion against Spain, saw the military value of the instrument and endeavoured to keep it secret.

He reckoned without Galileo, however. Hearing rumours of the invention of a far-seeing glass, Galileo, knowing no more than that it was made with lenses, soon discovered the principle and built his own telescope; his was completed within six months after Lippershey's.

By rearranging the lenses of his telescope, Galileo found that he could magnify close objects, so that it was in effect a 'microscope'. Over the next decades several scientists built microscopes. An Italian naturalist named Francesco Stelluti studied insect anatomy with one; Malpighi discovered the capillaries; and Hooke discovered the cells in cork.

But the importance of the microscope was not really appreciated until Anton van Leeuwenhoek, a merchant in the city of Delft, took it up (see p. 90). Some of van Leeuwenhoek's lenses could enlarge up to 200 times.

Van Leeuwenhoek looked at all sorts of objects quite indiscriminately, describing what he saw in lengthy detail in letters to the Royal Society in London. It was rather a triumph for the democracy of science that the tradesman was elected a fellow of the gentlemanly Royal Society. Before he died, the Queen of England and Peter the Great, Tsar of all the Russias, visited the humble microscope-maker of Delft.

Through his lenses van Leeuwenhoek discovered sperm cells and red blood cells, and actually saw blood moving through capillaries in the tail of a tadpole. More important, he was the first to see living creatures too small to be seen by the unaided eye. He discovered these 'animalcules' in stagnant water in 1675. He also resolved the tiny cells of yeast, and, at the limit of his lenses' magnifying power, he finally, in 1676, came upon 'germs', which today we know as bacteria.

Microscopes improved only slowly, and it took a century and a half before objects the size of germs could be studied with ease. For instance, it was not until 1830 that the English optician Joseph Jackson Lister devised an 'achromatic microscope', which eliminated the rings of colour that limited the sharpness of the image. Lister found that red blood corpuscles (first detected as featureless blobs by the Dutch physician Jan Swammerdam, in 1658) were biconcave discs – like tiny doughnuts with dents instead of a hole. The achromatic microscope was a great advance, and in 1878 the German physicist Ernst Abbe began a series of improvements that resulted in what might be called the modern optical microscope.

The members of the new world of microscopic life gradually received names. Van Leeuwenhoek's 'animalcules' actually were animals, feeding on small particles and moving about by means of small whips (flagellae) or hair-like cilia or advancing streams of protoplasm (pseudopods). These animals were given the name 'protozoa' (Greek for 'first animals'), and the German zoologist Karl Theodor Ernst Siebold identified them as single-celled creatures.

'Germs' were something else: much smaller than protozoa and much simpler. Although some could move about, most lay quiescent and merely grew and multiplied. Except for their lack of chlorophyll, they showed none of the properties associated with animals. For that reason they were usually classified among the fungi – plants that lack chlorophyll and live on organic matter. Nowadays most biologists tend to consider them as neither plant nor animal but put them in a class by themselves. 'Germ' is a misleading name for them. The same term may apply to the living part of a seed (e.g., the 'wheat germ'), or to sex cells ('germ cells'), or to embryonic organs ('germ layers'), or, in fact, to any small object possessing the potentiality of life.

The Danish microscopist Otto Frederik Müller managed to see the little creatures well enough in 1773 to distinguish two types: 'bacilli' (from a Latin word meaning 'little rods') and 'spirilla' (for their spiral shape). With the advent of achromatic microscopes, the Austrian surgeon Theodor Billroth saw still smaller varieties to which he applied the term 'coccus' (from the Greek word for 'berry'). It was the German botanist Ferdinand Julius Cohn who finally coined the name 'bacterium' (also from a Latin word meaning 'little rod').

Pasteur popularized the general term 'microbe' ('small life') for all forms of microscopic life – plant, animal, and bacterial. But this word was soon adopted for the bacteria, just then coming into notoriety. Today the general term for microscopic forms of life is 'micro-organisms'.

It was Pasteur who first definitely connected micro-organisms with disease, thus founding the modern science of 'bacteriology' or, to use a more general term, 'microbiology'. This came about through Pasteur's concern with something that seemed an industrial problem rather than a medical one. In the 1860s, the French silk industry was being ruined by a disease of the silkworms. Pasteur, having already rescued France's wine makers, was put to work on this problem, too. Again making inspired use of the microscope, as he had in studying asymmetric crystals and varieties of yeast cells, Pasteur found micro-organisms infecting the sick silkworms and the mulberry leaves on which they fed. He recommended that all infected worms and leaves be destroyed and a fresh start be made with the uninfected worms and leaves that remained. This drastic step was taken, and it worked.

Pasteur did more with these researches than merely to revive the silk industry. He generalized his conclusions and enunciated the 'germ theory of disease' – without question one of the greatest single medical discoveries ever made (and it was made, not by a physician, but by a chemist, as chemists delight in pointing out).

Before Pasteur, doctors had been able to do little more for their patients than recommend rest, good food, fresh air, and clean surroundings, and, occasionally, handle a few types of emergency. This much had been advocated by the Greek physician Hippocrates of Cos (the 'father of medicine') as long ago as 400 B.C. It was Hippocrates who introduced the rational view of medicine, turning away from the arrows of Apollo and demonic possession to proclaim that even epilepsy, called the 'sacred disease', was not the result of being affected by some god's influence, but was a mere physical disorder to be treated as such. The lesson was never entirely forgotten by later generations.

However, medicine progressed surprisingly little in the next two millennia. Doctors could lance boils, set broken bones, and prescribe a few specific remedies that were simply products of folk wisdom: such drugs as quinine from the bark of the cinchona tree (originally chewed by the Peruvian Indians to cure themselves of malaria) and digitalis from the plant called foxglove (an old herb-women's remedy to stimulate the heart). Aside from these few treatments (and the smallpox vaccine, which I shall discuss later), many of the medicines and treatments dispensed by physicians after Hippocrates tended to heighten the death rate rather than lower it.

One of the interesting advances made in the first two and a half centuries of the Age of Science was the invention, in 1819, of the stethoscope by the French physician, René Théophile Hyacinthe Laennec. In its original form, it was little more than a wooden tube designed to help the doctor hear and interpret the sounds of the beating heart. Improvements since then have made it as characteristic and inevitable an accompaniment of the physician as the slide rule is of an engineer.

It is not surprising, then, that up to the nineteenth century

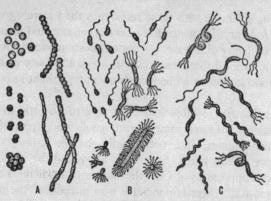

Types of bacteria: cocci (A), bacilli (B), and spirilla (C). Each type has a number of varieties.

even the most civilized countries were periodically swept by plagues, some of which had a profound effect on history. The plague in Athens that killed Pericles, at the time of the Peloponnesian War, was the first step in the ultimate ruin of Greece. Rome's downfall probably began with the plagues that fell upon the empire during the reign of Marcus Aurelius. The Black Death of the fourteenth century is estimated to have killed off a fourth of the population of Europe; this plague and gunpowder combined to destroy the social structure of the Middle Ages.

To be sure, plagues did not end when Pasteur discovered that infectious diseases were caused and spread by micro-organisms. In India cholera is still endemic, and other underdeveloped countries suffer severely from epidemics. Disease has remained a major hazard of wartime. Virulent new organisms arise from time to time and sweep over the world; indeed, the influenza pandemic of 1918 killed an estimated 15 million people, a larger number of people than died in any other plague in the history of mankind, and nearly twice as many as were killed in the then just-completed First World War.

Nevertheless, Pasteur's discovery was a great turning point. The death rate in Europe and the United States began to fall markedly, and life expectancy steadily rose. Thanks to the scientific study of disease and its treatment, which began with Pasteur, men and women in the more advanced regions of the world can now expect to live an average of 70 years, whereas before Pasteur the average was only 40 years under the most favourable conditions and perhaps only 25 years under unfavourable conditions. Since the Second World War, life expectancy has been zooming upwards even in the less advanced regions of the world.

Even before Pasteur advanced the germ theory in 1865, a Viennese physician named Ignaz Philipp Semmelweiss had made the first effective attack on bacteria, without, of course, knowing what he was fighting. He was working in the maternity ward of one of Vienna's hospitals, where 12 per cent or more of the new mothers died of something called 'puerperal' fever (in plain English, 'childbed fever'). Semmelweiss noted uneasily that women who bore their babies at home with only the services of ignorant midwives practically never got puerperal fever. His suspicions were further aroused by the death of a doctor in the hospital with symptoms that strongly resembled those of puerperal fever, after the doctor had cut himself while dissecting a corpse. Were the doctors and students who came in from the dissection wards somehow transmitting this disease to the women whose delivery they attended? Semmelweiss insisted that the doctors wash their hands in a solution of chlorinated lime. Within a year the death rate in the maternity wards fell from 12 per cent to 1·5 per cent.

But the veteran doctors were livid. Resentful of the implication that they had been murderers, and humiliated by all the handwashing, they drove Semmelweiss out of the hospital. (In this they were helped by the fact that he was a Hungarian and Hungary was in revolt against the Austrian rulers.) Semmelweiss went to Budapest, where he reduced the maternal death rate, while in Vienna the hospitals reverted to death traps for another decade or so. But Semmelweiss himself died of puerperal fever from an accidental infection (at the age of forty-seven) in 1865 – just too soon to see the scientific vindication of his suspicions about the transmission of disease. That was the year that Pasteur discovered micro–organisms in the diseased silkworms, and an English surgeon named Joseph Lister (the son of the inventor of the achromatic microscope) independently introduced the chemical attack upon germs.

Lister resorted to the drastic substance phenol (carbolic acid). He used it first in dressings for a patient with a compound fracture. Up to that time, any serious wound almost invariably led to infection. Of course, Lister's phenol killed the tissues around the wound, but it did kill the bacteria. The patient made a remarkably untroubled recovery.

Lister followed up this success with the practice of spraying the operating room with phenol. It must have been hard on those who had to breathe it, but it began to save lives. As in Semmelweiss's case, there was opposition, but Pasteur's experiments had created a rationale for antisepsis, and Lister easily won the day.

Pasteur himself had somewhat harder going in France (unlike Lister, he lacked the union label of the M.D.), but he prevailed on surgeons to boil their instruments and steam their bandages. Sterilization with steam *à la Pasteur* replaced Lister's unpleasant phenol spray. Milder antiseptics, which could kill bacteria without unduly damaging tissue, were sought and found. The French physician Casimir Joseph Davaine reported on the antiseptic properties of iodine in 1873, and 'tincture of iodine' (i.e., iodine dissolved in a mixture of alcohol and water) was at one time in common use in the home. It and similar products were automatically applied to every scratch. The number of infections prevented in this way was undoubtedly enormous.

In fact, the search for protection against infection leaned more and more in the direction of preventing germ entry ('asepsis') rather than of destroying them after they had gained a foothold, as was implied in antisepsis. The American surgeon William Stewart Halstead introduced the practice of using sterilized rubber gloves during operations in 1890; by 1900, the British physician William Hunter had added the gauze mask to protect the patient against the germ content of the physician's breath.

Meanwhile the German physician Robert Koch had begun to

identify the specific bacteria responsible for various diseases. To do this he introduced a vital improvement in the nature of culture media, i.e., the food supply in which bacteria were grown. Where Pasteur used liquid media, Koch introduced solid media. He planted isolated samples on gelatin (for which agar, a gelatin-like substance obtained from seaweed, was substituted later). If a single bacterium was deposited (with a fine needle) in a spot on this medium, a pure colony would grow round the spot, because on the solid surface of the agar the bacteria lacked the ability to move or drift away from the original parent, as they would have done in a liquid. An assistant of Koch, Julius Richard Petri, introduced the use of shallow glass dishes with covers, to pro-tect the cultures from contamination by bacterial spores floating in air; such 'Petri dishes' have been used for the purpose ever since.

In this way, individual bacteria would give rise to colonies which could then be cultured separately and tested to see what disease they would produce in an experimental animal. The technique not only made it possible to identify a given infection but also permitted experiments with various possible treatments to kill specific bacteria.

With his new techniques, Koch isolated a bacillus that caused anthrax and, in 1882, another that caused tuberculosis. In 1884, he also isolated the bacterium that caused cholera. Others followed in Koch's path. In 1883, for instance, the German pathologist Edwin Klebs isolated the bacterium that caused diphtheria. In 1905, Koch received the Nobel Prize in medicine and physiology.

Once bacteria had been identified, the next task was to find drugs that would kill a bacterium without killing the patient as well. To such a search, the German physician and bacteriologist Paul Ehrlich, who had worked with Koch, now addressed him-self. He thought of the task as looking for a 'magic bullet'which would not harm the body but strike only the bacteria.

Ehrlich was interested in dyes that stained bacteria. This had an important relationship to cell research. The cell in its natural state is colourless and transparent so that little detail within it could be seen. Early microscopists had tried to use dyes to colour the cells, but it was only after Perkin's discovery of aniline dyes (see p. 22) that the technique became practical. Though Ehrlich was not the first to use synthetic dyes in staining, he worked out the techniques in detail in the late 1870s, and it was this that led the way to Flemming's study of mitosis and Feulgen's study of DNA in the chromosomes (see p. 159).

But Ehrlich had other game in mind, too. He turned to these dyes as possible bactericides. A stain that reacted with bacteria more strongly than with other cells might well kill the bacteria, even when it was injected into the blood in a concentration low enough not to harm the cells of the patient. By 1907, Ehrlich had discovered a dye, called 'Trypan red', which would stain trypanosomes, the organisms responsible for the dreaded African sleeping sickness, transmitted via the tsetse fly. Trypan red, when injected into the blood in proper doses, could kill trypanosomes without killing the patient.

Ehrlich was not satisfied: he wanted a surer kill of the microorganisms. Assuming that the toxic part of the trypan-red molecule was the 'azo' combination – that is, a pair of nitrogen atoms ($-N=N-$) – he wondered what a similar combination of arsenic atoms ($-As=As-$) might accomplish. Arsenic is chemically similar to nitrogen but much more toxic. Ehrlich began to test arsenic compounds one after the other almost indiscriminately, numbering them methodically as he went. In 1909, a Japanese student of Ehrlich's, Sahachiro Hata, tested compound 606, which had failed against the trypanosomes, on the bacterium that caused syphilis. It proved deadly against this microbe (called a 'spirochete' because it is spiral-shaped).

At once Ehrlich realized he had stumbled on something more important than a cure for trypanosomiasis, which after all was a limited disease confined to the tropics. Syphilis had been a hidden scourge of Europe for more than 400 years, ever since Columbus's time. (Columbus's men are supposed to have brought it back from the Caribbean Indians; in return, Europe donated smallpox to the Indians.) Not only was there no cure for syphilis, but prudishness had clothed the disease in a curtain of silence that let it spread unchecked.

Ehrlich devoted the rest of his life (he died in 1915) to the attempt to combat syphilis with compound 606, or, as he called it, 'Salvarsan' – 'safe arsenic'. (Its chemical name is arsphenamine.) It could cure the disease, but its use was not without risk, and Ehrlich had to bully hospitals into using it correctly.

With Ehrlich, a new phase of chemotherapy came into being. Pharmacology, the study of the action of chemicals other than foods (that is 'drugs') upon organisms, finally came into its own as a twentieth-century adjunct of medicine. Arsphenamine was the first synthetic drug, as opposed to the plant remedies such as quinine or the mineral remedies of Paracelsus and those who imitated him.

Naturally, the hope at once arose that every disease might be fought with a little tailored antidote all its own. But for a quarter of a century after Ehrlich's discovery the concocters of new drugs had little luck. About the only success of any sort was the synthesis by German chemists of 'plasmochin' in 1924 and 'atabrine' in 1930; they could be used as substitutes for quinine against malaria. (They were very helpful to Western troops in jungle areas during the Second World War, when the Japanese held Java, the source of the world supply of quinine, which, like rubber, had moved from South America to south-east Asia.)

In 1932 came a breakthrough. A German chemist named Gerhard Domagk had been injecting various dyes into infected mice. He tried a new red dye called 'Prontosil' on mice infected with the deadly haemolytic streptococcus. The mice survived! He used it on his own daughter, who was dying of streptococcal blood poisoning. She survived also. Within three years Prontosil had gained worldwide renown as a drug that could stop the strep infection in man.

Oddly, Prontosil did not kill streptococci in the test tube – only in the body. At the Pasteur Institute in Paris, Jacques Trefouël and his co-workers decided that the body must change Prontosil into some other substance that took effect on the bacteria. They proceeded to break down Prontosil to the effective fragment, named 'sulphanilamide'. This compound had been synthesized in 1908, reported perfunctorily, and forgotten. Sulphanilamide's structure is:

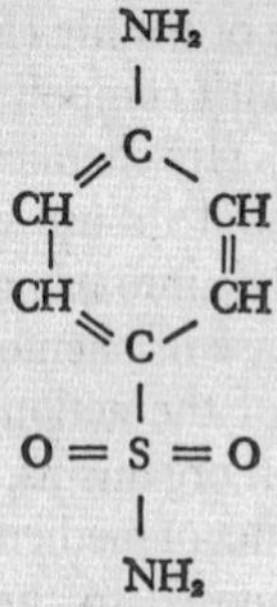

$$NH_2$$
$$|$$
$$C$$

CH   CH

CH   CH

$$C$$
$$|$$
$$O = S = O$$
$$|$$
$$NH_2$$

It was the first of the 'wonder drugs'. One after another bacterium fell before it. Chemists found that by substituting various groups for one of the hydrogen atoms on the sulphur-containing group, they could obtain a series of compounds, each of which had slightly different antibacterial properties. 'Sulphapyridine' was introduced in 1937, 'sulphathiazole' in 1939, and 'sulphadiazine' in 1941. Physicians now could choose from a whole platoon of 'sulpha drugs' for various infections. In the medically advanced countries, the death rates from bacterial diseases, notably, pneumococcal pneumonia, dropped dramatically.

Domagk was awarded the Nobel Prize in medicine and physiology in 1939. When he wrote the usual letter of acceptance, he was promptly arrested by the Gestapo; the Nazi government, for peculiar reasons of its own, refused to have anything to do with the Nobel prizes. Domagk felt it the better part of valour to refuse the prize. After the Second World War, when he was at last free to accept the honour, Domagk went to Stockholm to receive it officially.

The sulpha drugs had only a brief period of glory, for they were soon put in the shade by the discovery of a far more potent kind of antibacterial weapon – the antibiotics.

All living matter (including man) eventually returns to the soil to decay and decompose. With the dead matter and the wastes of living creatures go the germs of the many diseases that infect those creatures. Why is it, then, that the soil is usually so remarkably clean of infectious germs? Very few of them (the anthrax bacillus is one of the few) survive in the soil. A number of years ago bacteriologists began to suspect that the soil harboured micro-organisms or substances that destroyed bacteria. As early as 1877, for instance, Pasteur had noticed that some bacteria died in the presence of others. And if this is so, the soil offers a large variety of organisms within which to search for death to others of their kind. It is estimated that each acre of soil contains about 2,000 pounds of moulds, 1,000 pounds of bacteria, 200 pounds of protozoa, 100 pounds of algae, and 100 pounds of yeast.

One of those who conducted a deliberate search for bactericides in the soil was the French-American microbiologist René Jules Dubos. In 1939, he isolated from a soil micro-organism called *Bacillus brevis* a substance, 'tyrothricin', from which he isolated two bacteria-killing compounds that he named 'gramicidin' and 'tyrocidin'. They turned out to be peptides containing D-amino acids – the mirror images of the ordinary L-amino acids that make up most natural proteins.

Gramicidin and tyrocidin were the first antibiotics produced as such. But an antibiotic which was to prove immeasurably more important had been discovered – and merely noted in a scientific paper – twelve years earlier.

The British bacteriologist Alexander Fleming one morning found that some cultures of staphylococcus (the common pus-forming bacterium), which he had left on a bench, were contaminated with something that had killed the bacteria. There were little clear circles where the staphylococci had been destroyed in the culture dishes. Fleming, being interested in antisepsis (he had discovered that an enzyme in tears, called 'lysozome', had antiseptic properties), at once investigated to see what had killed the bacteria, and he discovered that it was a common bread mould, *Penicillium notatum*. Some substance, which he named 'penicillin', produced by the mould was lethal to germs. Fleming dutifully published his results in 1929, but no one paid much attention at the time.

Ten years later the British biochemist Howard Walter Florey and his German-born associate Ernst Boris Chain became intrigued by the almost forgotten discovery and set out to try to isolate the antibacterial substance. By 1941, they had obtained an extract which proved effective clinically against a number of 'gram-positive' bacteria (bacteria that retain a dye developed in 1884 by the Danish bacteriologist Hans Christian Joachim Gram).

Because wartime Britain was in no position to produce the drug, Florey went to the United States and helped to launch a programme which developed methods of purifying penicillin and speeding up its production by the mould. In 1943, five hundred cases were treated with penicillin and, by the war's end, large-scale production and use of penicillin were under way. Not only did penicillin pretty much supplant the sulpha drugs, but it became (and still is) one of the most important drugs in the entire practice of medicine. It is effective against a wide range of infections, including pneumonia, gonorrhoea, syphilis, puerperal fever, scarlet fever, and meningitis. (The range of effectivity is called the 'antibiotic spectrum'.) Furthermore, it has practically no toxicity or undesirable side effects, except in penicillin-sensitive individuals.

In 1945, Fleming, Florey, and Chain shared the Nobel Prize in medicine and physiology.

Penicillin set off an almost unbelievably elaborate hunt for other antibiotics. (The word was coined in 1942 by the Rutgers University bacteriologist Selman Abraham Waksman.)

In 1943, Waksman isolated from a soil mould of the genus *Streptomyces* the antibiotic known as 'streptomycin'. Streptomycin hit the 'gram-negative' bacteria (those that easily lose the Gram stain). Its greatest triumph was against the tubercle bacillus. But streptomycin, unlike penicillin, is rather toxic, and it must be used with caution.

For the discovery of streptomycin, Waksman received the Nobel Prize in medicine and physiology in 1952.

Another antibiotic, chloramphenicol, was isolated from moulds of the genus *Streptomyces* in 1947. It attacks not only gram-positive and gram-negative bacteria but also certain smaller organisms, notably those causing typhus fever and psittacosis ('parrot fever'). But its toxicity calls for care in its use.

Then came a whole series of 'broad-spectrum' antibiotics, found after painstaking examination of many thousands of soil samples – Aureomycin, Terramycin, Achromycin, and so on. The first of these, Aureomycin, was isolated by Benjamin Minge Duggar and his co-workers in 1944 and was placed on the market in 1948. These antibiotics are called 'tetracyclines', because in each case the molecule is composed of four rings side by side. They are effective against a wide range of micro-organisms and are particularly valuable because they are relatively non-toxic. One of their annoying side effects is that, by disrupting the balance of useful bacteria in the digestive tract, they upset the natural course of intestinal events and sometimes cause diarrhoea.

Next to penicillin (which is much less expensive), the tetra-cyclines are now the most commonly used prescription drugs for infection. Thanks to the antibiotics in general, the death rates for many of the infectious diseases have fallen to cheeringly low levels. (Of course human beings left alive by the continuing mastery of man over infectious disease have a much greater chance of succumbing to a metabolic disorder. Thus, in the last eighty years, the incidence of diabetes, the most common such disorder, has increased tenfold.)

The chief disappointment in the development of chemotherapy has been the speedy rise of resistant strains of bacteria. In 1939, for instance, all cases of meningitis and pneumococcal pneu-monia showed a favourable response to the administration of sulpha drugs. Twenty years later, only half the cases did. The various antibiotics also became less effective with time. It is not that the bacteria 'learn' to resist but that resistant mutants among them flourish and multiply when the 'normal' strains are killed off. This danger is greatest in hospitals, where antibiotics are used constantly and where the patients naturally have below-normal resistance to infection. Certain new strains of staphylo-cocci resist antibiotics with particular stubbornness. This 'hospital staph' is now a serious worry in maternity wards, for instance, and attained headline fame in 1961, when an attack of pneumonia, sparked by such resistant bacteria, nearly ended the life of screen star Elizabeth Taylor.

Fortunately, where one antibiotic fails another may still attack a resistant strain. New antibiotics, and synthetic modifications of the old, may hold the line in the contest against mutations. The ideal thing would be to find an antibiotic to which no mutants are immune. Then there would be no survivors of that particular bacterium to multiply. A number of such candidates have been produced. For instance, a modified penicillin, called 'Staphcyllin', was developed in 1960. It is partly synthetic, and because its structure is strange to bacteria, its molecule is not split and its activity ruined by enzymes such as 'penicillinase' (first discovery by Chain), which resistant strains use against ordinary penicillin. Consequently, Staphcyllin is death to otherwise resistant strains; it was used to save Miss Taylor's life, for instance. Yet strains of staphylococcus, resistant to synthetic penicillins, have also turned up. Presumably, the merry-go-round will go on forever.

Additional allies against resistant strains are various other new antibiotics and modified versions of old ones. One can only hope that the stubborn versatility of chemical science will manage to keep the upper hand over the stubborn versatility of the disease germs.

The same problem of the development of resistant strains arises in man's battle with his larger enemies, the insects, which not only compete dangerously for food, but which also spread disease. The modern chemical defences against insects arose in 1939, with the development by the Swiss chemist Paul Müller of the chemical, 'dichlorodiphenyltrichloroethane', better known by its initials, 'DDT'. Müller was awarded the Nobel Prize in medicine and physiology for this feat in 1948.

By then DDT had come into large-scale use and, already, resistant strains of houseflies had developed. Newer 'insecticides' (or, to use a more general term that will cover chemicals used against rats or against weeds, 'pesticides') must continually be developed.

In addition, there are critics of the over-chemicalization of man's battle against other forms of life. There are some who are concerned lest mankind make it possible for an increasingly large segment of the population to remain alive only through the grace of chemistry; they fear that if ever man's technological organization falters, even temporarily, great carnage will result as populations fall prey to the infections and diseases to which they lack natural resistance.

As for the pesticides, the American science-writer, Rachel Louise Carson, published a book, *Silent Spring*, in 1962 that dramatically brought to the fore the possibility that, by indiscriminate use of chemicals, mankind might kill harmless and even useful species along with those he was actually attempting to destroy. Furthermore, Miss Carson maintained that to destroy living things without due consideration might lead to a serious upsetting of the intricate system whereby one species depended on another and, in the end, hurt man more than it helps him. The study of this interlinking of species is termed 'ecology', and there is no question but that Miss Carson's book encourages a new hard look at this branch of biology.

The answer, of course, is not to abandon technology and give up all attempts to control insects (the price in disease and starvation would be too high) but to find methods that are more specific and less damaging to the ecological structure generally.

For instance, insects have their enemies. Those enemies, whether insect-parasites or insect-eaters, might be encouraged. Sounds and odours might be used to repel insects or to lure them to their death. Insects might be sterilized through radiation. In each case, every effort should be made to zero in on the insect being fought.

One hopeful line of attack led by the American biologist Carroll Milton Williams, is to make use of the insects' own hormones. Insects moult periodically and pass through two or three well-defined stages: larva, pupa, and adult. The transitions are complex and are controlled by hormones. Thus, one called 'juvenile hormone' prevents formation of the adult stage until an appropriate time. By isolating and applying the juvenile hormone, the adult stage is held back to the point where the insect is killed. Each insect has its own juvenile hormone and is affected only by its own. A particular juvenile hormone might thus be used to attack one particular species of insect without affecting any other organism in the world. Guided by the structure of the hormone, biologists may even prepare synthetic substitutes that will be much cheaper and do the job as well.

In short, the answer to the fact that scientific advance may sometimes have damaging side-effects is not to abandon scientific advance, but to substitute still more advance – intelligently and cautiously applied.

As to how the chemotherapeutic agents work, the best guess seems to be that each drug inhibits some key enzyme in the micro-organism in a competitive way. This is best established in the case of the sulpha drugs. They are very similar to 'para amino-benzoic acid' (generally written *p*-aminobenzoic acid), which has this structure:

$$
\begin{array}{c}
NH_2 \\
| \\
C \\
CH \quad\quad CH \\
| \quad\quad\quad || \\
CH \quad\quad CH \\
C \\
| \\
C = O \\
| \\
OH
\end{array}
$$

P-aminobenzoic acid is necessary for the synthesis of 'folic acid', a key substance in the metabolism of bacteria as well as other cells. A bacterium that picks up a sulphanilamide molecule instead of *p*-aminobenzoic acid can no longer produce folic acid, because the enzyme needed for the process is put out of action. Consequently, the bacterium ceases to grow and multiply. The cells of the human patient, on the other hand, are not disturbed; they obtain folic acid from food and do not have to synthesize it. There are no enzymes in human cells to be inhibited by moderate concentrations of the sulpha drugs in this fashion.

Even where a bacterium and the human cell possess similar enzymes, there are other ways of attacking the bacterium select-ively. The bacterial enzyme may be more sensitive to a given drug than the human enzyme is, so that a certain dose will kill the bacterium without seriously disturbing the human cells. Or a drug of the proper design may be able to penetrate the bacterial cell membrane but not the human cell membrane. Penicillin, for instance, interferes with the manufacture of cell walls, which bacteria possess but animal cells do not.

Do the antibiotics also work by competitive inhibition of enzymes? Here the answer is less clear. But there is good ground for believing that at least some of them do.

Gramicidin and tyrocidin, as I mentioned earlier, contain the 'unnatural' D-amino acids. Perhaps these jam up the enzymes that form compounds from the natural L-amino acids. Another peptide antibiotic, bacitracin, contains ornithine; this may inhibit enzymes from making use of arginine, which ornithine resembles. There is a similar situation in streptomycin: its molecule contains an odd variety of sugar which may interfere with some enzyme acting on one of the normal sugars of living cells. Again, chloramphenicol resembles the amino acid phenylalanine; likewise, part of the penicillin molecule resembles the amino acid cysteine. In both of these cases the possibility of competitive inhibition is strong.

The clearest evidence of competitive action by an antibiotic turned up so far involves 'puromycin', a substance produced by a *Streptomyces* mould. This compound has a structure much like that of nucleotides (the building units of nucleic acids), and Michael Yarmolinsky and his co-workers at Johns Hopkins University have shown that puromycin, competing with transfer-RNA, interferes with the synthesis of proteins. Again, streptomycin interferes with transfer-RNA, forcing the misreading of the genetic code and the formation of useless protein. Unfortunately, this form of interference makes it toxic to other cells besides bacteria, because it prevents their normal production of necessary proteins. Thus puromycin is too dangerous a drug to use, and streptomycin is nearly so.

Viruses

To most people it may seem mystifying that the 'wonder drugs' have had so much success against the bacterial diseases and so little success against the virus diseases. Since viruses, after all, can cause disease only if they reproduce themselves, why should it not be possible to jam the virus's machinery just as we jam the bacterium's machinery? The answer is quite simple, and indeed obvious, once you realize how a virus reproduces itself. As a complete parasite, incapable of multiplying anywhere except inside a living cell, the virus has very little, if any, metabolic machinery of its own. To make copies of itself, it depends entirely on materials supplied by the cell it invades. This, however, it can do with great efficiency. One virus within a cell can become 200 in twenty-five minutes. And it is therefore difficult to deprive it of those materials or jam the machinery without destroying the cell itself.

Biologists discovered the viruses only recently, after a series of encounters with increasingly simple forms of life. Perhaps as good a place as any to start this story is the discovery of the cause of malaria.

Malaria has, year in and year out, probably killed more people in the world than any other infectious ailment, since until recently about 10 per cent of the world's population suffered from the disease, with three million deaths a year caused by it. Until 1880, it was thought to be caused by the bad air (*mala aria* in Italian) of swampy regions. Then a French bacteriologist, Charles Louis Alphonse Laveran, discovered that the red blood cells of malaria-stricken individuals were infested with parasitic protozoa of the genus *Plasmodium*. (For this discovery, Laveran was awarded the Nobel Prize in medicine and physiology in 1907.)

In the early 1890s a British physician named Patrick Manson, who had conducted a missionary hospital in Hong Kong, pointed out that swampy regions harboured mosquitoes as well as dank air, and he suggested that mosquitoes might have something to do with the spread of malaria. A British physician in India,

Ronald Ross, pursued this idea, and he was able to show that the malarial parasite did indeed pass part of its life-cycle in mosquitoes of the genus *Anopheles*. The mosquito picked up the parasite in sucking the blood of an infected person and then would pass it on to any person it bit.

For his work, bringing to light for the first time the transmission of a disease by an insect 'vector', Ross received the Nobel Prize in medicine and physiology in 1902. It was a crucial discovery of modern medicine, for it showed that a disease might be stamped out by killing off the insect carrier. Drain the swamps that breed mosquitoes; eliminate stagnant water; destroy the mosquitoes with insecticides, and you can stop the disease. Since the Second World War, large areas of the world have been freed of malaria in just this way, and the total number of deaths from malaria has declined by at least one third from its maximum.

Malaria was the first infectious disease traced to a non-bacterial micro-organism (a protozoan in this case). Very shortly afterwards, another non-bacterial disease was tracked down to a similar cause. It was the deadly yellow fever, which as late as 1898, during an epidemic in Rio de Janeiro, killed nearly 95 per cent of those it struck. In 1899, when an epidemic of yellow fever broke out in Cuba, a United States board of inquiry, headed by the bacteriologist Walter Reed, went to Cuba to investigate the causes of the disease.

Reed suspected a mosquito vector, such as had just been exposed as the transmitter of malaria. He first established that the disease could not be transmitted by direct contact between the patients and doctors or by way of the patient's clothing or bedding. Then some of the doctors deliberately let themselves be bitten by mosquitoes that had previously bitten a man sick with yellow fever. They got the disease, and one of the courageous investigators, Jesse William Lazear, died. But the culprit was identified as the *Aedes aegypti* mosquito. The epidemic in Cuba was checked, and yellow fever is no longer a serious disease in the medically advanced parts of the world.

As a third example of a non-bacterial disease, there is typhus fever. This infection is endemic in North Africa and was brought into Europe via Spain during the long struggle of the Spaniards against the Moors of North Africa. Commonly known as 'plague', it is very contagious and has devastated nations. In the First World War the Austrian armies were driven out of Serbia by the typhus when the Serbian army itself was unequal to the task. The ravages of typhus in Poland and Russia during that war and its aftermath (some three million persons died of the disease) did as much as military action to ruin those nations.

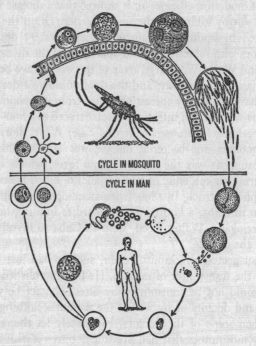

CYCLE IN MOSQUITO

CYCLE IN MAN

Life-cycle of the malarial micro-organism.

At the turn of the twentieth century the French bacteriologist Charles Nicolle, then in charge of the Pasteur Institute in Tunis, noticed that although typhus was rife in the city, no one caught it in the hospital. The doctors and nurses were in daily contact with typhus-ridden patients and the hospital was crowded, yet there was no spread of the disease there. Nicolle considered what happened when a patient came into the hospital, and it struck him that the most significant change was a thorough washing of the patient and removal of his lice-infested clothing. Nicolle decided that the body louse must be the vector of typhus. He proved the correctness of his guess by experiments. He received the Nobel Prize in medicine and physiology in 1928 for his discovery. Thanks to his finding, and the discovery of DDT, typhus fever did not repeat its deadly carnage in the Second World War. In January 1944, DDT was brought into play against the body louse. The population of Naples was sprayed *en masse* and the lice died. For the first time in history, a winter epidemic of typhus (when the multiplicity of clothes, not removed very often, made louse-infestation almost certain and almost universal) was stopped in its tracks. A similar epidemic was stopped in Japan in late 1945 after the American occupation. The Second World War became almost unique among history's wars by possessing the dubious merit of killing fewer people by disease than by guns and bombs.

Typhus, like yellow fever, is caused by an agent smaller than a bacterium, and we must now enter the strange and wonderful realm populated by sub-bacterial organisms.

To get some idea of the dimensions of objects in this world, let us look at them in order of decreasing size. The human ovum is about 100 microns (100 millionths of a metre) in diameter, and it is just barely visible to the naked eye. The paramecium, a large protozoan which in bright light can be seen moving about in a drop of water, is about the same size. An ordinary human cell is only 1/10 as large (about 10 microns in diameter), and it is quite invisible without a microscope. Smaller still is the red blood corpuscle – some 7 microns in maximum diameter. The bacteria, starting with species as large as ordinary cells, drop down to a tinier level: the average rod-shaped bacterium is only 2 microns long, and the smallest bacteria are spheres perhaps no more than 4/10 of a micron in diameter. They can barely be seen in ordinary microscopes.

At this level, organisms apparently have reached the smallest possible volume into which all the metabolic machinery necessary for an independent life can be crowded. Any smaller organism cannot be a self-sufficient cell and must live as a parasite. It must shed most of the enzymatic machinery as excess baggage, so to speak. It is unable to grow or multiply on any artificial supply of food, however ample; hence it cannot be cultured, as bacteria can, in the test tube. The only place it can grow is in a living cell, which supplies the enzymes that it lacks. Such a parasite grows and multiplies, naturally, at the expense of the host cell.

The first sub-bacteria were discovered by a young American pathologist named Howard Taylor Ricketts. In 1909, he was studying a disease called Rocky Mountain spotted fever, which is spread by ticks (blood-sucking arthropods, related to the spiders rather than to insects). Within the cells' infected hosts he found 'inclusion bodies' that turned out to be very tiny organisms, now called 'rickettsia' in his honour. Ricketts and others soon found that typhus also was a rickettsial disease. In the process of

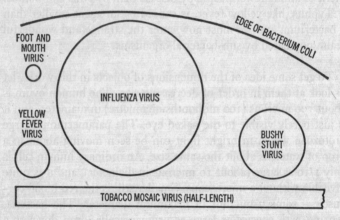

FOOT AND MOUTH VIRUS

EDGE OF BACTERIUM COLI

INFLUENZA VIRUS

YELLOW FEVER VIRUS

BUSHY STUNT VIRUS

TOBACCO MOSAIC VIRUS (HALF-LENGTH)

◻ HAEMOGLOBIN MOLECULE
• SUGAR MOLECULE
• SINGLE AMINO ACID MOLECULE

Relative sizes of simple substances and proteins and of various particles and bacteria. (An inch and a half on this scale = 1/10,000 of a millimetre in life.)

establishing a proof of this fact, Ricketts himself caught typhus, and he died in 1910 at the age of thirty-nine.

The rickettsia are still big enough to be attacked by antibiotics such as chloramphenicol and the tetracyclines. They range from about four fifths of a micron to one fifth of a micron in diameter. Apparently they possess enough metabolic machinery of their own to differ from the host cells in their reaction to drugs. Antibiotic therapy has therefore considerably reduced the danger of rickettsial diseases.

At the lowest end of the scale, finally, come the viruses. They overlap the rickettsia in size; in fact, there is no actual dividing line between rickettsia and viruses. But the smallest viruses are small indeed. The virus of yellow fever, for instance, is only 1/50 of a micron in diameter. The viruses are much too small to be detected in a cell or to be seen under any optical microscope. The average virus is only 1/1,000 the size of the average bacterium.

A virus is stripped practically clean of metabolic machinery. It depends almost entirely upon the enzyme equipment of the host cell. Some of the largest viruses are affected by certain antibiotics, but against the run-of-the-mill viruses drugs are helpless.

The existence of viruses was suspected many decades before they were finally seen. Pasteur, in his studies of hydrophobia, could find no organism in the body that could reasonably be suspected of causing the disease. Rather than decide that his germ theory of disease was wrong, Pasteur suggested that the germ in this case was simply too small to be seen. He was right.

In 1892, a Russian bacteriologist, Dmitri Ivanovski, was studying 'tobacco mosaic disease', a disease that gave the leaves of the tobacco plant a mottled appearance. He found that the juice of infected leaves could transmit the disease when placed on the leaves of healthy plants. In an effort to trap the germs, he passed the juice through porcelain filters with holes so fine that not even the smallest bacterium could pass through. Yet the filtered juice still infected tobacco plants. Ivanovski decided that his filters must be defective and were actually letting bacteria through.

A Dutch bacteriologist, Martinus Willem Beijerinck, repeated the experiment in 1897, and he came to the decision that the agent of the disease was small enough to pass through the filter. Since he could see nothing in the clear, infective fluid under any microscope, and was unable to grow anything from it in a test-tube culture, he thought the infective agent might be a small molecule, perhaps about the size of a sugar molecule. Beijerinck called the infective agent a 'filtrable virus' (virus being a Latin word meaning 'poison').

In the same year a German bacteriologist, Friedrich August Johannes Löffler, found that the agent causing foot-and-mouth disease in cattle could also pass through a filter. And, in 1901, Walter Reed, in the course of his yellow-fever researches, found that the infective agent of that disease also was a filtrable virus. In 1914, the German bacteriologist Walther Kruse demonstrated the common cold to be virus-produced.

By 1931, some forty diseases (including measles, mumps, chickenpox, influenza, smallpox, poliomyelitis, and hydrophobia) were known to be caused by viruses, but the nature of viruses was still a mystery. Then an English bacteriologist, William Joseph Elford, finally began to trap some in filters and to prove that at least they were material particles of some kind. He used fine collodion membranes, graded to keep out smaller and smaller particles, and he worked his way down to membranes fine enough to remove the infectious agent from a liquid. From the fineness of the membrane that could filter out the agent of a given disease, he was able to judge the size of that virus. He found that Beijerinck had been wrong: even the smallest virus was larger than most molecules. The largest viruses approached the rickettsia in size.

For some years afterwards, biologists debated whether viruses were living or dead particles. Their ability to multiply and transmit disease certainly suggested that they were alive. But in 1935 the American biochemist Wendell Meredith Stanley produced a piece of evidence which seemed to speak forcefully in favour of 'dead'. He mashed up tobacco leaves heavily infected with the tobacco mosaic virus and set out to isolate the virus in as pure and concentrated a form as he could get, using protein-separation techniques for the purpose. Stanley succeeded beyond his expectations, for he obtained the virus in crystalline form! His preparation was just as crystalline as a crystallized molecule, yet the virus evidently was still intact; when he redissolved it in liquid, it was just as infectious as before.

For his crystallization of the virus, Stanley shared the 1946 Nobel Prize in chemistry with Sumner and Northrop, the crystallizers of enzymes (see p. 93).

Still, for twenty years after Stanley's feat, the only viruses that could be crystallized were the very simple 'plant viruses' (those infesting plant cells). Not until 1955 was the first 'animal virus' crystallized. In that year, Carlton E. Schwerdt and Frederick L. Schaffer crystallized the poliomyelitis virus.

The fact that viruses could be crystallized seemed to many, including Stanley himself, to be proof that they were merely dead protein. Nothing living had ever been crystallized, and life and crystallinity just seemed to be mutually contradictory. Life was flexible, changeable, dynamic; a crystal was rigid, fixed, strictly ordered.

Yet the fact remained that viruses were infective, that they could grow and multiply even after having been crystallized. And growth and reproduction had always been considered the essence of life.

The turning point came when two British biochemists, Frederick Charles Bawden and Norman W. Pirie, showed that the tobacco mosaic virus contained ribonucleic acid! Not much, to be sure; the virus was 94 per cent protein and only 6 per cent RNA. But it was nonetheless definitely a nucleoprotein. Furthermore, all other viruses proved to be nucleoprotein, containing RNA or DNA or both.

The difference between being nucleoprotein and being merely protein is practically the difference between being alive and dead. Viruses turned out to be composed of the same stuff as genes, and the genes are the very essence of life. The larger viruses give every appearance of being chromosomes on the loose, so to speak. Some contain as many as 75 genes, each of which controls the formation of some aspect of its structure – a fibre here, a folding there. By producing mutations in the nucleic acid, one gene or another may be made defective, and through this means its function and even its location can be determined. The total gene analysis (both structural and functional) of a virus is within reach, though of course this represents but a small step towards a similar total analysis for cellular organisms, with their much more elaborate genic equipment.

We can picture viruses in the cell as raiders that, pushing aside the supervising genes, take over the chemistry of the cell in their own interests, often causing the death of the cell or of the entire host organism in the process. Sometimes, a virus may even replace a gene, or series of genes, with its own, introducing new characteristics that can be passed along to daughter cells. This phenomenon is called 'transduction'.

If the genes carry the 'living' properties of a cell, then viruses are living things. Of course, a lot depends on how one defines life. I, myself, think it fair to consider any nucleoprotein molecule that is capable of replication to be living. By that definition, viruses are as alive as elephants and human beings.

No amount of indirect evidence of the existence of viruses is as good as seeing one, of course. Apparently the first man to lay eyes on a virus was a Scottish physician named John Brown Buist. In 1887, he reported that in the fluid from a vaccination blister he had managed to make out some tiny dots under the microscope. Presumably they were the cowpox virus, the largest known virus.

To get a good look – or any look at all – at a typical virus, something better than an ordinary microscope was needed. The something better was finally invented in the late 1930s: the electron microscope, which could reach magnifications as high as 100,000 and resolve objects as small as 1/1,000 of a micron in diameter.

The electron microscope has its drawbacks. The object has to be placed in a vacuum, and the inevitable dehydration may change its shape. An object such as a cell must be sliced extremely thin. The image is only two-dimensional; furthermore, the electrons tend to go right through a biological material, so that it does not stand out against the background.

In 1944, the American astronomer and physicist Robley Cook Williams and the electron microscopist Ralph Walter Graystone

Wyckoff jointly worked out an ingenious solution of these last difficulties. It occurred to Williams, as an astronomer, that just as the craters and mountains of the moon are brought into relief by shadows when the sun's light falls on them obliquely, so viruses might be seen in three dimensions in the electron microscope if they could somehow be made to cast shadows. The solution the experimenters hit upon was to blow vaporized metal obliquely across the virus particles set up on the stage of the microscope. The metal stream left a clear space – a 'shadow' – behind each virus particle. The length of the shadow indicated the height of the blocking particle. And the metal, condensing as a thin film, also defined the virus particles sharply against the background.

The shadow picture of various viruses then disclosed their shapes. The cowpox virus was found to be shaped something like a barrel. It turned out to be about 0·25 of a micron thick – about the size of the smallest rickettsia. The tobacco mosaic virus proved to be a thin rod 0·28 micron long by 0·015 micron thick. The smallest viruses, such as those of poliomyelitis, yellow fever, and foot-and-mouth disease, were tiny spheres ranging in diameter from 0·025 down to 0·020 micron. This is considerably smaller than the estimated size of a single human gene. The weight of these viruses is only about 100 times that of an average protein molecule. The brome grass mosaic virus, the smallest yet characterized, has a molecular weight of 4·5 million. It is only one tenth the size of the tobacco mosaic virus and may, perhaps, bear off the prize as 'smallest living thing'.

In 1959, the Finnish cytologist Alvar P. Wilska designed an electron microscope using comparatively low-speed electrons. Because they are less penetrating than high-speed electrons, they can define some of the internal detail in the structure of viruses. And in 1961, the French cytologist Gaston DuPouy devised a way of placing bacteria in air-filled capsules and taking electron microscope views of living cells in this way. In the absence of metal-shadowing, however, detail was lacking.

Virologists have actually begun to take viruses apart and put them together again. For instance, at the University of California, the German-American biochemist Heinz Fraenkel-Conrat, working with Robley Williams, found that gentle chemical treatment broke down the protein of the tobacco mosaic virus into some 2,200 fragments, consisting of peptide chains made up of 158 amino acids apiece, and individual molecular weights of 18,000. The exact amino-acid constitution of these virus–protein units was completely worked out in 1960. When such units are dissolved, they tend to coalesce once more into the long, hollow rod (in which form they exist in the intact virus). The units are held together by calcium and magnesium atoms.

In general, virus–protein units make up geometric patterns when they combine. Those of tobacco mosaic virus, just discussed, form segments of a helix. The sixty sub-units of the protein of the poliomyelitis virus are arranged in twelve pentagons. The twenty sub-units of the *Tipula* iridescent virus are arranged in a regular twenty-sided solid, an icosahedron.

The protein of the virus is hollow. The protein helix of tobacco mosaic virus, for instance, is made up of 130 turns of the peptide chain, producing a long, straight cavity within. Inside the protein cavity is the nucleic-acid portion of the virus. This may be DNA or RNA, but, in either case, it is made up of about 6,000 nucleotides, although Sol Spiegelman has detected an RNA molecule with as few as 470 nucleotides that is capable of replication.

Fraenkel-Conrat separated the nucleic acid and protein portions of tobacco mosaic viruses and tried to find out whether each portion alone could infect a cell. It developed that separately they could not, as far as he could tell. But when he mixed the protein and nucleic acid together again, as much as 50 per cent of the original infectiousness of the virus sample could eventually be restored!

What had happened? The separated virus protein and nucleic acid had seemed dead, to all intents and purposes; yet, mixed together again, some at least of the material seemed to come to life. The public press hailed Fraenkel-Conrat's experiment as the creation of a living organism from non-living matter. The stories were mistaken, as we shall see in a moment.

Apparently, some recombination of protein and nucleic acid had taken place. Each, it seemed, had a role to play in infection. What were the respective roles of the protein and the nucleic acid, and which was more important?

Fraenkel-Conrat performed a neat experiment that answered the question. He mixed the protein part of one strain of the virus with the nucleic-acid portion of another strain. The two parts combined to form an infectious virus with a mixture of properties! In virulence (i.e., the degree of its power to infect tobacco plants) it was the same as the strain of virus that had contributed the protein; in the particular disease produced (i.e., the nature of the mosaic pattern on the leaf), it was identical with the strain of virus that had supplied the nucleic acid.

This finding fitted well with what virologists already suspected about the respective functions of the protein and the nucleic acid. It seems that when a virus attacks a cell, its protein shell, or coat, serves to attach itself to the cell and to break open an entrance into the cell. Its nucleic acid then invades the cell and engineers the production of virus particles.

After Fraenkel-Conrat's hybrid virus had infected a tobacco leaf, the new generation of virus that it bred in the leaf's cells turned out to be not a hybrid but just a replica of the strain that had contributed the nucleic acid. It copied that strain in degree of infectiousness as well as in the pattern of disease produced. In other words, the nucleic acid had dictated the construction of the new virus's protein coat. It had produced the protein of its own strain, not that of the strain with which it had been combined in the hybrid.

This reinforced the evidence that the nucleic acid is the 'live' part of a virus, or, for that matter, of any nucleoprotein. Actually, Fraenkel-Conrat found in further experiments that pure virus nucleic acid alone could produce a little infection in a tobacco leaf – about 0·1 per cent as much as the intact virus. Apparently once in a while the nucleic acid somehow managed to breach an entrance into a cell all by itself.

So putting virus nucleic acid and protein together to form a virus is not creating life from non-life; the life is already there, in the shape of the nucleic acid. The protein merely serves to protect the nucleic acid against the action of hydrolysing enzymes ('nucleases') in the environment and to help it go about the business of infection and reproduction more efficiently. We might compare the nucleic-acid fraction to a man and the protein fraction to a motor car. The combination makes easy work of travelling from one place to another. The car by itself could never make the trip. The man could make it on foot (and occasionally does), but the car is a big help.

The clearest and most detailed information about the mechanism by which viruses infect a cell has come from studies of the viruses called bacteriophages, first discovered by the English bacteriologist Frederick William Twort in 1915 and, independently, by the Canadian bacteriologist Félix Hubert d'Hérelle in 1917. Oddly enough, these viruses are germs that prey on germs – namely, bacteria. D'Hérelle gave them the name 'bacteriophage', from Greek words meaning 'bacteria-eater'.

The bacteriophages are beautifully convenient things to study, because they can be cultured with their hosts in a test tube. The process of infection and multiplication goes about as follows.

A typical bacteriophage (usually called 'phage' by the workers with the beast) is shaped like a tiny tadpole, with a blunt head and a tail. Under the electron microscope investigators have been able to see that the phage first lays hold of the surface of a bacterium with its tail. The best guess as to how it does this is that the pattern of electric charge on the tip of the tail (determined by charged amino acids) just fits the charge pattern on certain portions of the bacterium's surface. The configurations of the opposite, and attracting, charges on the tail and on the bacterial surface match so neatly that they come together with something like a click of perfectly meshing gear teeth. Once the virus has attached itself to its victim by the tip of its tail, it cuts a tiny opening in the cell wall, perhaps by means of an enzyme that cleaves the molecules at that point. As far as the electron-microscope pictures show, nothing whatever is happening. The phage, or at least its visible shell, remains attached to the outside of the bacterium. Inside the bacterial cell there is no visible activity. But, within half an hour the cell bursts open and hundreds of full-grown viruses pour out.

Evidently only the protein shell of the attacking virus stays outside the cell. The nucleic acid within the virus's shell must pour into the bacterium through the hole in its wall made by the protein. That the invading material is just nucleic acid, without any detectable admixture of protein, was proved by the American

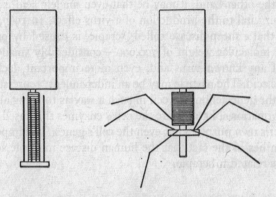

Model of T-2 bacteriophage, a tadpole-shaped virus that preys on other germs, in 'untriggered' (*left*) and 'triggered' (*right*) forms.

bacteriologist Alfred Day Hershey by means of radioactive tracers. He tagged phages with radioactive phosphorus and radioactive sulphur atoms (by growing them in bacteria that had incorporated these radio-isotopes from their nutritive medium). Now phosphorus occurs both in proteins and in nucleic acids, but sulphur will turn up only in proteins, because there is no sulphur in a nucleic acid. Therefore if a phage labelled with both tracers invaded a bacterium and its progeny turned up with radio-phosphorus but no radio-sulphur, the experiment would indicate that the parent virus's nucleic acid had entered the cell but its protein had not. The absence of radio-sulphur would suggest that all the protein in the virus progeny was supplied by the host bacterium. The experiment, in fact, turned out just this way: the new viruses contained radio-phosphorus (contributed by the parent) but no radio-sulphur.

Once more, the dominant role of nucleic acid in the living process was demonstrated. Apparently, only the phage's nucleic acid went into the bacterium, and there it superintended the construction of new viruses – protein and all – from the material in the cell. Indeed, the potato spindle tuber virus, an unusually small one, seems to be nucleic acid without a protein coat at all.

On the other hand, it may be that even nucleic acid is not altogether vital to the production of a virus effect. In 1967, it was found that a sheep disease called 'scrapie' is caused by particles with a molecular weight of 700,000 – considerably smaller than that of any known virus and, even more important, lacking in nucleic acid. The particle may be an independent 'repressor' that alters the gene action of a cell in such a way as to bring about its own formation. Not only are the cell's enzymes thus used for the invader's own purpose, but even the cell's genes. The importance to man lies in the fact that the human disease multiple sclerosis may be related to scrapie.

Immunity

Viruses are man's most formidable living enemy (except man himself). By virtue of their intimate association with the body's own cells, viruses have been all but invulnerable to attack by drugs or any other artificial weapon. And yet man has been able to hold his own against them, even under the most unfavourable conditions. The human organism is endowed with impressive natural defences against disease.

Consider the Black Death, the great plague of the fourteenth century. It attacked a Europe living in appalling filth, without any modern conception of cleanliness and hygiene, without plumbing, without any form of reasonable medical treatment – a crowded and helpless population. To be sure, people could flee from the infected villages, but the fugitive sick only spread the epidemics faster and farther. Notwithstanding all this, three fourths of the population successfully resisted the infections. Under the circumstances, the marvel is not that one out of four died; the marvel is that three out of four survived.

There is clearly such a thing as natural resistance to any given disease. Of a number of people exposed to a serious contagious disease, some will have a relatively mild case, some will be very ill, some will die. There is also such a thing as complete immunity – sometimes inborn, sometimes acquired. A single attack of measles, mumps, or chickenpox, for instance, will usually make a person immune to that particular disease for the rest of his life.

All three of these diseases, as it happens, are caused by viruses. Yet they are comparatively minor infections, seldom fatal. Measles usually produces only mild symptoms, at least in a child. How does the body fight off these viruses, and then fortify itself so that the virus it has defeated never troubles it again? The answer to that question forms a thrilling episode in modern medical science, and for the beginning of the story we must go back to the conquest of smallpox.

Up to the end of the eighteenth century, smallpox was a particularly dreaded disease, not only because it was often fatal but also because those who recovered were permanently disfigured. A light case would leave the skin pitted; a severe attack could destroy all traces of beauty and almost of humanity. A very large proportion of the population bore the marks of smallpox on their faces. And those who had not yet caught it lived in fear of when it might strike.

In the seventeenth century, people in Turkey began to infect themselves deliberately with mild forms of smallpox, with the hope of making themselves immune to severe attack. They would have themselves scratched with the serum from blisters of a person who had a mild case. Sometimes they developed only a light infection; sometimes they suffered the very disfigurement or death they had sought to avoid. It was a risky business, but it is a measure of the horror of the disease that people were willing to risk the horror itself in order to escape from it.

In 1718, the famous beauty Lady Mary Wortley Montagu

learned about this practice when she went to Turkey with her husband, sent there briefly as the British ambassador, and she had her own children inoculated. They escaped without harm. But the idea did not catch on in England, perhaps partly because Lady Montagu was considered a notorious eccentric. A similar case, across the ocean, was that of Zabdiel Boylston, an American physician. During a smallpox epidemic in Boston, he inoculated 241 people, of whom six died. He underwent considerable criticism for this.

Certain country folk in Gloucestershire had their own idea about how to avoid smallpox. They believed that a case of cowpox, a disease that attacked cows and sometimes people, would make a person immune to both cowpox and smallpox. This was wonderful, if true, for cowpox produced hardly any blisters and left hardly any marks. A Gloucestershire doctor, Edward Jenner, decided that there might be some truth in this folk 'superstition'. Milkmaids, he noticed, were particularly prone to catch cowpox and apparently also particularly prone not to be pockmarked by smallpox. (Perhaps the eighteenth-century vogue of romanticizing the beautiful milkmaid was based on the fact that milkmaids, having clear complexions, were indeed beautiful in a pockmarked world.)

Was it possible that cowpox and smallpox were so alike that a defence formed by the body against cowpox would also protect against smallpox? Very cautiously Dr Jenner began to test this notion (probably experimenting on his own family first). In 1796, he decided to chance the supreme test. First he inoculated an eight-year-old boy named James Phipps with cowpox, using fluid from a cowpox blister on a milkmaid's hand. Two months later came the crucial and desperate part of the test. Jenner deliberately inoculated young James with smallpox itself.

The boy did not catch the disease. He was immune.

Jenner called the process 'vaccination', from *vaccinia*, the Latin name for cowpox. Vaccination spread through Europe like wildfire. It is one of the rare cases of a revolution in medicine that was adopted easily and almost at once – a true measure of the deadly fear inspired by smallpox and the eagerness of the public to try anything that promised escape. Even the medical profession put up only weak opposition to vaccination – though its leaders put up such stumbling blocks as they could. When Jenner was proposed for election to the Royal College of Physicians in London in 1813, he was refused admission, on the ground that he was not sufficiently up on Hippocrates and Galen.

Today smallpox has practically been wiped out in civilized countries, though its terrors as a disease are still as strong as ever. A report of a single case in any large city is sufficient to send virtually the entire population running to doctors' surgeries for renewed vaccination.

Attempts to discover similar inoculations for other severe diseases got nowhere for more than a century and a half. It was Pasteur who made the next big step forward. He discovered, more or less by accident, that he could change a severe disease into a mild one by weakening the microbe that produced it.

Pasteur was working with a bacterium that caused cholera in chickens. He concentrated a preparation so virulent that a little injected under the skin of a chicken would kill it within a day. On one occasion he used a culture that had been standing for a week. This time the chickens became only slightly sick and recovered. Pasteur decided that the culture was spoiled and prepared a virulent new batch. But his fresh culture failed to kill the chickens that had recovered from the dose of 'spoiled' bacteria. Clearly, the infection with the weakened bacteria had equipped the chickens with a defence against the fully potent ones.

In a sense, Pasteur had produced an artificial 'cowpox' for this particular 'smallpox'. He recognized the philosophical debt he owed to Jenner by calling his procedure vaccination, too, although it had nothing to do with *vaccinia*. Since then, the term has been used quite generally to mean inoculations against any disease, and the preparation used for the purpose is called a 'vaccine'.

Pasteur developed other methods of weakening (or 'attenuating') disease agents. For instance, he found that culturing anthrax bacteria at a high temperature produced a weakened strain that would immunize animals against the disease. Until then, anthrax had been so hopelessly fatal and contagious that as soon as one member of a herd came down with it, the whole herd had to be slaughtered and burned.

Pasteur's most famous victory, however, was over the virus disease called hydrophobia, or 'rabies' (from a Latin word meaning 'to rave', because the disease attacked the nervous system and produced symptoms akin to madness). A person bitten by a rabid dog would, after an incubation period of a month or two, be seized by violent symptoms and almost invariably die an agonizing death.

Pasteur could find no visible microbe as the agent of the disease (of course, he knew nothing of viruses), so he had to use living animals to cultivate it. He would inject the infectious fluid into the brain of a rabbit, let it incubate, mash up the rabbit's spinal cord, inject the extract into the brain of another rabbit, and so on. Pasteur attenuated his preparations by aging and testing them continuously until the extract could no longer cause the disease in a rabbit. He then injected the attenuated virus into a dog, which survived. After a time, he infected the dog with hydrophobia in full strength and found the animal immune.

In 1885, Pasteur got his chance to try the cure on a human being. A nine-year-old boy, Joseph Meister, who had been severely bitten by a rabid dog, was brought to him. With considerable hesitation and anxiety, Pasteur treated the boy with inoculations of successively less and less attenuated virus, hoping to build up resistance before the incubation period had elapsed. He succeeded. At least, the boy survived. (Meister became the gatekeeper of the Pasteur Institute, and in 1940 he committed suicide when the Nazi army in Paris ordered him to open Pasteur's crypt.)

In 1890, a German army doctor named Emil von Behring, working in Koch's laboratory, tried another idea. Why take the risk of injecting the microbe itself, even in attenuated form, into a human being? Assuming that the disease agent caused the body to manufacture some defensive substance, would it not serve just as well to infect an animal with the agent, extract the defence substance that it produced, and inject that substance into the human patient?

Von Behring found that this scheme did indeed work. The defensive substance turned up in the blood serum, and von Behring called it 'antitoxin'. He caused animals to produce anti-toxins against tetanus and diphtheria. His first use of the diph-theria antitoxin on a child with the disease was so dramatically successful that the treatment was adopted immediately and pro-ceeded to cut the death rate from diphtheria drastically.

Paul Ehrlich (who later was to discover the 'magic bullet' for syphilis) worked with von Behring, and it was probably he who calculated the appropriate antitoxin dosages. Later he broke with von Behring (Ehrlich was an irascible individual who found it easy to break with anyone), and alone he went on to work out the rationale of serum therapy in detail. Von Behring received the Nobel Prize in medicine and physiology in 1901, the first year in which it was awarded. Ehrlich also was awarded that Nobel Prize, sharing it with a Russian biologist in 1908.

The immunity conferred by an antitoxin lasts only as long as the antitoxin remains in the blood. But the French bacteriologist Gaston Ramon found that by treating the toxin of diphtheria or tetanus with formaldehyde or heat he was able to change its struc-ture in such a way that the new substance (called 'toxoid') could safely be injected into a human patient. The antitoxin then made by the patient himself lasts longer than that from an animal; furthermore, new doses of the toxoid can be injected when necessary to renew immunity. After toxoid was introduced in 1925, diphtheria lost most of its terrors.

Serum reactions were also used to detect the presence of dis-ease. The best-known example of this is the 'Wassermann test', introduced by the German bacteriologist August von Wasser-mann, in 1906, for the detection of syphilis. This was based on techniques first developed by a Belgian bacteriologist, Jules Bordet, who worked with serum fractions that came to be called 'complement'. For his work, Bordet received the Nobel Prize in medicine and physiology in 1919.

Pasteur's laborious wrestle with the virus of rabies showed the difficulties of dealing with viruses. Bacteria can be cultured, manipulated, and attenuated on artificial media in the test tube.

229

Viruses cannot; they can be grown only in living tissue. In the case of smallpox, the living hosts for the experimental material (the cowpox virus) were cows and milkmaids. In the case of rabies, Pasteur used rabbits. But living animals are, at best, an awkward, expensive, and time-consuming type of medium for culturing micro-organisms.

In the first quarter of this century the French biologist Alexis Carrel won considerable fame with a feat which was to prove immensely valuable to medical research – keeping bits of tissue alive in the test tube. Carrel had become interested in this sort of thing through his work as a surgeon. He had developed new methods of transplanting animals' blood vessels and organs, for which he received the Nobel Prize in medicine and physiology in 1912. Naturally, he had to keep the excised organ alive while he was getting ready to transplant it. He worked out a way to nourish it, which consisted in perfusing the tissue with blood and supplying the various extracts and ions. As an incidental dividend, Carrel, with the help of Charles Augustus Lindbergh, developed a crude 'mechanical heart' to pump the blood through the tissue.

Carrel's devices were good enough to keep a piece of embryonic chicken heart alive for thirty-four years – much longer than a chicken's lifetime. Carrel even tried to use his tissue cultures to grow viruses – and he succeeded in a way. The only trouble was that bacteria also grew in the tissues, and in order to keep the virus pure, such tedious aseptic precautions had to be taken that it was easier to use animals.

The chicken-embryo idea, however, was in the right court, so to speak. Better than just a piece of tissue would be the whole thing – the chick embryo itself. A chick embryo is a self-contained organism, protected by the egg shell, equipped with its own natural defences against bacteria, and cheap and easy to come by in quantity. And, in 1931, the pathologist Ernest William Goodpasture and his co-workers at Vanderbilt University succeeded in transplanting a virus into a chick embryo. For the first time, pure viruses could be cultured almost as easily as bacteria.

The first great medical victory by means of the culture of viruses in fertile eggs came in 1937. At the Rockefeller Institute, bacteriologists were still hunting for further protection against the yellow-fever virus. It was impossible to eradicate the mosquito completely, after all, and infected monkeys maintained a constantly threatening reservoir of the disease in the tropics. The South-African bacteriologist Max Theiler at the Institute set out to produce an attenuated yellow-fever virus. He passed the virus through 200 mice and 100 chick embryos until he had a mutant that caused only mild symptoms yet gave rise to complete immunity against yellow fever. For this achievement Theiler received the 1951 Nobel Prize in medicine and physiology.

When all is said and done, nothing can beat culture in glassware for speed, control of the conditions, and efficiency. In the late 1940s John Franklin Enders, Thomas Huckle Weller, and Frederick Chapman Robbins at the Harvard Medical School went back to Carrel's approach. (He had died in 1944 and was not to see their success.) This time they had a new and powerful weapon against bacteria contaminating the tissue culture – the antibiotics. They added penicillin and streptomycin to the supply of blood that kept the tissues alive, and they found that they could grow viruses without trouble. On impulse, they tried the poliomyelitis virus. To their delight, it flourished in this medium. It was the breakthrough that was to conquer polio, and the three men received the Nobel Prize in medicine and physiology in 1954.

The poliomyelitis virus could now be bred in the test tube, instead of solely in monkeys (which are expensive and temperamental laboratory subjects). Large-scale experimentation with the virus became possible. Thanks to the tissue-culture technique, Jonas Edward Salk of the University of Pittsburgh was able to experiment with chemical treatment of the virus, to learn that polio viruses killed by formaldehyde could still produce immune reactions in the body, and to develop his now-famous Salk vaccine.

Polio's sizable death rate, its dreaded paralysis, its partiality for children (so that it has the alternative name of 'infantile paralysis'), the fact that it seems to be a modern scourge, with no epidemics on record prior to 1840, and particularly the interest attracted to the disease by its eminent victim, Franklin Delano Roosevelt, made its conquest one of the most celebrated victories over a disease in all human history. Probably no medical announcement ever received such a Hollywood-première-type reception as did the report in 1955, of the evaluating committee that found the Salk vaccine effective. Of course, the event merited such a celebration – more than do most of the performances that arouse people to throw ticker tape and trample one another. But science does not thrive on furore or wild publicity. The rush to respond to the public pressure for the vaccine apparently resulted in a few defective, disease-producing samples of the vaccine slipping through, and the subsequent counter-furore set back the vaccination programme against the disease.

The set-back was, however, made up, and the Salk vaccine was found effective and, properly prepared, safe. In 1957, the Polish-American microbiologist Albert Bruce Sabin went a step further. He did not make use of dead virus (which, when not entirely dead, could be dangerous) but of strains of living virus, incapable of producing the disease itself, but capable of bringing about the production of appropriate antibodies. Such a 'Sabin vaccine' could be taken by mouth, moreover, and did not require the hypodermic. The Sabin vaccine gained popularity first in the Soviet Union and then in the east European countries; but by 1960 it came into use in the United States as well, and the fear of poliomyelitis has lifted.

What does a vaccine do, exactly? The answer to this question may some day give us the chemical key to immunity.

For more than half a century biologists have known the body's main defences against infection as 'antibodies'. (Of course, there are also the white blood cells called 'phagocytes', which devour bacteria. This was discovered in 1883 by the Russian biologist Ilya Mechnikov, who later succeeded Pasteur as the head of the Pasteur Institute in Paris and shared the 1908 Nobel Prize in medicine and physiology with Ehrlich. But phagocytes are no help against viruses and seem not to be involved in the immunity process we are considering.) A virus, or indeed almost any foreign substance entering into the body's chemistry, is called an 'antigen'. The antibody is a substance manufactured by the body to fight the specific antigen. It puts the antigen out of action by combining with it.

Long before the chemists actually ran down an antibody, they were pretty sure the antibodies must be proteins. For one thing, the best-known antigens were proteins, and presumably it would take a protein to catch a protein. Only a protein could have the subtlety of structure necessary to single out and combine with a particular antigen.

Early in the 1920s Landsteiner (the discoverer of blood groups) carried out a series of experiments which clearly showed that antibodies were very specific indeed. The substances he used to generate antibodies were not antigens but much simpler compounds whose structure was well known. They were arsenic-containing compounds called 'arsanilic acids'. In combination with a simple protein, such as the albumin of egg white, an arsanilic acid acted as an antigen: when injected into an animal, it gave rise to an antibody in the blood serum. Furthermore, this antibody was specific for the arsanilic acid; the blood serum of the animal would clump only the arsanilic–albumin combination, not albumin alone. Indeed, sometimes the antibody could be made to react with just an arsanilic acid, not combined with albumin. Landsteiner also showed that very small changes in the structure of the arsanilic acid would be reflected in the antibody. An antibody evoked by one variety of arsanilic acid would not react with a slightly altered variety.

Landsteiner coined the name 'haptens' (from a Greek word meaning 'to bind') for compounds, such as the arsanilic acids, that can give rise to antibodies when they are combined with protein. Presumably, each natural antigen has a specific region in its molecule that acts as a hapten. On that theory, a germ or virus that can serve as a vaccine is one that has had its structure changed sufficiently to reduce its ability to damage cells but still has its hapten group intact, so that it can cause the formation of a specific antibody.

It would be interesting to learn the chemical nature of the natural haptens. If that could be determined, it might be possible to use a hapten, perhaps in combination with some harmless protein, to serve as a vaccine giving rise to antibodies for a specific antigen. That would avoid the necessity of resorting to toxins or attenuated viruses, which always carries some small risk.

Just how an antigen evokes an antibody has not been determined. Ehrlich believed that the body normally contains a small supply of all the antibodies it may need, and that when an invading antigen reacts with the appropriate antibody, this stimulates the body to produce an extra supply of that particular antibody. Some immunologists still adhere to that theory or to modifications of it. Yet it seems highly unlikely that the body is prepared with specific antibodies for all the possible antigens, including unnatural substances such as the arsanilic acids.

The alternate suggestion is that the body has some generalized protein molecule which can be moulded to fit any antigen. The antigen, then, acts as a template to shape the specific antibody formed in response to it. Pauling proposed such a theory in 1940. He suggested that the specific antibodies are varying versions of the same basic molecule, merely folded in different ways. In other words, the antibody is moulded to fit its antigen as a glove fits the hand.

By 1969, however, the advance of protein analysis had made it possible for a team under Gerald M. Edelman to work out the amino-acid structure of a typical antibody made up of well over a thousand amino acids. No doubt this will pave the way for determining the manner of working of such molecules with considerably more subtlety than had hitherto been possible.

The very specificity of antibodies is a disadvantage in some ways. Suppose a virus mutates so that its protein has a slightly different structure. The old antibody for the virus often will not fit the new structure. It follows that immunity against one strain of virus is no safeguard against another strain. The virus of influenza and of the common cold are particularly prone to minor mutations, and that is one reason why we are plagued by frequent recurrences of these diseases. Influenza, in particular, will occasionally develop a mutant of extraordinary virulence, which may then sweep a surprised and non-immune world. This happened in 1918 and, with much less fatal result, in the 'Asian flu' pandemic of 1957.

A still more annoying effect of the body's over-sharp efficiency in forming antibodies is its tendency to produce them even against a harmless protein that happens to enter the body. The body then becomes 'sensitized' to that protein, and it may react violently to any later incursion of the original innocent protein. The reaction may take the form of itching, tears, production of mucus in the nose and throat, asthma, and so on. Such 'allergic reactions' are evoked by the pollen of certain plants (causing hay fever), by certain foods, by the fur or dandruff of animals, and so on. An allergic reaction may be acute enough to cause serious disablement, or even death. The discovery of such 'anaphylactic shock' won for the French physiologist Charles Robert Richet the Nobel Prize in medicine and physiology in 1913.

In a sense, every human being is more or less allergic to every other human being. A transplant, or graft, from one individual to another will not take, because the receiver's body treats the transplanted tissue as foreign protein and manufactures antibodies against it. The person-to-person graft that will work best is from one identical twin to the other. Since their identical heredity gives them exactly the same proteins, they can exchange tissues or even a whole organ, such as a kidney.

The first successful kidney transplant took place in December 1954 in Boston, from one identical twin to another. The receiver died in 1962 at the age of thirty of coronary artery disease. Since then, hundreds of individuals have lived for months and even years with kidney transplanted from *other* than identical twins.

Attempts at transplanting other organs, such as the lungs or the liver, have been made, but that which most caught the public fancy was the heart transplant. The first reasonably successful heart transplants were conducted in December 1967 by the South African surgeon Christiaan Barnard. The fortunate receiver – Philip Blaiberg, a retired South African dentist – lived on for many months on someone else's heart.

For a while after that, heart transplants became the rage, but the furore by late 1969 had died down. Few receivers lived very long, for the problems of tissue rejection seemed mountainous, despite massive attempts to solve the reluctance of the body to incorporate any tissue but its own.

The Australian bacteriologist Macfarlane Burnet had suggested that embryonic tissues might be 'immunized' to foreign tissues and that the free-living animal might then tolerate grafts of that tissue. The British biologist Peter Medawar demonstrated this to be so, using mouse embryos. The two men shared in the 1960 Nobel Prize in medicine and physiology as a result.

In 1962, a French-Australian immunologist, Jacques Francis Albert Pierre Miller, working in England, went even further and discovered what may be the reason for this ability to work with embryos in order to make future toleration possible. He discovered that the thymus gland (a piece of tissue which until then had had no known use) was the tissue capable of forming antibodies. If the thymus gland was removed from mice at birth, those mice died after three or four months out of sheer incapacity to protect themselves against the environment. If the thymus was allowed to remain in the mice for three weeks, it already had time to bring about the development of antibody-producing cells in the body, and the thymus gland might then be removed without harm. Embryos in which the thymus has not yet done its work may be so treated as to 'learn' to tolerate foreign tissue; the day may yet come when, by the way of the thymus, we may improve tissue toleration, when that is desirable, perhaps even in adults.

And yet, even if the problem of tissue rejection were surmounted, there would remain serious problems. After all, every person who receives a living organ must receive it from someone who is giving it up, and the question arises as to when the prospective donor may be considered dead enough to yield up his organs.

In that respect it might prove better if mechanical organs were prepared which would involve neither tissue rejection nor knotty ethical issues. Artificial kidneys became practical in the 1940s, and it is possible for patients without natural kidney function to visit a hospital once or twice a week and have their blood cleansed of wastes. It makes for a restricted life even for those fortunate enough to be serviced, but it is preferable to death.

In the 1940s, researchers found that allergic reactions are brought about by the liberation of small quantities of a substance called 'histamine' into the bloodstream. This led to the successful search for neutralizing 'anti-histamines', which can relieve the allergic symptoms but, of course, do not remove the allergy. The first successful anti-histamine was produced at the Pasteur Institute in Paris in 1937 by the Swiss-born chemist Daniel Bovet, who for this and subsequent researches in chemotherapy was awarded the Nobel Prize in physiology and medicine in 1957.

Noting that sniffling and other allergic symptoms were much like those of the common cold, pharmaceutical firms decided that what worked for one ought to work for the other, and in 1949 and 1950 they flooded the country with anti-histamine tablets. (The tablets turned out to do little or nothing for colds, and their vogue diminished.)

In 1937, thanks to the protein-isolating techniques of electrophoresis, biologists finally tracked down the physical location of antibodies in the blood. The antibodies were located in the blood fraction called 'gamma globulin'.

Physicians have long been aware that some children are unable to form antibodies and therefore are easy prey to infection. In 1951, doctors at the Walter Reed Hospital in Washington made an electrophoretic analysis of the plasma of an eight-year-old boy suffering from a serious septicaemia ('blood poisoning'), and to their astonishment they discovered that his blood had no gamma globulin at all. Other cases were quickly discovered. Investigators established that this lack is due to an inborn defect of metabolism which deprives the person of the ability to make gamma globulin; it is called 'agammaglobulinaemia'. Such persons cannot develop immunity to bacteria. They can now be kept alive, however, by antibiotics. Surprisingly enough, they *are* able to become immune to virus infections, such as measles and chickenpox, after having the disease once. Apparently, antibodies are not the body's only defence against viruses.

In 1957, a group of British bacteriologists, headed by Alick

Isaacs, showed that cells, under the stimulus of a virus invasion, liberated a protein that had broad anti-viral properties. It countered not only the virus involved in the immediate infection but others as well. This protein, named 'interferon', is produced more quickly than antibodies are and may explain the anti-virus defences of those with agammaglobulinaemia. Apparently its production is stimulated by the presence of RNA in the double-stranded variety found in viruses. Interferon seems to direct the synthesis of a messenger-RNA that produces an anti-virus protein that inhibits production of virus protein but not of other forms of protein. Interferon seems to be as potent as antibiotics and doesn't activate resistance. It is, however, fairly species-specific. Only interferon from humans and from other primates will work on human beings.

Cancer

As the danger of infectious diseases diminishes, the incidence of other types of disease increases. Many people who a century ago would have died young of tuberculosis or diphtheria or pneumonia or typhus now live long enough to die of heart disease or cancer. That is one reason why heart disease and cancer have become, respectively, the number one and the number two killers in the Western world. Cancer, in fact, has succeeded plague and smallpox as the great fear of man. It is a nightmare hanging over all of us, ready to strike anyone without warning or mercy. Three hundred thousand Americans die of it each year while ten thousand new cases are recorded each week. The incidence has risen 50 per cent since 1900.

Cancer is actually a group of many diseases (about 200 types are known), affecting various parts of the body in various fashions. But the primary disorder is always the same: disorganization and uncontrolled growth of the affected tissues. The name cancer (the Latin word for 'crab') comes from the fact that Hippocrates and

Galen fancied the disease spreading its ravages through diseased veins like the crooked, outstretched claws of a crab.

'Tumour' (from the Latin word meaning 'grow') is by no means synonymous with cancer; it applies to harmless growths such as warts and moles ('benign tumours') as well as to cancers ('malignant tumours'). The cancers are variously named according to the tissues affected. Cancers of the skin or the intestinal linings (the most common malignancies) are called 'carcinomas' (from the Greek word for 'crab'); cancers of the connective tissues are 'sarcomas'; of the liver, 'hepatoma'; of glands generally, 'adenomas'; of the white blood cells, 'leukaemia'; and so on.

Rudolf Virchow of Germany, the first to study cancer tissue under the microscope, believed that cancer was caused by the irritations and shocks of the outer environment. This is a natural thought, for it is just those parts of the body most exposed to the outer world that are most subject to cancer. But when the germ theory of disease became popular, pathologists began to look for some microbe as the cause of cancer. Virchow, a staunch opponent of the germ theory of disease, stubbornly insisted on the irritation theory. (He left pathology for archaeology and politics when it turned out that the germ theory of disease was going to win out. Few scientists in history have gone down with the ship of mistaken belief in quite so drastic a fashion.)

If Virchow was stubborn for the wrong reason, he may have been so in the right cause. There has been increasing evidence that some environments are particularly conducive to cancer. In the eighteenth century, chimney sweeps were found to be more prone to cancer of the scrotum than other people were. After the coal-tar dyes were developed, workers in the dye industries showed an above-average incidence of cancers of the skin or bladder. It seemed that something in soot and in the aniline dyes must be capable of causing cancer. Then, in 1915, two Japanese scientists, K. Yamagiwa and K. Ichikawa, discovered that a certain coal-tar fraction could produce cancer in rabbits when it was applied to the rabbits' ears for long periods. In 1930, two British chemists induced cancer in animals with a synthetic chemical called 'dibenzanthracene' (a hydrocarbon with a molecule made up of five benzene rings). This does not occur in coal tar, but three years later it was discovered that 'benzpyrene' (also containing five benzene rings but in a different arrangement), a chemical that *does* occur in coal tar, can cause cancer.

Quite a number of 'carcinogens' (cancer-producers) have now been identified. Many are hydrocarbons made up of numerous benzene rings, like the first two discovered. Some are molecules related to the aniline dyes. In fact, one of the chief concerns about using artificial dyes in foods is the possibility that in the long run such dyes may be carcinogenic.

Many biologists believe that man has introduced a number of new cancer-producing factors into his environment within the last two or three centuries. There is the increased use of coal; there is the burning of oil on a large scale, particularly in petrol engines; there is the growing use of synthetic chemicals in food, cosmetics, and so on. The most dramatic of the suspects, of course, is cigarette-smoking, which, statistically at least, seems to be accompanied by a relatively high rate of incidence of lung cancer.

One environmental factor that is certainly carcinogenic is energetic radiation, and man has been exposed to such radiation in increasing measure since 1895.

On 5 November 1895, the German physicist Wilhelm Konrad Roentgen performed an experiment to study the luminescence produced by cathode rays. The better to see the effect, he darkened the room. His cathode-ray tube was enclosed in a black cardboard box. When he turned on the cathode-ray tube, he was startled to catch a flash of light from something across the room. The flash came from a sheet of paper coated with barium platino-cyanide, a luminescent chemical. Was it possible that radiation from the closed box had made it glow? Roentgen turned off his cathode-ray tube, and the glow stopped. He turned it on again – the glow returned. He took the paper into the next room, and it still glowed. Clearly, the cathode-ray tube was producing some form of radiation which could penetrate cardboard and walls.

Roentgen, having no idea what kind of radiation this might be, called it simply 'X-rays'. Other scientists tried to change the name to 'Roentgen rays', but this was so hard for anyone but Germans to pronounce that 'X-rays' stuck. (We now know that the speeding electrons making up the cathode rays are strongly decelerated on striking a metal barrier. The kinetic energy lost is converted into radiation that is called 'Bremsstrahlung' – German for 'braking radiation'. X-rays are an example of such radiation.)

The X-rays revolutionized physics. They captured the imagination of physicists, started a typhoon of experiments, led within a few months to the discovery of radioactivity, and opened up the inner world of the atom. When the award of Nobel Prizes began in 1901, Roentgen was the first to receive the prize in physics.

The hard X-radiation also started something else – exposure of human beings to intensities of energetic radiation such as man had never experienced before. Four days after the news of Roentgen's discovery reached the United States, X-rays were used to locate a bullet in a patient's leg. They were a wonderful means of exploring the interior of the body. X-rays pass easily through the soft tissues (consisting chiefly of elements of low atomic weight) and tend to be stopped by elements of higher atomic weight, such as make up the bones (composed largely of phosphorus and calcium). On a photographic plate placed behind the body, bone shows up as a cloudy white, in contrast to the black areas where X-rays have come through in greater intensity because they have been much less absorbed by the soft tissues. A lead bullet shows up as pure white; it stops the X-rays completely.

X-rays are obviously useful for showing bone fractures, calcified joints, cavities in the teeth, foreign objects in the body, and so on. But it is also a simple matter to outline the soft tissues by introducing an insoluble salt of a heavy element. Barium sulphate, when swallowed, will outline the stomach or intestines. An iodine compound injected into the blood will travel to the kidneys and the ureter and outline those organs, for iodine has a high atomic weight and therefore is opaque to X-rays.

Even before X-rays were discovered, a Danish physician, Niels Ryberg Finsen, had found that high-energy radiation could kill micro-organisms; he used ultra-violet light to destroy the bacteria causing lupus vulgaris, a skin disease. (For this he was awarded the Nobel Prize in physiology and medicine in 1903.) The X-rays turned out to be far more deadly. They could kill the fungus of ringworm. They could damage or destroy human cells, and were eventually used to kill cancer cells beyond reach of the surgeon's knife.

What was also discovered – the hard way – was that high-energy radiation could *cause* cancer. At least one hundred of the early workers with X-rays and radioactive materials died of cancer, the first death taking place in 1902. As a matter of fact, both Marie Curie and her daughter, Irène Joliot-Curie, died of leukaemia, and it is easy to believe that radiation was a contributing cause in both cases. In 1928, a British physician, George William Marshall Findlay, found that even ultra-violet radiation was energetic enough to cause skin cancer in mice.

It is certainly reasonable to suspect that man's increasing exposure to energetic radiation (in the form of medical X-rays and so on) may be responsible for part of the increased incidence of cancer. And the future will tell whether the accumulation in our bones of traces of strontium 90 from fall-out will increase the incidence of bone cancer and leukaemia.

What can all the various carcinogens – chemicals, radiation, and so on – possibly have in common? One reasonable thought is that all of them may cause genetic mutations, and that cancer may be the result of mutations in body cells.

Suppose that some gene is changed so that it no longer can produce a key enzyme needed in the process that controls the growth of cells. When a cell with such a defective gene divides, it will pass on the defect. With the control mechanism not functioning, further division of these cells may continue indefinitely, without regard to the needs of the body as a whole or even to the needs of the tissue involved (for example, the specialization of cells in an organ). The tissue is disorganized. It is, so to speak, a case of anarchy in the body.

That energetic radiation can produce mutations is well established. What about the chemical carcinogens? Well, mutation by chemicals also has been demonstrated. The 'nitrogen mustards' are a good example. These compounds, like the 'mustard gas' of the First World War, produce burns and blisters on the skin resembling those caused by X-rays. They can also damage the chromosomes and increase the mutation rate. Moreover, a number of other chemicals have been found to imitate energetic radiation in the same way.

The chemicals that can induce mutations are called 'mutagens'. Not all mutagens have been shown to be carcinogens, and not all carcinogens have been shown to be mutagens. But there are enough cases of compounds that are both carcinogenic and mutagenic to arouse suspicion that the coincidence is not accidental.

Meanwhile, the notion that micro-organisms may have something to do with cancer is far from dead. With the discovery of viruses, this suggestion of the Pasteur era was revived. In 1903, the French bacteriologist Amédée Borrel suggested that cancer might be a virus disease, and, in 1908, two Danes, Wilhelm Ellerman and Olaf Bang, showed that fowl leukaemia was indeed caused by a virus. However, leukaemia was not at the time recognized as a form of cancer, and the issue hung fire. In 1909, however, the American physician Francis Peyton Rous ground up a chicken tumour, filtered it, and injected the clear filtrate into other chickens. Some of them developed tumours. The finer the filter, the fewer the tumours. This certainly looked as if particles of some kind were responsible for the initiation of tumours, and it seemed that these particles were the size of viruses.

The 'tumour viruses' have had a rocky history. At first, the tumours pinned down to viruses turned out to be uniformly benign; for instance, viruses were shown to cause such things as rabbits' papillomas (similar to warts). In 1936, John Joseph Bittner, working in the famous mouse-breeding laboratory at Bar Harbor, Maine, came on something more exciting. Maude Slye of the same laboratory had bred strains of mice that seemed to have an inborn resistance to cancer and other strains that seemed cancer-prone. The mice of some strains rarely developed cancer; those of other strains almost invariably did, after reaching maturity. Bittner tried the experiment of switching mothers on the newborn mice so that they would suckle at the opposite strain. He discovered that when baby mice of a 'cancer-resistant' strain suckled at mothers of a 'cancer-prone' strain, they usually developed cancer. On the other hand, supposedly cancer-prone baby mice that were fed by cancer-resistant mothers did not develop cancer. Bittner concluded that the cancer cause, whatever it was, was not inborn but was transmitted in the mothers' milk. He called it the 'milk factor'.

Naturally, Bittner's milk factor was suspected to be a virus. Eventually the Columbia University biochemist Samuel Graff identified the factor as a particle containing nucleic acids. Other tumour viruses, causing certain types of mouse tumours, and animal leukaemias, have been found, and all of them contain nucleic acids. No viruses have been detected in connection with human cancers, but research on human cancer is obviously limited.

Now the mutation and virus theories of cancer begin to converge. Perhaps the seeming contradiction between the two notions is not a contradiction after all. Viruses and genes have a very important thing in common: the key to the behaviour of both lies in their nucleic acids. Indeed, in 1959, G. A. di Mayorca and co-workers at the Sloan-Kettering Institute and the National Institutes of Health isolated DNA from a mouse-tumour virus and found that the DNA alone could induce cancers in mice just as effectively as the virus did.

Thus the difference between the mutation theory and the virus theory boils down to whether the cancer-causing nucleic acid arises by a mutation in a gene within the cell or is introduced by a virus invasion from outside the cell. These ideas are not mutually exclusive; cancer may come about in both ways.

It was not until 1966, however, that the virus hypothesis was deemed fruitful enough to be worth a Nobel Prize. Fortunately, Peyton Rous, who had made his discovery fifty-five years before, was still alive and received a share of the 1966 Nobel Prize for medicine and physiology. (He lived on to 1970, dying at the age of ninety, active in research nearly to the end.)

What goes wrong in the metabolic machinery when cells grow unrestrainedly? This question has as yet received no answer. But strong suspicion rests on some of the hormones, especially the sex hormones.

For one thing, the sex hormones are known to stimulate rapid, localized growth in the body (as in the breasts of an adolescent girl). For another, the tissues of sexual organs – the breasts, cervix, and ovaries in a woman, the testes and prostate in a man – are particularly prone to cancer. Strongest of all is the chemical evidence. In 1933, the German biochemist Heinrich Wieland (who had won the Nobel Prize in chemistry in 1927 for his work with bile acids) managed to convert a bile acid into a complex hydrocarbon called 'methylcholanthrene', a powerful carcinogen. Now methylcholanthrene (like the bile acids) has the four-ring structure of a steroid, and it so happens that all the sex hormones are steroids. Could a misshapen sex-hormone molecule act as a carcinogen? Or might even a correctly shaped hormone be mistaken for a carcinogen, so to speak, by a distorted gene pattern in a cell, and so stimulate uncontrolled growth? It is anyone's guess, but these are interesting speculations.

Curiously enough, changing the supply of sex hormones sometimes checks cancerous growth. For instance, castration, to reduce the body's manufacture of male sex hormone, or the administration of neutralizing female sex hormone, has a mitigating effect on cancer of the prostate. As a treatment, this is scarcely something to shout about, and it is a measure of the desperation regarding cancer that such devices are resorted to.

The main line of attack against cancer still is surgery. And its limitations are still what they have always been: sometimes the cancer cannot be cut out without killing the patient; often the knife frees bits of malignant tissue (since the disorganized cancer tissue has a tendency to fragment), which are then carried by the bloodstream to other parts of the body where they take root and grow.

The use of energetic radiation to kill the cancer tissue also has its drawbacks. Artificial radioactivity has added new weapons to the traditional X-rays and radium. One of them is cobalt 60, which yields high-energy gamma rays and is much less expensive than radium; another is a solution of radioactive iodine (the 'atomic cocktail'), which concentrates in the thyroid gland and thus attacks a thyroid cancer. But the body's tolerance of radiation is limited, and there is always the danger that the radiation will start more cancers than it stops.

Still, surgery and radiation are the best we have, and they have saved, or at least prolonged, many a life. They will perforce be man's main reliance against cancer until biologists find what they are seeking: a 'magic bullet' that, without harming normal cells, will search out the cancer cells and either destroy them or stop their wild division in its tracks.

A great deal of work is going on along two principal routes. One is to find out everything possible about how cells divide. The other is to learn in the greatest possible detail exactly how cells conduct metabolism, with the hope of finding some decisive difference between cancer cells and normal cells. Differences have been found, but they are pretty minor – so far.

Meanwhile a stupendous sifting of chemicals by trial and error is being carried out. About 50,000 new drugs a year are tested. For a time the nitrogen mustards looked hopeful, on the theory that they would mimic radiation in killing cancer cells. Some of the drugs of this type do seem to help against certain types of cancer, at least to the extent of prolonging life, but they are obviously only a stopgap.

More hope lies in the direction of the nucleic acids themselves. There must be some difference between the nucleic acids in cancer cells and those in normal cells. The object, then, is to find a way to interfere with the chemical workings of one and not the other. Then again, perhaps the disorganized cancer cells are less efficient than normal cells in manufacturing nucleic acids. If so, throwing a few grams of sand into the machinery might cripple the less efficient cancer cells without seriously disturbing the more efficient normal cells.

For instance, one substance that is vital to the production of nucleic acid is folic acid. It plays a major role in the formation of the purines and pyrimidines, the building blocks for nucleic acid. Now a compound resembling folic acid might, by competitive inhibition, slow things up just enough to prevent cancer cells from making nucleic acid while allowing normal cells to produce it at an adequate rate. And, of course, without nucleic acid the cancer cells could not multiply. There are, in fact, 'folic-acid antagonists' of this sort. One of them, called 'amethopterin', has shown some effect against leukaemia.

There is a still more direct attack. Why not inject competitive substitutes for the purines and pyrimidines themselves? The most hopeful candidate is '6-mercaptopurine'. This compound is just like adenine except that it has an —SH group in place of adenine's —NH$_2$.

Even the possibility of treatment of just one of the cancer group of diseases is not to be ignored. The malignant cells of certain types of leukaemias require an outside source of the substance asparagine, something healthy cells can manufacture for themselves. Treatment with the enzyme asparaginase, which catalyses the breakdown of asparagine, reduces the body's supply and starves the malignant cells while the normals manage to survive.

The world-wide research attack upon cancer is keen, resourceful, and, in comparison with other biological research, handsomely financed. Treatment has reached the point where one out of three cancer victims will survive and live out a normal life span. But a cure will not be found easily, for the secret of cancer is as subtle as the secret of life itself.

5 The Body

Food

Perhaps the first great advance in medical science was the recognition by physicians that good health called for a simple, balanced diet. The Greek philosophers recommended moderation in eating and drinking, not only for philosophical reasons but also because those who followed this rule were more comfortable and lived longer. That was a good start, but biologists eventually learned that moderation alone was not enough. Even if someone has the good fortune to avoid eating too little and the good sense to avoid eating too much, he will still do poorly if his diet happens to be shy of certain essential ingredients, as is actually the case for large numbers of people in some parts of the world.

The human body is rather specialized (as organisms go) in its dietary needs. A plant can live on just carbon dioxide, water, and certain inorganic ions. Some of the micro-organisms likewise get along without any organic food; they are called 'autotrophic' ('self-feeding'), which means that they can grow in environments in which there is no other living thing. The bread mould *Neurospora* begins to get a little more complicated: in addition to inorganic substances it has to have sugar and the vitamin biotin. And as the forms of life become more and more complex, they seem to become more and more dependent on their diet to supply the organic building blocks necessary for building living tissue. The reason is simply that they have lost some of the enzymes that primitive organisms possess. A green plant has a complete supply of enzymes for making all the necessary amino acids, proteins, fats, and carbohydrates from inorganic materials. *Neurospora* has all the enzymes except one or more of those needed to make sugar and biotin. By the time we get to man, we find that he lacks the enzymes required to make many of the amino acids, the vitamins, and various other necessities, and he must get these ready-made in his food.

This may seem a kind of degeneration – a growing dependence on the environment which puts the organism at a disadvantage. Not so. If the environment supplies the building blocks, why carry the elaborate enzymatic machinery needed to make them? By dispensing with this machinery, the cell can use its energy and space for more refined and specialized purposes.

It was the English physician William Prout (the same Prout who was a century ahead of his time in suggesting that all the elements were built from hydrogen) who first suggested that the organic foods could be divided into three types of substances, later named carbohydrates, fats, and proteins.

The chemists and biologists of the nineteenth century, notably Justus von Liebig of Germany, gradually worked out the nutritive properties of these foods. Protein, they found, is the most essential, and the organism could get along on it alone. The body cannot make protein from carbohydrate and fat, because those substances have no nitrogen, but it could make the necessary carbohydrates and fats from materials supplied by protein. Since protein is comparatively scarce in the environment, however, it would be wasteful to live on an all-protein diet – like stoking the fire with furniture when firewood is available.

Under favourable circumstances the human body's daily requirement of proteins, by the way, is surprisingly low. The Food and Nutrition Board of the National Research Council in its 1958 chart of recommendations suggested that the minimum for adults is one gram of protein per kilogram of body weight per day, which amounts to a little more than two ounces for the average grown man. About two quarts of milk can supply that amount. Children and pregnant or nursing mothers need somewhat more protein.

Of course, a lot depends on what proteins you choose. Nineteenth-century experimenters tried to find out whether the population could get along, in times of famine, on gelatin – a protein material obtained by heating bones, tendons, and otherwise inedible parts of animals. But the French physiologist François Magendie demonstrated that dogs lost weight and died when gelatin was their sole source of protein. This does not mean there is anything wrong with gelatin as a food, but it simply does not supply all the necessary building blocks when it is the only protein in the diet. The key to the usefulness of a protein lies in the efficiency with which the body can use the nitrogen it supplies. In 1854 the English agriculturists John Bennet Lawes and Joseph Henry Gilbert fed pigs protein in two forms – lentil meal and barley meal. They found that the pigs retained much more of the nitrogen in barley than of that in lentils. These were the first 'nitrogen balance' experiments.

A growing organism gradually accumulates nitrogen from the food it ingests ('positive nitrogen balance'). If it is starving or suffering a wasting disease, and gelatin is the sole source of protein, the body continues to starve or waste away, from a nitrogen-balance standpoint (a situation called 'negative nitrogen balance'). It keeps losing more nitrogen than it takes in, regardless of how much gelatin it is fed.

Why so? The nineteenth-century chemists eventually discovered that gelatin is an unusually simple protein. It lacks tryptophan and other amino acids present in most proteins. Without these building blocks, the body cannot build the proteins it needs for its own substance. Therefore, unless it gets other protein in its food as well, the amino acids that do occur in the gelatin are useless and have to be excreted. It is as if housebuilders found themselves with plenty of timber but no nails. Not only could they not build the house, but the timber would just be in the way and eventually would have to be disposed of. Attempts were made in the 1890s to make gelatin a more efficient article of diet by adding some of those amino acids in which it was deficient, but without success.

Better luck was obtained with proteins not as drastically limited as gelatin.

In 1906, the English biochemists Frederick Gowland Hopkins and E. G. Willcock fed mice on a diet in which the only protein was 'zein', found in corn. They knew that this protein had very little of the amino acid tryptophan. The mice died in about fourteen days. The experimenters then tried mice on zein plus tryptophan. This time the mice survived twice as long. It was the first hard evidence that amino acids, rather than protein, might be the essential components of the diet. (Although the mice still died prematurely, this was probably due mainly to a lack of certain vitamins not known at the time.)

In the 1930s, the American nutritionist William Cumming Rose got to the bottom of the amino-acid problem. By that time the major vitamins were known, so he could supply the animals with those needs and focus on the amino acids. Rose fed rats a mixture of amino acids instead of protein. The rats did not live long on this diet. But when he fed rats on the milk protein casein, they did well. Apparently there was something in casein – some undiscovered amino acid, in all probability – which was not present in the amino-acid mixture he was using. Rose broke down the casein and tried adding various of its molecular fragments to his amino-acid mixture. In this way he tracked down the amino acid 'threonine', the last of the major amino acids to be discovered. When he added the threonine from casein to his amino-acid mixture, the rats grew satisfactorily, without any intact protein in the diet.

Rose proceeded to remove the amino acids from their diet one at a time. By this method he eventually identified ten amino acids as indispensable items in the diet of the rat: lysine, tryptophan, histidine, phenylalanine, leucine, isoleucine, threonine, methionine, valine, and arginine. If supplied with ample quantities of these, the rat could manufacture all it needed of the others, such as glycine, proline, aspartic acid, alanine, and so on.

In the 1940s, Rose turned his attention to man's requirements of amino acids. He persuaded graduate students to submit to controlled diets in which a mixture of amino acids was the only source of nitrogen. By 1949, he was able to announce that the adult male required only eight amino acids in the diet: phenylalanine, leucine, isoleucine, methionine, valine, lysine, tryptophan, and threonine. Since arginine and histidine, indispensable to the rat, are dispensable in the human diet, it would seem that in this respect man is less specialized than the rat, or, indeed, than any other mammal that has been tested in detail.

Potentially a person could get along on the eight dietarily essential amino acids; given enough of these, he could make not only all the other amino acids he needs but also all the carbohydrates and fats. Actually a diet made up only of amino acids would be much too expensive, to say nothing of its flatness and monotony. But it is enormously helpful to have a complete blueprint of our amino-acid needs so that we can reinforce natural proteins when necessary for maximum efficiency in absorbing and utilizing nitrogen.

Vitamins

Food fads and superstitions unhappily still delude too many people – and spawn too many cure-everything best sellers – even in these enlightened times. In fact, it is perhaps because these times are enlightened that food faddism is possible. Through most of man's history, his food consisted of whatever could be produced in the vicinity, of which there usually was not very much. It was eat what there was to eat or starve; no one could afford to be choosy, and without choosiness there can be no food faddism.

Modern transportation has made it possible to ship food from any part of the earth to any other, particularly since the use of large-scale refrigeration has arisen. This reduced the threat of famine, which, before modern times, was invariably local, with neighbouring provinces loaded with food that could not be transported to the famine area.

Home storage of a variety of foods became possible as early

man learned to preserve foods by drying, salting, increasing the sugar content, fermenting, and so on. It became possible to preserve food in states closer to the original when methods of storing cooked food in vacuum were developed. (The cooking kills micro-organisms and the vacuum prevents others from growing and reproducing.) Vacuum storage was first made practical by a French chef, François Appert, who developed the technique in response to a prize offered by Napoleon I for a way of preserving food for his armies. Appert made use of glass jars, but nowadays tin-lined steel cans (inappropriately called 'tin cans' or just 'tins') are used for the purpose. Since the Second World War, fresh-frozen food has become popular and the growing number of home freezers has further increased the general availability and variety of fresh foods. Each broadening of food availability has increased the practicality of food faddism.

All this is not to say that a shrewd choice of food may not be useful. There are certain cases in which specific foods will definitely cure a particular disease. In every instance, these are 'deficiency diseases', diseases produced by the lack in the diet of some substance essential to the body's chemical machinery. These arise almost invariably when a person is deprived of a normal, balanced diet – one containing a wide variety of foods.

To be sure, the value of a balanced and variegated diet was understood by a number of medical practitioners of the nineteenth century and before, when the chemistry of food was still a mystery. A famous example is that of Florence Nightingale, the heroic English nurse of the Crimean War who pioneered the adequate feeding of soldiers, as well as decent medical care. And yet 'dietetics' (the systematic study of diet) had to await the end of the century and the discovery of trace substances in food, essential to life.

The ancient world was well acquainted with scurvy, a disease in which the capillaries become increasingly fragile, gums bleed and teeth loosen, wounds heal with difficulty if at all, and the patient grows weak and eventually dies. It was particularly prevalent in besieged cities and on long ocean voyages. (Magellan's crew suffered more from scurvy than from general undernourishment.) Ships on long voyages, lacking refrigeration, had to carry non-spoilable food, which meant hardtack and salt pork. Nevertheless, physicians for many centuries failed to connect scurvy with diet.

In 1536, while the French explorer Jacques Cartier was wintering in Canada, 110 of his men were stricken with scurvy. The native Indians knew and suggested a remedy: drinking water in which pine needles had been soaked. Cartier's men in desperation followed this seemingly childish suggestion. It cured them of their scurvy.

Two centuries later, in 1747, the Scottish physician James Lind took note of several incidents of this kind and experimented with fresh fruits and vegetables as a cure. Trying his treatments on scurvy-ridden sailors, he found that oranges and lemons brought about improvement most quickly. Captain Cook, on a voyage of exploration across the Pacific from 1772 to 1775 kept his crew scurvy-free by enforcing the regular eating of sauerkraut. Nevertheless, it was not until 1795 that the brass hats of the British Navy were sufficiently impressed by Lind's experiments (and by the fact that a scurvy-ridden flotilla could lose a naval engagement with scarcely a fight) to order daily rations of lime juice for British sailors. (They have been called 'limeys' ever since, and the Thames area in London where the crates of limes were stored is still called 'Limehouse'.) Thanks to the lime juice, scurvy disappeared from the British Navy.

A century later, in 1891, Admiral Takaki of the Japanese Navy similarly introduced a broader diet into the rice monotony of his ships. The scourge of a disease known as 'beri-beri' came to an end in the Japanese Navy as a result.

In spite of occasional dietary victories of this kind (which no one could explain), nineteenth-century biologists refused to believe that a disease could be cured by diet, particularly after Pasteur's germ theory of disease came into its own. In 1896, however, a Dutch physician named Christiaan Eijkman convinced them almost against his own will.

Eijkman was sent to the Dutch East Indies to investigate beri-beri, which was endemic in those regions (and which, even today, when medicine knows its cause and cure, still kills 100,000 people a year). Takaki had stopped beri-beri by dietary measures, but the West, apparently, placed no stock in what might have seemed merely the mystic lore of the Orient.

Supposing that beri-beri was a germ disease, Eijkman took along some chickens as experimental animals in which to establish the germ. A highly fortunate piece of skulduggery upset his plans. Without warning, most of his chickens came down with a paralytic disease from which some died, but after about four months those still surviving regained their health. Eijkman, mystified by failing to find any germ responsible for the attack, finally investigated the chickens' diet. He discovered that the person originally in charge of feeding the chickens had economized (and no doubt profited) by using scraps of left-over food, mostly polished rice, from the wards of the military hospital. It happened that after a few months a new cook had arrived and taken over the feeding of the chickens; he had put a stop to the petty graft and supplied the animals with the usual chicken feed, containing unhulled rice. It was then that the chickens had recovered.

Eijkman experimented. He put chickens on a polished-rice diet, and they fell sick. Back on the unhulled rice, they recovered. It was the first case of a deliberately produced dietary-deficiency disease. Eijkman decided that this 'polyneuritis' that afflicted fowls was similar in symptoms to human beri-beri. Did human beings get beri-beri because they ate only polished rice?

For human consumption, rice was stripped of its hulls mainly so that it would keep better, for the rice germ removed with the hulls contains oils that go rancid easily. Eijkman and a co-worker, Gerrit Grijns, set out to see what it was in rice hulls that prevented beri-beri. They succeeded in dissolving the crucial factor out of the hulls with water, and they found that it would pass through membranes which would not pass proteins. Evidently the substance in question must be a fairly small molecule. They could not, however, identify it.

Meanwhile other investigators were coming across other mysterious factors that seemed to be essential for life. In 1905, a

Dutch nutritionist, C. A. Pekelharing, found that all his mice died within a month on an artificial diet which seemed ample as far as fats, carbohydrates, and proteins were concerned. But mice did fine when he added a few drops of milk to this diet. And in England the biochemist Frederick Hopkins, who was demonstrating the importance of amino acids in the diet, carried out a series of experiments in which he, too, showed that something in the casein of milk would support growth if added to an artificial diet. This something was soluble in water. Even better than casein as the dietary supplement was a small amount of a yeast extract.

For their pioneer work in establishing that trace substances in the diet were essential to life, Eijkman and Hopkins shared the Nobel Prize in medicine and physiology in 1929.

The next task was to isolate these vital trace factors in food. By 1912, three Japanese biochemists, Umetaro Suzuki, T. Shimamura, and S. Ohdake, had extracted from rice hulls a compound which was very potent in combating beri-beri. Doses of five to ten milligrams sufficed to effect a cure in fowl. In the same year the Polish-born biochemist Casimir Funk (then working in England and later to come to the United States) prepared the same compound from yeast.

Because the compound proved to be an amine (that is, one containing the amine group, NH_2), Funk called it a 'vitamine', Latin for 'life amine'. He made the guess that beri-beri, scurvy, pellagra, and rickets all arose from deficiencies of 'vitamines'. Funk's guess was correct as far as his identification of these diseases as dietary-deficiency diseases was concerned. But it turned out that not all 'vitamines' were amines.

In 1913, two American biochemists, Elmer Vernon McCollum and Marguerite Davis, discovered another trace factor vital to health in butter and egg yolk. This one was soluble in fatty substances instead of water. McCollum called it 'fat-soluble A', to contrast it with 'water-soluble B', which was the name he applied to the anti-beri-beri factor. In the absence of chemical information as to the nature of the factors, this seemed fair enough, and it started the custom of naming them by letters. In 1920, the

British biochemist Jack Cecil Drummond changed the names to 'vitamin A' and 'vitamin B', dropping the final *e* of 'vitamine' as a gesture towards taking 'amine' out of the name. He also suggested that the anti-scurvy factor was still a third such substance, which he named 'vitamin C'.

Vitamin A was quickly identified as a food factor required to prevent the development of abnormal dryness of the membranes around the eye, called 'xerophthalmia', from Greek words meaning 'dry eyes'. In 1920, McCollum and his associates found that a substance in cod-liver oil, which was effective in curing both xerophthalmia and a bone disease called 'rickets', could be so treated as to cure rickets only. They decided the anti-rickets factor must represent a fourth vitamin, which they named vitamin D. Vitamins D and A are fat-soluble; C and B are water-soluble.

By 1930, it had become clear that vitamin B was not a simple substance but a mixture of compounds with different properties. The food factor that cured beri-beri was named vitamin B_1, a second factor was called vitamin B_2, and so on. Some of the reports of new factors turned out to be false alarms, so that one does not hear of B_3, B_4, or B_5 any longer. However, the numbers worked their way up to B_{14}. The whole group of vitamins (all water-soluble) is frequently referred to as the 'B-vitamin complex'.

New letters also were added. Of these, vitamins E and K (both fat-soluble) remain as veritable vitamins, but 'vitamin F' turned out to be not a vitamin and 'vitamin H' turned out to be one of the B-complex vitamins.

Nowadays, with their chemistry identified, the letters of even the true vitamins are going by the board, and most of them are known by their chemical names, though the fat-soluble vitamins, for some reason, have held on to their letter designations more tenaciously than the water-soluble ones.

It was not easy to work out the chemical composition and structure of the vitamins, for these substances occur only in minute amounts. For instance, a ton of rice hulls contains only about five grams (a little less than a fifth of an ounce) of vitamin B_1. Not until 1926 did anyone extract enough of the reasonably pure vitamin to analyse it chemically. Two Dutch biochemists, Barend Coenraad Petrus Jansen and William Frederick Donath, worked up a composition for vitamin B, from a tiny sample, but it turned out to be wrong. In 1932, Ohdake tried again on a slightly larger sample and got it almost right. He was the first to detect a sulphur atom in a vitamin molecule.

Finally in 1934 Robert R. Williams, then director of chemistry at the Bell Telephone Laboratories, climaxed twenty years of research by painstakingly separating vitamin B_1 from tons of rice hulls until he had enough to work out a complete structural formula. The formula follows:

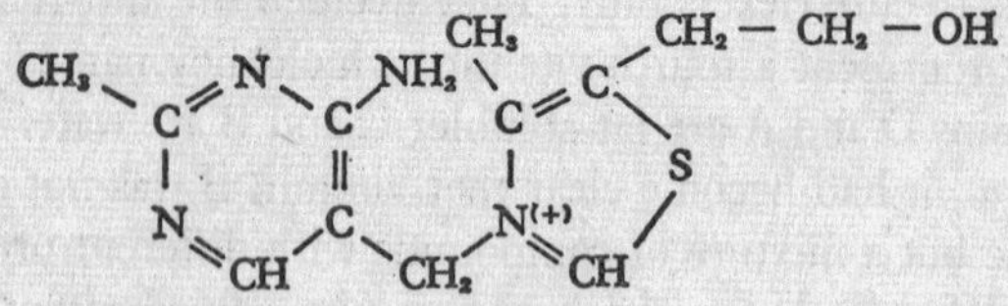

Since the most unexpected feature of the molecule was the atom of sulphur ('theion' in Greek), the vitamin was named 'thiamine'.

Vitamin C was a different sort of problem. Citrus fruits furnish a comparatively rich source of this material, but one difficulty was finding an experimental animal that did not make its own vitamin C. Most mammals, aside from man and the other primates, have retained the capacity to form this vitamin. Without a cheap and simple experimental animal that would develop scurvy, it was difficult to follow the location of vitamin C among the various fractions into which the fruit juice was broken down chemically.

In 1918 the American biochemists B. Cohen and Lafayette Benedict Mendel solved this problem by discovering that guinea pigs could not form the vitamin. In fact, guinea pigs developed scurvy much more easily than men did. But another difficulty remained. Vitamin C was found to be very unstable (it is the most unstable of the vitamins), so it was easily lost in chemical procedures to isolate it. A number of research workers ardently pursued the vitamin without success.

As it happened, vitamin C was finally isolated by someone who was not particularly looking for it. In 1928, the Hungarian-born biochemist Albert Szent-Györgi, then working in London in Hopkins' laboratory and interested mainly in finding out how tissues made use of oxygen, isolated from cabbages a substance which helped transfer hydrogen atoms from one compound to another. Shortly afterwards Charles Glen King and his co-workers at the University of Pittsburgh, who *were* looking for vitamin C, prepared some of the substance from cabbages and found that it was strongly protective against scurvy. Furthermore, they found it identical with crystals they had obtained from lemon juice. King determined its structure in 1933, and it turned out to be a sugar molecule of six carbons, belonging to the L-series instead of the D-series:

$$O = C \begin{matrix} O \\ CH - CH - CH_2OH \\ C = C \quad OH \\ OH \quad OH \end{matrix}$$

It was named 'ascorbic acid' (from Greek words meaning 'no scurvy').

As for vitamin A, the first hint as to its structure came from the observation that the foods rich in vitamin A were often yellow or orange (butter, egg yolk, carrots, fish-liver oil, and so on). The substance largely responsible for this colour was found to be a hydrocarbon named 'carotene', and in 1929 the British biochemist Thomas Moore demonstrated that rats fed on diets containing carotene stored vitamin A in the liver. The vitamin itself was not coloured yellow, so that deduction was that though carotene was not itself vitamin A, the liver converted it into something which was vitamin A. (Carotene is now considered an example of a 'provitamin'.)

In 1937, the American chemists Harry Nicholls Holmes and Ruth Elizabeth Corbet isolated vitamin A as crystals from fish-liver oil. It turned out to be a 20-carbon compound – half of the carotene molecule with a hydroxyl group added:

$$CH_3 \quad CH_3$$

(Chemical structure diagram of vitamin D precursor chain with methyl groups and conjugated double bonds, ending in $-CH_2-OH$)

The chemists hunting for vitamin D found their best chemical clue by means of sunlight. As early as 1921, the McCollum group (who first demonstrated the existence of the vitamin) showed that rats did not develop rickets on a diet lacking vitamin D if they were exposed to sunlight. Biochemists guessed that the energy of sunlight converted some provitamin in the body into vitamin D. Since vitamin D was fat-soluble, they went searching for the provitamin in the fatty substances of food.

By breaking down fats into fractions and exposing each fragment separately to sunlight, they determined that the provitamin that sunlight converted into vitamin D was a steroid. What steroid? They tested cholesterol, the most common steroid of the body, and that was not it. Then, in 1926, the British biochemists Otto Rosenheim and T. A. Webster found that sunlight would convert a closely related sterol, 'ergosterol' (so named from the fact that it was first isolated from ergot-infested rye), into vitamin D. The German chemist Adolf Windaus discovered this independently at about the same time. For this and other work in steroids, Windaus received the Nobel Prize in chemistry in 1928.

The difficulty in producing vitamin D from ergosterol rested on the fact that ergosterol did not occur in animals. Eventually the human pro-vitamin was identified as '7-dehydrocholesterol', which differs from cholesterol only in having two hydrogen atoms fewer in its molecule. The vitamin D formed from it has this formula:

(Chemical structure diagram of vitamin D steroid ring system with side chains)

Vitamin D in one of its forms is called 'calciferol', from Latin words meaning 'calcium-carrying', because it is essential to the proper laying down of bone structure.

Not all the vitamins show their absence by producing an acute disease. In 1922, Herbert McLean Evans and K. J. Scott at the University of California implicated a vitamin as a cause of sterility in animals. Evans and his group did not succeed in isolating this one, vitamin E, until 1936. It was then given the name 'tocopherol' (from Greek words meaning 'to bear children').

Unfortunately, whether human beings need vitamin E, or how much, is not yet known. Obviously, dietary experiments designed to bring about sterility cannot be tried on human subjects. And even in animals, the fact that they can be made sterile by withholding vitamin E does not necessarily mean that natural sterility arises in this way.

In the 1930s, the Danish biochemist Carl Peter Henrik Dam discovered by experiments on chickens that a vitamin was involved in the clotting of blood. He named it 'Koagulations-vitamine', and this was eventually shortened to vitamin K. Edward Doisy and his associates at St Louis University then isolated vitamin K and determined its structure. Dam and Doisy shared the Nobel Prize in medicine and physiology in 1943.

Vitamin K is not a major vitamin nor a nutritional problem. Normally a more than adequate supply of this vitamin is manufactured by the bacteria in the intestines. In fact, they make so much of it that the faeces may be richer in vitamin K than the food is. Newborn infants are the most likely to run a danger of poor blood clotting and consequent haemorrhage because of vitamin deficiency. In the hygienic modern hospital it takes infants three days to accumulate a reasonable supply of intestinal bacteria, and they are protected by injections of the vitamin into themselves directly or into the mother shortly before birth. In the old days, the infants picked up the bacteria almost at once, and though they might die of various infections and disease, they were at least safe from the dangers of haemorrhage.

In fact, one might wonder whether organisms could live at all in the complete absence of intestinal bacteria, or whether the symbiosis had not become too intimate to abandon. However, germ-free animals have been grown from birth under completely sterile conditions and have even been allowed to reproduce under such conditions. Mice have been carried through twelve generations in this fashion. Experiments of this sort have been conducted at the University of Notre Dame since 1928.

During the late 1930s and early 1940s, biochemists identified several additional B vitamins, which now go under the names of biotin, pantothenic acid, pyridoxine, folic acid, and cyanocobalamine. These vitamins are all made by intestinal bacteria; moreover, they are present so universally in foodstuffs that no cases of deficiency diseases have appeared. In fact, investigators have had to feed animals an artificial diet deliberately excluding them, and even to add 'anti-vitamins' to neutralize those made by the intestinal bacteria, in order to see what the deficiency symptoms are. (Anti-vitamins are substances similar to the vitamin in structure. They immobilize the enzyme making use of the vitamin by means of competitive inhibition.)

The determination of the structure of each of the various vitamins was usually followed speedily (or even preceded) by synthesis of the vitamin. For instance, Williams and his group synthesized thiamine in 1937, three years after they had deduced its structure. The Polish-born Swiss biochemist Tadeus Reichstein and his group synthesized ascorbic acid in 1933, somewhat before the structure was completely determined by King. Vitamin A, for another example, was synthesized in 1936 (again somewhat before the structure was completely determined) by two different groups of chemists.

The use of synthetic vitamins has made it possible to fortify food (milk was first vitamin-fortified as early as 1924) and to prepare vitamin mixtures at reasonable prices and sell them over the chemist's counter. The need for vitamin pills varies with individual cases. Of all the vitamins, the one most apt to be deficient in supply is vitamin D. Young children in northern climates, where sunlight is weak in winter time, run the danger of rickets, so they may require irradiated foods or vitamin supplements. But the dosage of vitamin D (and of vitamin A) should be carefully controlled, because an overdose of these vitamins can be harmful. As for the B vitamins, anyone eating an ordinary, rounded diet does not need to take pills for them. The same is true of vitamin C, which in any case should not present a problem, for there are few people who do not enjoy orange juice or who do not drink it regularly in these vitamin-conscious times.

On the whole, the wholesale use of vitamin pills, while redounding chiefly to the profit of drug houses, usually does people no harm and may be partly responsible for the fact that the current generation of Americans is taller and heavier than previous generations.

Biochemists naturally were curious to find out how the vitamins, present in the body in such tiny quantities, exerted such important effects on the body chemistry. The obvious guess was that they had something to do with enzymes, also present in small quantities.

The answer finally came from detailed studies of the chemistry of enzymes. Protein chemists had known for a long time that some proteins were not made up solely of amino acids, and that non-amino-acid prosthetic groups might exist, such as the haem in haemoglobin (see p. 68). In general, these prosthetic groups tended to be tightly bound to the rest of the molecule. With enzymes, however, there were in some cases non-amino-acid portions that were quite loosely bound and might be removed with little trouble.

This was first discovered in 1904 by Arthur Harden (who was soon to discover phosphorus-containing intermediates; see p. 100). Harden worked with a yeast extract capable of bringing about the fermentation of sugar. He placed it in a bag made of a semi-permeable membrane and placed that bag in fresh water. Small molecules could penetrate the membrane, but the large protein molecule could not. After this 'dialysis' had progressed for a while, Harden found that the activity of the extract was lost. Neither the fluid within nor that outside the bag would ferment sugar. If the two fluids were combined, activity was regained.

Apparently, the enzyme was made up not only of a large protein molecule, but also of a 'coenzyme' molecule, small enough to pass through the pores of the membrane. The coenzyme was essential to enzyme activity (it was the 'cutting edge', so to speak).

Chemists at once tackled the problem of determining the structure of this coenzyme (and of similar adjuncts to other enzymes). The German-Swedish chemist Hans Karl August Simon von Euler-Chelpin was the first to make real progress in this respect. As a result, he and Harden shared the Nobel Prize in chemistry in 1929.

The coenzyme of the yeast enzyme studied by Harden proved to consist of a combination of an adenine molecule, two ribose molecules, two phosphate groups, and a molecule of 'nicotinamide'. Now this last was an unusual kind of thing to find in living tissue, and interest naturally centred on the nicotinamide. (It is called 'nicotinamide' because it contains an amide group, $CONH_2$, and can be formed easily from nicotinic acid. Nicotinic acid is structurally related to the tobacco alkaloid 'nicotine', but they are utterly different in properties; for one thing, nicotinic acid is necessary to life, whereas nicotine is a deadly poison.) The formulas of nicotinamide and nicotinic acid are:

nicotinic acid   nicotinamide

Once the formula of Harden's coenzyme was worked out, it was promptly renamed 'diphosphopyridine nucleotide' (DPN) – 'nucleotide' from the characteristic arrangement of the adenine, ribose, and phosphate, similar to that of the nucleotides making up nucleic acid, and 'pyridine' from the name given to the combination of atoms making up the ring in the nicotinamide formula.

Soon a very similar coenzyme was found, differing from DPN only in the fact that it contained three phosphate groups rather than two. This, naturally, was named 'triphosphopyridine nucleotide' (TPN). Both DPN and TPN proved to be coenzymes for a number of enzymes in the body, all serving to transfer hydrogen atoms from one molecule to another. (Such enzymes are called 'dehydrogenases'.) It was the coenzyme that did the actual job of hydrogen transfer; the enzyme proper in each case selected the particular substrate on which the operation was to be performed. The enzyme and the coenzyme each had a vital function, and if either was deficient in supply the release of energy from foodstuffs via hydrogen transfer would slow to a limp.

What was immediately striking about all this was that the nicotinamide group represented the only part of the enzyme the body cannot manufacture itself. It can make all the protein it needs and all the ingredients of DPN and TPN except the nicotinamide; that it must find ready-made (or at least in the form of nicotinic acid) in the diet. If not, then the manufacture of DPN and TPN stops and all the hydrogen-transfer reactions they control slow down.

Was nicotinamide or nicotinic acid a vitamin? As it happened, Funk (who coined the word 'vitamine') had isolated nicotinic acid from rice hulls. Nicotinic acid was not the substance that cured beri-beri, so he had ignored it. But on the strength of nicotinic acid's appearance in connection with coenzymes, the University of Wisconsin biochemist Conrad Arnold Elvehjem and his co-workers tried it on another deficiency disease.

In the 1920s, the American physician Joseph Goldberger had studied pellagra (sometimes called Italian leprosy), a disease endemic in the Mediterranean area and almost epidemic in the southern United States in the early part of this century. Pellagra's most noticeable symptoms are a dry, scaly skin, diarrhoea, and an inflamed tongue; it sometimes leads to mental disorders. Goldberger noticed that the disease struck people who lived on a limited diet (e.g., mainly cornmeal) and spared families that owned a milch cow. He began to experiment with artificial diets, feeding them to animals and inmates of gaols (where pellagra seemed to blossom). He succeeded in producing 'blacktongue' (a disease analogous to pellagra) in dogs, and in curing this disease with a yeast extract. He found he could cure gaol inmates of pellagra by adding milk to their diet. Goldberger decided that a vitamin must be involved, and he named it the P-P ('pellagra-preventive') factor.

It was pellagra, then, that Elvehjem chose for the test of nicotinic acid. He fed a tiny dose to a dog with blacktongue, and the dog responded with a remarkable improvement. A few more doses cured him. Nicotinic acid was a vitamin, all right; it was the P-P factor.

The American Medical Association, worried that the public might get the impression there were vitamins in tobacco, urged that the vitamin not be called nicotinic acid and suggested instead the name 'niacin' (an abbreviation of *ni*cotine *a*cid) or 'niacinamide'. Niacin has caught on fairly well.

Gradually, it became clear that the various vitamins were merely portions of coenzymes, each consisting of a molecular group an animal or a human being cannot make for itself. In 1932, Warburg had found a yellow coenzyme that catalysed the transfer of hydrogen atoms. The Austrian chemist Richard Kuhn and his associates shortly afterwards isolated vitamin B_2, which proved to be yellow, and worked out its structure:

$$
\begin{array}{c}
CH_2-OH \\
| \\
HO-CH \\
| \\
HO-CH \\
| \\
HO-CH \\
| \\
CH_2
\end{array}
$$

The carbon chain attached to the middle ring is like a molecule called 'ribitol', so vitamin B_2 was named 'riboflavin', 'flavin' coming from a Latin word meaning 'yellow'. Since examination of its spectrum showed that riboflavin was very similar in colour to Warburg's yellow coenzyme, Kuhn tested the coenzyme for riboflavin activity in 1935 and found such activity to be there. In the same year the Swedish biochemist Hugo Theorell worked out the structure of Warburg's yellow coenzyme and showed that it was riboflavin with a phosphate group added. (In 1954, a second and more complicated coenzyme also was shown to have riboflavin as part of its molecule.)

Kuhn was awarded the 1938 Nobel Prize in chemistry, and Theorell received the 1955 Nobel Prize in medicine and physiology. Kuhn, however, was unfortunate enough to be selected for his prize shortly after Austria had been absorbed by Nazi Germany, and (like Gerhard Domagk) he was compelled to refuse it.

Riboflavin was synthesized, independently, by the Swiss chemist Paul Karrer. For this and other work on vitamins, Karrer was awarded a share of the 1937 Nobel Prize in chemistry. (He shared it with the English chemist Walter Norman Haworth, who had worked on the structure of carbohydrate molecules.)

In 1937, the German biochemists K. Lohmann and P. Schuster discovered an important coenzyme that contained thiamine as part of its structure. Through the 1940s other connections were found between B vitamins and coenzymes. Pyridoxine, pantothenic acid, folic acid, biotin – each in turn was found to be tied to one or more groups of enzymes.

The vitamins beautifully illustrate the economy of the human body's chemical machinery. The human cell can dispense with making them because they serve only one special function, and the cell can take the reasonable risk of finding the necessary supply in the diet. There are many other vital substances that the body needs only in trace amounts but must make for itself. ATP, for instance, is formed from much the same building blocks that make up the indispensable nucleic acids. It is inconceivable that any organism could lose any enzyme necessary for nucleic-acid synthesis and remain alive, for nucleic acid is needed in such quantities that the organism dare not trust to the diet for its supply of the necessary building blocks. And to be able to make nucleic acid automatically implies the ability to make ATP. Consequently, no organism is known that is incapable of manufacturing its own ATP, and in all probability no such organism will ever be found.

To make such special products as vitamins would be like setting up a special machine next to an assembly line to turn out nuts and bolts for the motor cars. The nuts and bolts can be obtained more efficiently from a parts supplier, without any loss to the apparatus for assembling the automobiles; by the same token the organism can obtain vitamins in its diet, with a saving in space and material.

The vitamins illustrate another important fact of life. As far as is known, all living cells require the B vitamins. The coenzymes are an essential part of the cell machinery of every cell alive – plant, animal, or bacterial. Whether the cell gets the B vitamins from its diet or makes them itself, it must have them if it is to live and grow. This universal need for a particular group of substances is an impressive piece of evidence for the essential unity of all life and its descent (possibly) from a single original scrap of life formed in the primeval ocean.

While the roles of the B vitamins are now well known, the chemical functions of the other vitamins have proved rather hard nuts to crack. The only one on which any real advance has been made is vitamin A.

In 1925, the American physiologists L. S. Fridericia and E. Holm found that rats fed on a diet deficient in vitamin A had difficulty performing tasks in dim light. An analysis of their retinas showed that they were deficient in a substance called 'visual purple'.

There are two kinds of cell in the retina of the eye – 'rods' and 'cones'. The rods specialize in vision in dim light, and they contain the visual purple. A shortage of visual purple therefore hampers only vision in dim light, and it results in what is known as 'night blindness'.

In 1938, the Harvard biologist George Wald began to work out the chemistry of vision in dim light. He showed that light causes visual purple, or 'rhodopsin', to separate into two components: the protein 'opsin' and a non-protein called 'retinene'. Retinene proved to be very similar in structure to vitamin A.

The retinene always re-combines with the opsin to form rhodopsin in the dark. But during its separation from opsin in the light, a small percentage of it breaks down, because it is unstable. However, the supply of retinene is replenished from vitamin A, which is converted to retinene by the removal of two hydrogen atoms with the aid of enzymes. Thus vitamin A acts as a stable reserve for retinene. If vitamin A is lacking in the diet, eventually the retinene supply and the amount of visual purple decline, and night blindness is the result. For his work in this field, Wald shared in the 1967 Nobel Prize for medicine and physiology.

Vitamin A must have other functions as well, for a deficiency causes dryness of the mucous membranes and other symptoms which cannot very well be traced to troubles in the retina of the eye. But the other functions are still unknown.

The same has to be said about the chemical functions of vitamins C, D, E, and K. In 1970, Linus Pauling created a stir by maintaining that massive doses of vitamin C would reduce the incidence of colds. The public stripped chemists' shelves of the vitamin at once, and no doubt the contention will forthwith receive a thorough testing.

Minerals

It is natural to suppose that the materials making up anything as wonderful as living tissue must themselves be something pretty exotic. Wonderful the proteins and nucleic acids certainly are, but it is a little humbling to realize that the elements making up the human body are as common as dirt, and the whole lot could be bought for a few pounds. (It used to be pence, but inflation has raised the price of everything.)

In the early nineteenth century, when chemists were beginning to analyse organic compounds, it became quite clear that living tissue was made up, in the main, of carbon, hydrogen, oxygen, and nitrogen. These four elements alone constituted about 96 per cent of the weight of the human body. Then there was also a little sulphur in the body. If you burned off these five elements, you were left with a bit of white ash, mostly the residue from the bones. The ash was a collection of minerals.

It was not surprising to find common salt, sodium chloride, in the ash. After all, salt is not a mere condiment to improve the taste of food – as dispensable as, say, basil, rosemary, or thyme. It is a matter of life and death. You need only taste blood to realize that salt is a basic component of the body. Herbivorous animals, which presumably lack sophistication as far as the delicacies of food preparation are concerned, will undergo much danger and privation to reach a 'salt lick', where they can make up the lack of salt in their diet of grass and leaves.

As early as the mid eighteenth century, the Swedish chemist Johann Gottlieb Gahn had shown that bones were made up largely of calcium phosphate, and an Italian scientist, V. Menghini, had established that the blood contained iron. In 1847, Justus von Liebig found potassium and magnesium in the tissues. By the mid nineteenth century, then, the mineral constituents of the body were known to include calcium, phosphorus, sodium, potassium, chlorine, magnesium, and iron. Furthermore, these were as active in life processes as any of the elements usually associated with organic compounds.

The case of iron is the clearest. If it is lacking in the diet, the blood becomes deficient in haemoglobin and transports less oxygen from the lungs to the cells. The condition is known as 'iron-deficiency anaemia'. The patient is pale for lack of the red pigment and tired for lack of oxygen.

In 1882 the English physician Sidney Ringer found that a frog's heart could be kept alive and beating outside its body in a solution (called 'Ringer's solution' to this day) containing, among other things, sodium, potassium, and calcium in about the proportions found in the frog's blood. Each was essential for functioning of muscle. An excess of calcium caused the muscle to lock in permanent contraction ('calcium rigor'), whereas an excess of potassium caused it to unlock in permanent relaxation ('potassium inhibition'). Calcium, moreover, was vital to blood clotting. In its absence blood would not clot, and no other element could substitute for calcium in this respect.

Of all the minerals, phosphorus was eventually discovered to perform the most varied and crucial functions in the chemical machinery of life (see Chapter 3).

Calcium, a major component of bone, makes up 2 per cent of the body; phosphorus, 1 per cent. The other minerals I have mentioned come in smaller proportions, down to iron, which makes up only 0·004 per cent of the body. (That still leaves the average adult male 1/10 of an ounce of iron in his tissues.) But we are not at the end of the list; there are other minerals that, though present in tissue only in barely detectable quantities, are yet essential to life.

The mere presence of an element is not necessarily significant; it may be just an impurity. In our food we take in at least traces of every element in our environment, and some small amount of each finds its way into our tissues. But elements such as silicon and aluminium, for instance, contribute nothing. On the other hand, zinc is vital. How does one distinguish an essential mineral from an accidental impurity?

The best way is to show that some necessary enzyme contains the trace element as an essential component. (Why an enzyme? Because in no other way can any trace component possibly play an important role.) In 1939, David Keilin and T. Mann of England showed that zinc was an integral part of the enzyme carbonic anhydrase. Now carbonic anhydrase is essential to the body's handling of carbon dioxide, and the proper handling of that important waste material in turn is essential to life. It follows in theory that zinc is indispensable to life, and experiment shows that it actually is. Rats fed on a diet low in zinc stop growing, lose hair, suffer scaliness of the skin, and die prematurely for lack of zinc as surely as for lack of a vitamin.

In the same way it has been shown that copper, manganese, cobalt, and molybdenum are essential to animal life. Their absence from the diet gives rise to deficiency diseases. Molybdenum, the latest of the essential trace elements to be identified (in 1954), is a constituent of an enzyme called 'xanthine oxidase'. The importance of molybdenum was first noticed in the 1940s in connection with plants, when soil scientists found that plants would not grow well in soils deficient in the element. It seems that molybdenum is a compound of certain enzymes in soil micro-organisms that catalyse the conversion of the nitrogen of the air into nitrogen-containing compounds. Plants depend on this help from micro-organisms because they cannot themselves take nitrogen from the air. (This is only one of an enormous number of examples of the close interdependence of all life on our planet. The living world is a long and intricate chain which may suffer hardship or even disaster if any link is broken.)

Minerals Necessary to Life

Sodium (Na)	Zinc (Zn)
Potassium (K)	Copper (Cu)
Calcium (Ca)	Manganese (Mn)
Phosphorus (P)	Cobalt (Co)
Chlorine (Cl)	Molybdenum (Mo)
Magnesium (Mg)	Iodine (I)
Iron (Fe)	

Not all 'trace elements' are universally essential. Boron seems to be essential in traces to plant life, but not, apparently, to animals. Certain tunicates gather vanadium from sea water and use it in their oxygen-transporting compound, but few, if any, other animals require vanadium for any reason.

It is now realized that there are trace-element deserts, just as there are waterless deserts; the two usually go together but not always. In Australia soil scientists have found that an ounce of molybdenum in the form of some appropriate compound spread over sixteen acres of molybdenum-deficient land results in a considerable increase in fertility. A survey of American farmland in 1960 showed areas of boron deficiency in forty-one states, of zinc deficiency in twenty-nine states, and of molybdenum deficiency in twenty-one states. The dosage of trace elements is crucial.

Too much is as bad as too little, for some substances that are essential for life in small quantities (e.g., copper) become poisonous in large quantities.

This, of course, carries to its logical extreme the much older custom of using 'fertilizers' for soil. Until modern times, fertilization was through the use of animal excreta, manure or guano, which restored nitrogen and phosphorus to the soil. While this worked, it was accompanied by foul odours and by the ever-present possibility of infection. The substitution of chemical fertilizers, clean and odour-free, was through the work of Justus von Liebig in the early nineteenth century.

One of the most dramatic episodes in the discovery of mineral deficiencies has to do with cobalt. It involves the once incurably fatal disease called 'pernicious anaemia'.

In the early 1920s, the University of Rochester pathologist George Hoyt Whipple was experimenting on the replenishment of haemoglobin by means of various food substances. He would bleed dogs to induce anaemia and then feed them various diets to see which would permit them to replace the lost haemoglobin most rapidly. He did this not because he was interested in pernicious anaemia, or in any kind of anaemia, but because he was investigating bile pigments, compounds produced by the body from haemoglobin. Whipple discovered that the food that enabled the dogs to make haemoglobin most quickly was liver.

In 1926, two Boston physicians, George Richards Minot and William Parry Murphy, considering Whipple's results, decided to try liver as a treatment for pernicious-anaemia patients. The treatment worked. The incurable disease was cured, so long as the patients ate liver as an important portion of their diet. Whipple, Minot, and Murphy shared the Nobel Prize in physiology and medicine in 1934.

Unfortunately liver, although it is a great delicacy when properly cooked, then chopped, and lovingly mixed with such things as eggs, onions, and chicken fat, becomes wearing as a steady diet. (After a while, a patient might be tempted to think pernicious anaemia was preferable.) Biochemists began to search for the curative substance in liver, and by 1930 Edwin Joseph Cohn and his co-workers at the Harvard Medical School had prepared a concentrate a hundred times as potent as liver itself. To isolate the active factor, however, further purification was needed. Fortunately, chemists at the Merck Laboratories discovered in the 1940s that the concentrate from liver could accelerate the growth of certain bacteria. This provided an easy test of the potency of any preparation from it, so the biochemists could proceed to break down the concentrate into fractions and test them in quick succession. Because the bacteria reacted to the liver substance in much the same way that they reacted to, say, thiamine or riboflavin, the investigators now suspected strongly that the factor they were hunting for was a B vitamin. They called it 'vitamin B_{12}'.

By 1948, using bacterial response and chromatography, Ernest Lester Smith in England and Karl August Folkers at Merck succeeded in isolating pure samples of vitamin B_{12}. The vitamin proved to be a red substance, and both scientists thought it resembled the colour of certain cobalt compounds. It was known by this time that a deficiency of cobalt caused severe anaemia in cattle and sheep. Both Smith and Folkers burned samples of vitamin B_{12}, analysed the ash, and found that it did indeed contain cobalt. The compound has now been named 'cyanocobalamine'. So far it is the only cobalt-containing compound that has been found in living tissue.

By breaking it up and examining the fragments, chemists quickly decided that vitamin B_{12} was an extremely complicated compound, and they worked out an empirical formula of $C_{63}H_{88}O_{14}N_{14}PCo$. Then a British chemist, Dorothy Crowfoot Hodgkin, determined its over-all structure by means of X-rays. The diffraction pattern given by crystals of the compound allowed her to build up a picture of the 'electron densities' along the molecule, that is, those regions where the probability of finding an electron is high and those where it is low. If lines are drawn through regions of equal probability, a kind of skeletal picture is built up of the shape of the molecule as a whole.

This is not as easy as it sounds. Complicated organic molecules can produce an X-ray scattering truly formidable in its complexity. The mathematical operations required to translate that scattering into electron densities are tedious in the extreme. By 1944, electronic computers had been called in to help work out the structural formula of penicillin. Vitamin B_{12} was much more complicated and Miss Hodgkin had to use a more advanced computer – the National Bureau of Standards Western Automatic Computer (SWAC) – and do some heavy spadework. It eventually earned for her, however, the 1964 Nobel Prize for chemistry.

The molecule of vitamin B_{12}, or cyanocobalamine, turned out to be a lopsided porphyrin ring, with one of the carbon bridges connecting two of the smaller pyrrole rings missing, and with complicated side chains on the pyrrole rings. It resembled the somewhat simpler haem molecule, with this key difference: where haem had an iron atom at the centre of the porphyrin ring, cyanocobalamine had a cobalt atom.

Cyanocobalamine is active in very small quantities when injected into the blood of pernicious-anaemia patients. The body can get along on only $1/1,000$ as much of this substance as it needs of the other B vitamins. Any diet, therefore, ought to have enough cyanocobalamine for our needs. Even if it did not, the bacteria in the intestines manufacture quite a bit of it. Why, then, should anyone ever have pernicious anaemia?

Apparently, the sufferers from this disease are simply unable to absorb enough of the vitamin into the body through the intestinal walls. Their faeces are actually rich in the vitamin (for want of which they are dying). From feedings of liver, providing a particularly abundant supply, such a patient manages to absorb enough cyanocobalamine to stay alive. But he needs 100 times as much of the vitamin if he takes it by mouth as he does when it is injected directly into the blood.

Something must be wrong with the patient's intestinal apparatus, preventing the passage of the vitamin through the walls of the intestines. It has been known since 1929, thanks to the researches of the American physician William Bosworth Castle, that the answer lies somehow in the gastric juice. Castle called the necessary component of gastric juice 'intrinsic factor'. And in

1954 investigators found a product, from the stomach linings of animals, which assists the absorption of the vitamin and proved to be Castle's intrinsic factor. Apparently this substance is missing in pernicious-anaemia patients. When a small amount of it is mixed with cyanocobalamine, the patient has no difficulty in absorbing the vitamin through the intestines. Just how this intrinsic factor helps absorption is still not known.

Getting back to the trace elements. . . . The first one discovered was not a metal; it was iodine, an element with properties like those of chlorine. This story begins with the thyroid gland.

In 1896, a German biochemist, Eugen Baumann, discovered that the thyroid was distinguished by containing iodine, practically absent from all other tissues. In 1905, a physician named David Marine, who had just set up practice in Cleveland, was amazed to find how widely prevalent goitre was in that area. Goitre is a conspicuous disease, sometimes producing grotesque enlargement of the thyroid and causing its victims to become either dull and listless, or nervous, over-active, and pop-eyed. For the development of surgical techniques in the treatment of abnormal thyroids with resulting relief from goitrous conditions, the Swiss physician Emil Theodor Kocher earned the 1909 Nobel Prize in medicine and physiology.

But Marine wondered whether the enlarged thyroid might not be the result of a deficiency of iodine, the one element in which the thyroid specialized, and whether goitre might not be treated more safely and expeditiously by chemicals rather than by the knife. Iodine deficiency and the prevalence of goitre in the Cleveland area might well go hand in hand, at that, for Cleveland, being inland, might lack the iodine that was so plentiful in the soil near the ocean and in the seafood that is an important article of diet there.

The doctor experimented on animals and after ten years felt sure enough of his ground to try feeding iodine-containing compounds to goitre patients. He was probably not too surprised to find that it worked. Marine then suggested that iodine-containing compounds be added to table salt and to the water supply of inland cities where the soil was poor in iodine. There was strong opposition to his proposal, however, and it took another ten years to get water iodination and iodized salt generally accepted. Once the iodine supplements became routine, simple goitre declined in importance as one of mankind's woes.

Today researchers (and the public) are engaged in studies and discussion of a similar health question – the fluoridation of water to prevent tooth decay. This issue is still a matter of bitter controversy in the non-scientific and political arena; so far the opposition has been far more stubborn and successful than in the case of iodine. Perhaps one reason is that cavities in the teeth do not seem nearly as serious as the disfigurement of goitre.

In the early decades of this century dentists noticed that people in certain areas in the United States (e.g., some localities in Arkansas) tended to have darkened teeth – a mottling of the enamel. Eventually this was traced to a higher-than-average content of fluorine compounds ('fluorides') in the natural drinking water of those areas. With the attention of researchers directed to fluoride in the water, another interesting discovery turned up. Where the fluoride content of the water was above average, the population has an unusually low rate of tooth decay. For instance, the town of Galesburg in Illinois, with fluoride in its water, had only one third as many cavities per youngster as the nearby town of Quincy, whose water contained practically no fluoride.

Tooth decay is no laughing matter, as anyone with a toothache will readily agree. It costs the people of the United States more than 150,000 million dollars a year in dental bills, and by the age of thirty-five two thirds of all Americans have lost at least some of their teeth. Dental researchers succeeded in getting support for large-scale studies to find out whether fluoridation of water would be safe and would really help to prevent tooth decay. They found that one part per million of fluoride in the drinking water, at an estimated cost of 5 to 10 cents per person per year, did not mottle teeth and yet showed an effect in decay prevention. They therefore adopted one part per million as a standard for testing the results of fluoridation of community water supplies.

277

The effect is, primarily, on those whose teeth are being formed; that is, on children. The presence of fluoride in the drinking water ensures the incorporation of tiny particles of fluoride into the tooth structure; it is this, apparently, that makes the tooth mineral unpalatable to bacteria. (The use of small quantities of fluoride in pill-form or in toothpaste has also shown some protective effect against tooth decay.)

The dental profession is now convinced, on the basis of a quarter of a century of research, that for a few pennies per person per year tooth decay can be reduced by about two thirds, with a saving of at least a thousand million dollars a year in dental costs and a relief of pain and of dental handicaps that cannot be measured in money. The nation's dental and medical organizations, the United States Public Health Service, and the state health agencies recommend fluoridation of public water supplies. And yet, in the realm of politics, fluoridation has lost a majority of its battles. A group called the National Committee Against Fluoridation has aroused community after community to vote down fluoridation and even to repeal it in some localities where it has been adopted.

Two chief arguments have been employed by the opponents with the greatest effect. One is that fluorine compounds are poisonous. So they are, but not in the doses used for fluoridation! The other is that fluoridation is compulsory medication, infringing the individual's freedom. That may be so, but it is questionable whether the individual in any society should have the freedom to expose others to preventable sickness. If compulsory medication is evil, then we have a quarrel not only with fluoridation but also with chlorination, iodination, and, for that matter, with all the forms of inoculation, including vaccination against smallpox, that are compulsory in most civilized countries today.

Hormones

Enzymes, vitamins, trace elements – how potently these sparse substances decide life-or-death issues for the organism! But there is a fourth group of substances that, in a way, are even more potent. They conduct the whole performance; they are like a master switch that awakens a city to activity, or the throttle that controls an engine, or the red cape that excites the bull.

At the turn of the century two English physiologists, William Maddock Bayliss and Ernest Henry Starling, became intrigued by a striking little performance in the digestive tract. The gland behind the stomach known as the pancreas releases its digestive fluid into the upper intestines at just the moment when food leaves the stomach and enters the intestine. How does it get the message? What tells the pancreas that the right moment has arrived? The obvious guess was that the information must be transmitted via the nervous system, which was then the only known means of communication in the body. Presumably, the entry of food into the intestines from the stomach stimulated nerve endings that relayed the message to the pancreas by way of the brain or the spinal cord.

To test this theory, Bayliss and Starling cut every nerve to the pancreas. Their manoeuvre failed! The pancreas still secreted juice at precisely the right moment.

The puzzled experimenters went hunting for an alternate signalling system. In 1902, they tracked down a 'chemical messenger'. It was a substance secreted by the walls of the intestine. When they injected this into an animal's blood, it stimulated the secretion of pancreatic juice even though the animal was not eating. Bayliss and Starling concluded that, in the normal course of events, food entering the intestines stimulates their linings to secrete the substance, which then travels via the bloodstream to the pancreas and triggers the gland to start giving forth pancreatic juice. The two investigators named the substance secreted by the intestines 'secretin', and they called it a 'hormone', from a Greek word meaning 'rouse to activity'. Secretin is now known to be a small protein molecule.

Several years earlier, physiologists had discovered that an extract of the adrenals (two small organs just above the kidneys) could raise blood pressure if injected into the body. The Japanese chemist Jokichi Takamine, working in the United States, isolated the responsible substance in 1901 and named it 'adrenalin'. (This later became a trade name; the scientific name for it now is 'epinephrine'.) Its structure proved to resemble that of the amino acid tyrosine, from which it is derived in the body.

Plainly, adrenalin, too, was a hormone. As the years went on, the physiologists found that a number of other 'glands' in the body secreted hormones. (The word 'gland' comes from the Greek word for acorn, and it was originally applied to any small lump of tissue in the body. But it became customary to give the name gland to any tissue that secreted a fluid, even large organs such as the liver and the mammaries. Small organs that did not secrete fluids gradually lost this name, so that the 'lymph glands', for instance, were renamed the 'lymph nodes'. Even so, when lymph nodes in the throat or the armpit become enlarged during infections, physicians and mothers alike still refer to them as 'enlarged glands'.)

Many of the glands, such as those along the alimentary canal, the sweat glands, and the salivary glands, discharge their fluids through ducts. Some, however, are 'ductless'; they release substances directly into the bloodstream, which then circulates the secretions through the body. It is the secretions of these ductless or 'endocrine' glands that contain hormones. The study of hormones is for this reason termed 'endocrinology'.

Naturally, biologists are most interested in hormones that control functions of the mammalian body and, in particular, that of man. However, I should like at least to mention the fact that there are 'plant hormones' that control and accelerate plant growth, 'insect hormones' that control pigmentation and moulting, and so on.

When biochemists found that iodine was concentrated in the thyroid gland, they made the reasonable guess that the element was part of a hormone. In 1915, Edward Calvin Kendall of the Mayo Foundation in Minnesota isolated from the thyroid an iodine-containing amino acid which behaved like a hormone, and he named it 'thyroxine'. Each molecule of thyroxine contained four atoms of iodine. Like adrenalin, thyroxine has a strong family resemblance to tyrosine and is manufactured from it in the body. (Many years later, in 1952, the biochemist Rosalind Pitt-Rivers and her associates isolated another thyroid hormone – 'tri-iodothyronine', so named because its molecule contains three atoms of iodine rather than four. It is less stable than thyroxine but three to five times as active.)

The thyroid hormones control the over-all rate of metabolism in the body: they arouse all the cells to activity. People with an under-active thyroid are sluggish, torpid, and after a time may become mentally retarded, because the various cells are running in low gear. Conversely, people with an over-active thyroid are nervous and jittery, because their cells are racing. Either an under-active or an over-active thyroid can produce goitre.

The thyroid controls the body's 'basal metabolism', that is, its rate of consumption of oxygen at complete rest in comfortable environmental conditions – the 'idling rate', so to speak. If a person's basal metabolism is above or below the norm, suspicion falls upon the thyroid gland. Measurement of the basal metabolism is a tedious affair, for the subject must fast for a period in advance and lie still for half an hour while the rate is measured, to say nothing of an even longer period beforehand. Instead of going through this troublesome procedure, why not go straight to the horse's mouth – that is, measure the amount of rate-controlling hormone that the thyroid is producing? In recent years researchers have developed a method of measuring the amount of 'protein-bound iodine' (PBI) in the bloodstream; this indicates the rate of thyroid-hormone production and so has provided a simple, quick blood test to replace the basal-metabolism determination.

The best-known hormone is insulin, the first protein whose structure was fully worked out (see p. 85). Its discovery was the culmination of a long chain of events.

Diabetes is the name of a whole group of diseases, all characterized by unusual thirst and, in consequence, an unusual output of urine. It is the most common of the inborn errors of metabolism. There are 1·5 million diabetics in the United States, 80 per cent of whom are over forty-five. It is one of the few diseases to which the female is more subject than the male; women diabetics outnumber men four to three.

The name comes from a Greek word meaning 'syphon' (apparently the coiner pictured water syphoning endlessly through the body). The most serious form of the disease is 'diabetes mellitus'. 'Mellitus' comes from the Greek word for 'honey', and it refers to the fact that in advanced stages of certain cases of the disease the urine has a sweet taste. (This may have been determined directly by some heroic physician, but the first indication of this was rather indirect. Diabetic urine tended to gather flies.) In 1815, the French chemist Michel Eugène Chevreul was able to show the sweetness was due to the presence of the simple sugar glucose. This waste of glucose plainly indicates that the body is not utilizing its food efficiently. In fact, the diabetic patient, despite an increase in appetite, may steadily lose weight as the disease advances. Up to a generation ago there was no helpful treatment for the disease.

In the nineteenth century, the German physiologists Joseph von Mering and Oscar Minkowski found that removal of the pancreas gland from a dog produced a condition just like human diabetes. After Bayliss and Starling discovered the hormone secretin, it began to appear that a hormone of the pancreas might be involved in diabetes. But the only known secretion from the pancreas was the digestive juice. Where did the hormone come from? A significant clue turned up. When the duct of the pancreas was tied off, so that it could not pour out its digestive secretions, the major part of the gland shrivelled, but the groups of cells known as the 'islets of Langerhans' (after the German physician Paul Langerhans, who had discovered them in 1869) remained intact.

In 1916, a Scottish physician, Albert Sharpey-Schafer, suggested, therefore, that the islets must be producing the antidiabetes hormone. He named the assumed hormone 'insulin', from the Latin word for 'island'.

Attempts to extract the hormone from the pancreas at first failed miserably. As we now know, insulin is a protein, and the protein-splitting enzymes of the pancreas destroyed it even while the chemists were trying to isolate it. In 1921, the Canadian physician Frederick Grant Banting and the physiologist Charles Herbert Best (working in the laboratories of John James Rickard MacLeod at the University of Toronto) tried a new approach. First they tied off the duct of the pancreas. The enzyme-producing portion of the gland shrivelled, the production of protein-splitting enzymes stopped, and the scientists were then able to extract the intact hormone from the islets. It proved indeed effective in countering diabetes, and it is estimated that in the next fifty years it saved the lives of some 20 to 30 million diabetics. Banting called the hormone 'isletin', but the older and more Latinized form proposed by Sharpey-Schafer won out. Insulin it became and still is.

In 1923, Banting and, for some reason, MacLeod (whose only service to the discovery of insulin was to allow the use of his laboratory over the summer while he was on vacation) received the Nobel Prize in physiology and medicine.

The effect of insulin within the body shows most clearly in connection with the level of glucose concentration in the blood. Ordinarily the body stores most of its glucose in the liver, in the form of a kind of starch called 'glycogen' (discovered in 1856 by the French physiologist Claude Bernard), keeping only a small quantity of glucose in the bloodstream to serve the immediate energy needs of the cells. If the glucose concentration in the blood rises too high, this stimulates the pancreas to increase its production of insulin, which pours into the bloodstream and brings about a lowering of the glucose level. On the other hand, when the glucose level falls too low, the lowered concentration inhibits the production of insulin by the pancreas, so that the sugar level rises. Thus a balance is achieved. The production of insulin lowers the level of glucose, which lowers the production of insulin, which raises the level of glucose, which raises the production of insulin, which lowers the level of glucose – and so on. This is an example of what is called 'feedback'. The thermostat that controls the heating of a house works in the same fashion.

Feedback is probably the customary device by which the body maintains a constant internal environment. Another example involves the hormone produced by the parathyroid glands, four small bodies embedded in the thyroid gland. The hormone 'parathormone' was finally purified in 1960 by the American biochemists Lyman Creighton Craig and Howard Rasmussen after five years of work.

The molecule of parathormone is somewhat larger than that of insulin, being made up of eighty-three amino acids and possessing a molecular weight of 9,500. The action of the hormone is to increase calcium absorption in the intestine and decrease calcium loss through the kidneys. Whenever calcium concentration in the blood falls slightly below normal, secretion of the hormone is stimulated. With more calcium coming in and less going out, the blood level soon rises; this rise inhibits the secretion of the hormone. This interplay between calcium concentration in the blood and parathyroid hormone flow keeps the calcium level close to the needed level at all times. (And a good thing, too, for even a small departure of the calcium concentration from the proper level can lead to death. Thus, removal of the parathyroids is fatal. At one time, doctors, in their anxiety to snip away sections of thyroid to relieve goitre, thought nothing of tossing away the much smaller and less prominent parathyroids. The death of the patient taught them better.)

At some times, the action of feedback is refined by the existence of two hormones working in opposite directions. In 1961, for instance, D. Harold Copp, at the University of British Columbia, demonstrated the presence of a thyroid hormone he called 'calcitonin', which acted to depress the level of calcium in the blood by encouraging the deposition of its ions in bone. With parathormone pulling in one direction and calcitonin in the other, the feedback produced by calcium levels in the blood can be all the more delicately controlled. (The calcitonin molecule is made up of a single polypeptide chain that is thirty-two amino acids long.)

Then, too, in the case of blood-sugar concentration, where insulin is involved, a second hormone, also secreted by the islets of Langerhans, cooperates. The islets are made up of two distinct kinds of cells, 'alpha' and 'beta'. The beta cells produce insulin, while the alpha cells produce 'glucagon'. The existence of glucagon was first suspected in 1923 and it was crystallized in 1955. Its molecule is made up of a single chain of twenty-nine acids, and, by 1958, its structure had been completely worked out.

Glucagon opposes the effect of insulin, so the two hormonal forces push in opposite directions, and the balance shifts very slightly this way and that under the stimulus of the glucose concentration in blood. Secretions from the pituitary gland (which I shall discuss shortly) also have a countering effect on insulin activity. For the discovery of this the Argentinian physiologist Bernardo Alberto Houssay shared in the 1947 Nobel Prize for medicine and physiology.

Now the trouble in diabetes is that the islets have lost the ability to turn out enough insulin. The glucose concentration in the blood therefore drifts upwards. When the level rises to about 50 per cent higher than normal, it crosses the 'renal threshold' – that is, glucose spills over into the urine. In a way this loss of glucose into the urine is the lesser of two evils, for if the glucose concentration were allowed to build up any higher, the resulting rise in viscosity of the blood would cause undue heartstrain. (The heart is designed to pump blood, not molasses.)

The classic way of checking for the presence of diabetes is to test the urine for sugar. For instance, a few drops of urine can be heated with 'Benedict's solution' (named after the American chemist Francis Gano Benedict). The solution contains copper sulphate, which gives it a deep blue colour. If glucose is not present in the urine, the solution remains blue. If glucose is present, the copper sulphate is converted to cuprous oxide. Cuprous oxide is a brick-red, insoluble substance. A reddish precipitate at the bottom of the test tube therefore is an unmistakable sign of sugar in the urine, which usually means diabetes.

Nowadays an even simpler method is available. Small paper strips about two inches long are impregnated with two enzymes, glucose dehydrogenase and peroxidase, plus an organic substance called 'orthotolidine'. The yellowish strip is dipped into a sample of the patient's urine and then exposed to the air. If glucose is present, it combines with oxygen from the air with the catalytic help of the glucose dehydrogenase. In the process, hydrogen peroxide is formed. The peroxidase in the paper then causes the hydrogen peroxide to combine with the orthotolidine to form a deep blue compound. In short, if the yellowish paper is dipped into urine and turns blue, diabetes can be strongly suspected.

Once glucose begins to appear in the urine, diabetes mellitus is fairly far along in its course. It is better to catch the disease earlier by checking the glucose level in the blood before it crosses the renal threshold. The 'glucose tolerance test', now in general use, measures the rate of fall of the glucose level in the blood after it has been raised by feeding the person glucose. Normally, the pancreas responds with a flood of insulin. In a healthy person the sugar level will drop to normal within two hours. If the level stays high for three hours or more, it shows a sluggish insulin response, and the person is likely to be in the early stages of diabetes.

It is possible that insulin has something to do with controlling appetite.

To begin with, we are all born with what some physiologists call an 'appestat', which regulates appetite as a thermostat regulates a furnace. If the appestat is set too high, the individual finds himself continually taking in more calories than he expends, unless he exerts a strenuous self-control which sooner or later wears him out.

In the early 1940s, a physiologist, Stephen Walter Ranson, showed that animals grew obese after destruction of a portion of the hypothalamus (located in the lower part of the brain). This seems to fix the location of the appestat. What controls its operation? 'Hunger pangs' spring to mind. An empty stomach contracts in waves, and the entry of food ends the contractions. Perhaps it is these contractions that signal to the appestat. Not so; surgical removal of the stomach has never interfered with appetite control.

The Harvard physiologist Jean Mayer has advanced a more

subtle suggestion. He believes that the appestat responds to the level of glucose in the blood. After food has been digested, the glucose level in the blood slowly drops. When it falls below a certain level, the appestat is turned on. If, in response to the consequent urgings of the appetite, the person eats, the glucose level in his blood momentarily rises, and the appestat is turned off.

The hormones I have discussed so far are all either proteins (as insulin, glucagon, secretin, parathormone) or modified amino acids (as thyroxine, triiodothyronine, adrenalin). We come now to an altogether different group – the steroid hormones.

The story of these begins in 1927, when two German physiologists, Bernhard Zondek and Selmar Aschheim, discovered that extracts of the urine of pregnant women, when injected into female mice or rats, aroused them to sexual heat. (Their discovery led to the first early test for pregnancy.) It was clear at once that they had found a hormone, specifically, a 'sex hormone'.

Within two years pure samples of the hormone were isolated by Adolf Butenandt in Germany and by Edward Adelbert Doisy at St Louis University. It was named 'oestrone', from 'oestrus', the term for sexual heat in females. Its structure was quickly found to be that of a steroid, with the four-ring structure of cholesterol. For his part in the discovery of sex hormones, Butenandt was awarded the Nobel Prize for chemistry in 1939. He, like Domagk and Kuhn, was forced to reject it and could only accept the honour in 1949 after the destruction of the Nazi tyranny.

Oestrone is now one of a group of known female sex hormones, called 'oestrogens' ('giving rise to oestrus'). In 1931, Butenandt isolated the first male sex hormone, or 'androgen' ('giving rise to maleness'). He called it 'androsterone'.

It is the production of sex hormones that governs the changes that take place during adolescence: the development of facial hair in the male and of enlarged breasts in the female, for instance. The complex menstrual cycle in females depends on the interplay of several oestrogens.

The female sex hormones are produced in large part in the ovaries, the male sex hormones in the testes.

The sex hormones are not the only steroid hormones. The first non-sexual chemical messenger of the steroid type was discovered in the adrenals. These, as a matter of fact, are double glands, consisting of an inner gland called the adrenal 'medulla' (the Latin word for 'marrow') and an outer gland called the adrenal 'cortex' (the Latin word for 'bark'). It is the medulla that produces adrenalin. In 1929, investigators found that extracts from the cortex could keep animals alive after their adrenal glands had been removed – a 100 per cent fatal operation. Naturally, a search immediately began for 'cortical hormones'.

The search had a practical medical reason behind it. The well-known affliction called 'Addison's disease' (first described by the English physician Thomas Addison in 1855) had symptoms like those resulting from the removal of the adrenals. Clearly the disease must be caused by a failure in hormone production by the adrenal cortex. Perhaps injections of cortical hormones might deal with Addison's disease as insulin dealt with diabetes.

Two men were outstanding in this search. One was Tadeus Reichstein (who was later to synthesize vitamin C); the other was

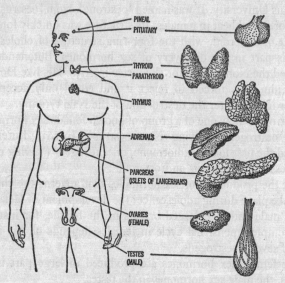

The endocrine glands.

Edward Kendall (who had first discovered the thyroid hormone nearly twenty years before). By the late 1930s, the researchers had isolated more than two dozen different compounds from the adrenal cortex. At least four showed hormonal activity. Kendall named the substances Compounds A, B, E, F, and so on. All the cortical hormones proved to be steroids.

Now the adrenals are very tiny glands, and it would take the glands of countless numbers of animals to provide enough cortical extracts for general use. Apparently, the only reasonable solution was to try to synthesize the hormones.

A false rumour drove cortical-hormone research forwards under full steam during the Second World War. It was reported that the Germans were buying up adrenal glands in Argentine slaughterhouses to manufacture cortical hormones that improved the efficiency of their aeroplane pilots in high-altitude flight. There was nothing to it, but the rumour had the effect of stimulating the United States Government to place a high priority on research into methods for the synthesis of the cortical hormones; the priority was even higher than that given to the synthesis of penicillin or the anti-malarials.

Compound A was synthesized by Kendall in 1944, and by the following year Merck & Co. had begun to produce it in substantial amounts. It proved of little value for Addison's disease, to the disappointment of all. After prodigious labour the Merck biochemist Lewis H. Sarrett then synthesized, by a process involving thirty-seven steps, Compound E, which was later to become known as 'cortisone'.

The synthesis of Compound E created little immediate stir in medical circles. The war was over; the rumour of cortical magic worked on German pilots had proved untrue; and Compound A had fizzled. Then, in an entirely unexpected quarter, Compound E suddenly came to life.

For twenty years, the Mayo Clinic physician Philip Showalter Hench had been studying rheumatoid arthritis, a painful, sometimes paralytic disease. Hench suspected that the body possessed natural mechanisms for countering this disease, because the arthritis was often relieved during pregnancy or during attacks of jaundice. He could not think of any biochemical factor that jaundice and pregnancy held in common. He tried injections of bile pigments (involved in jaundice) and sex hormones (involved in pregnancy) but neither helped his arthritic patients.

However, various bits of evidence pointed towards cortical hormones as a possible answer, and, in 1949, with cortisone available in reasonable quantity, Hench tried that. It worked! It did not cure the disease, any more than insulin cures diabetes, but it seemed to relieve the symptoms, and to an arthritic that alone is manna from heaven. What was more, cortisone later proved to be helpful as a treatment for Addison's disease, where Compound A had failed.

For their work on the cortical hormones, Kendall, Hench, and Reichstein shared the Nobel Prize in medicine and physiology in 1950.

Unfortunately, the influences of the cortical hormones on the body's workings are so multiplex that there are always side-effects, sometimes serious. Physicians are reluctant to use cortical-hormone therapy unless the need is clear and urgent. Synthetic substances related to cortical hormones (some with a fluorine atom inserted in the molecule) are being used in an attempt to avoid the worst of the side-effects, but nothing approaching a reasonable ideal has yet been found. One of the most active of the cortical hormones discovered so far is 'aldosterone', isolated in 1953 by Reichstein and his co-workers.

What controls all the varied and powerful hormones? All of them (including a number I have not mentioned) can exert more or less drastic effects in the body. Yet they are tuned together so harmoniously that they keep the body functioning smoothly without a break in the rhythm. Seemingly, there must be a conductor somewhere that directs their cooperation.

The nearest thing to an answer is the pituitary, a small gland suspended from the bottom of the brain (but not part of it). The name of the gland arose from an ancient notion that its function was to secrete phlegm, the Latin word for which is *pituita* (also the source of the word 'spit'). Because this notion is false, scientists have renamed the gland the 'hypophysis' (from Greek words meaning 'growing under' – i.e., under the brain), but pituitary is still the more common term.

The gland has three parts: the anterior lobe, the posterior lobe, and, in some organisms, a small bridge connecting the two. The anterior lobe is the most important, for it produces at least six hormones (all small-molecule proteins), which seem to act specifically upon other ductless glands. In other words, the anterior pituitary can be viewed as the orchestra leader that keeps the other glands playing in time and in tune. (It is interesting that the pituitary is located just about in the centre of the skull, as if deliberately placed in a spot of maximum security.)

One of the pituitary's messengers is the 'thyroid-stimulating hormone' (TSH). It stimulates the thyroid on a feedback basis. That is, it causes the thyroid to produce thyroid hormone; the rise in concentration of thyroid hormone in the blood in turn inhibits the formation of TSH by the pituitary; the fall of TSH in the blood in its turn reduces the thyroid's production; that stimulates the production of TSH by the pituitary, and so the cycle maintains a balance.

In the same way, the 'adrenal-cortical-stimulating hormone', or 'adrenocorticotropic hormone' (ACTH), maintains the level of cortical hormones. If extra ACTH is injected into the body, it will raise the level of these hormones, and thus it can serve the same purpose as the injection of cortisone itself. ACTH has therefore been used to treat rheumatoid arthritis.

Research into the structure of ACTH has proceeded with vigour because of this tie-in with arthritis. By the early 1950s, its molecular weight had been determined as 20,000, but it was easily broken down into smaller fragments ('corticotropins'), which possessed full activity. One of them, made up of a chain of thirty-nine amino acids, has had its structure worked out completely, and even shorter chains have been found effective.

ACTH has the ability of influencing the skin pigmentation of animals, and even man is affected. In diseases involving over-production of ACTH, human skin darkens. It is known that in lower animals, particularly the amphibians, special skin-darken-ing hormones exist. A hormone of this sort was finally detected among the pituitary products in the human being in 1955. It is called 'melanocyte-stimulating hormone' (melanocytes being the cells that produce skin pigment) and is usually abbreviated as 'MSH'.

The molecule of MSH has been largely worked out; it is interesting to note that MSH and ACTH share a seven amino-acid sequence in common. The indication that structure is allied to function (as, indeed, it must be) is unmistakable.

While on the subject of pigmentation, it might be well to mention the pineal gland, a conical body attached, like the pituitary, to the base of the brain and so-named because of its resemblance to a pine cone in shape. The pineal gland has seemed glandular in nature, but no hormone could be located until the late 1950s. Then the discoverers of MSH, working with 200,000 beef pineals, finally isolated a tiny quantity of substance that, on injection, lightened the skin of a tadpole. The hormone, named 'melatonin', does not, however, appear to have any effect on human melanocytes.

The list of pituitary hormones is not yet complete. A couple of pituitary hormones, ICSH ('interstitial cell-stimulating hormone') and FSH ('follicle-stimulating hormone'), control the growth of tissues involved in reproduction. There is also the 'lactogenic hormone', which stimulates milk production.

Lactogenic hormone stimulates other post-pregnancy activities. Young female rats injected with the hormone busy themselves with nest-building even though they have not given birth. On the other hand, mice whose pituitaries have been removed shortly before giving birth to young exhibit little interest in the baby mice. The newspapers at once termed lactogenic hormone the 'mother-love hormone'.

These pituitary hormones, associated with sexual tissues, are lumped together as the 'gonadotropins'. Another substance of this type is produced by the placenta (the organ that serves to transfer nourishment from the mother's blood to the blood of the developing infant and to transfer wastes in the opposite direction). The placental hormone is called 'human chorionic gonadotropin' and is abbreviated 'HCG'. As early as two to four weeks after the beginning of pregnancy, HCG is produced in appreciable quantities and makes its appearance in the urine. When extracts of the urine of a pregnant woman are injected into mice, frogs, or rabbits, recognizable effects can be detected. Pregnancy can be determined in this way at a very early stage.

The most spectacular of the anterior pituitary hormones is the 'somatotropic hormone' (STH), more popularly known as the 'growth hormone'. Its effect is a general one stimulating growth of the whole body. A child who cannot produce a sufficient supply of the hormone will become a dwarf; one who produces too much will turn into a circus giant. If the disorder that results in an over-supply of the growth hormone does not occur until after the person has matured (i.e., when his bones have been fully formed and hardened), only the extremities, such as the hands, feet, and chin, grow grotesquely large – a condition known as 'acromegaly' (Greek for 'large extremities'). It is this growth hormone that Li (who first determined its structure in 1966) synthesized in 1970.

As to how the hormones work, in chemical terms, investigators have so far made little headway.

It seems certain that the hormones do not act as enzymes. At least, no hormone has been found to catalyse a specific reaction directly. The next alternative is to suppose that a hormone, if not itself an enzyme, acts upon an enzyme – that it either promotes or inhibits an enzyme's activity. Insulin, the most thoroughly investigated of all the hormones, does seem to be definitely connected with an enzyme called 'glucokinase', which is essential for the conversion of glucose to glycogen. This enzyme is inhibited by extracts from the anterior pituitary and the adrenal cortex, and insulin can nullify that inhibition. Thus, insulin in the blood may serve to activate the enzyme and so speed up the conversion of glucose to glycogen. That would help to explain how insulin lowers the glucose concentration in the blood.

Yet the presence or absence of insulin affects metabolism at so many points that it is hard to see how this one action could bring about all the abnormalities that exist in the body chemistry of a diabetic. (The same is true for other hormones.) Some biochemists have therefore tended to look for grosser and more wholesale effects.

There is the suggestion that insulin somehow acts as an agent to get glucose into the cell. On this theory, a diabetic has a high glucose level in his blood for the simple reason that the sugar cannot get into his cells and therefore he cannot use it. (In explaining the insatiable appetite of a diabetic, Mayer, as I have already mentioned, suggested that glucose in the blood has difficulty in entering the cells of the appestat.)

If insulin assists glucose in entering the cell, then it must act on the cell membrane in some way. How? Cell membranes are composed of protein and fatty substances. We can speculate that insulin, as a protein molecule, may somehow change the arrangement of amino-acid side-chains in the protein of the membrane and thus open doors for glucose (and possibly many other substances).

If we are willing to be satisfied with generalities of this kind, we can go on to suppose that the other hormones also act on the cell membranes, each in its own fashion because each has its own specific amino-acid arrangement. Similarly, steroid hormones, as fatty substances, may act on the fatty molecules of the membrane, either opening or closing the door to certain substances. Clearly, by helping a given material to enter the cell or preventing it from doing so, a hormone could exert a drastic effect on what goes on in the cell. It could supply one enzyme with plenty of substrate to work on and deprive another of material, thus controlling what the cell produces. Assuming that a single hormone may decide the entrance or non-entrance of several different substances, we can see how the presence or absence of a hormone could profoundly influence metabolism, as in fact it does in the case of insulin.

The picture drawn above is attractive, but it is vague. Biochemists would much prefer to know the exact reactions that take place at the cell membrane under the influence of a hormone. The beginning of such knowledge came with the discovery in 1960 of a nucleotide like adenylic acid except that the phosphate group was attached to two different places in the sugar molecule. Its discoverers, Earl W. Sutherland and T. W. Rall, called it 'cyclic AMP'. It was 'cyclic' because the doubly attached phosphate-group formed a circle of atoms, and the 'AMP' stood for '*a*denine *mono*phosphate', an alternative name for adenylic acid.

Once discovered, cyclic AMP was found to be widely spread in tissue, and to have a pronounced effect on the activity of many different enzymes and cell processes. Cyclic AMP is produced from the universally occurring ATP by means of an enzyme named 'adenyl cyclase', which is located at the surface of cells. There may be several such enzymes, each geared for activity in the presence of a particular hormone. In other words, the surface activity of hormones serves to activate an adenyl cyclase that leads to the production of cyclic AMP, which alters the enzyme activity within the cell, producing many changes.

Undoubtedly, the details are enormously complex, and compounds other than cyclic AMP may be involved – but it is a beginning.

Death

The advances made by modern medicine in the battle against infection, against cancer, against nutritional disorders, have increased the probability that any given individual will live long enough to experience old age. Half the people born in this generation can be expected to reach the age of seventy (barring a nuclear war or some other prime catastrophe).

The rarity of survival to old age in earlier eras no doubt accounts in part for the extravagant respect paid to longevity in those times. The *Iliad*, for instance, makes much of 'old' Priam and 'old' Nestor. Nestor is described as having survived three generations of men, but at a time when the average length of life could not have been more than twenty to twenty-five, Nestor need not have been older than seventy to have survived three generations. That is old, yes, but not extraordinary by present standards.

Because Nestor's antiquity made such an impression on people in Homer's time, later mythologists supposed that he must have been something like 200 years old.

To take another example at random, Shakespeare's *Richard II* opens with the rolling words: 'Old John of Gaunt, time-honoured Lancaster'. John's own contemporaries, according to the chroniclers of the time, also considered him an old man. It comes as a slight shock to realize that John of Gaunt lived only to the age of fifty-nine. An interesting example from American history is that of Abraham Lincoln. Whether because of his beard, or his sad, lined face, or songs of the time that referred to him as 'Father Abraham', most people think of him as an old man at the time of his death. One could only wish that he had lived to be one. He was assassinated at the age of fifty-six.

All this is not to say that really old age was unknown in the days before modern medicine. In ancient Greece Sophocles, the play-wright, lived to be 90, and Isocrates, the orator, to 98. Flavius Cassiodorus of fifth-century Rome died at 95. Enrico Dandolo, the twelfth-century Doge of Venice, lived to be 97. Titian, the Renaissance painter, survived to 99. In the era of Louis XV, the Duc de Richelieu, grand-nephew of the famous cardinal, lived 92 years, and the French writer Bernard Le Bovier de Fontenelle managed to arrive at just 100 years.

This emphasizes the point that although the average life expectancy in medically advanced societies has risen greatly, the maximum life span has not. We expect very few men, even today, to attain or exceed the lifetime of an Isocrates or a Fontenelle. Nor do we expect modern nonagenarians to be able to participate in the business of life with any greater vigour. Sophocles was writing great plays in his nineties, and Isocrates was composing great orations. Titian painted to the last year of his life, Dandolo was the indomitable leader of a Venetian war against the Byzantine Empire at the age of 96. (Among comparably vigorous old men of our day, the best examples I can think of are George Bernard Shaw, who lived to 94, and the English mathematician and philosopher Bertrand Russell, who was still active in his ninety-eighth year, when he died.)

Although a far larger proportion of our population reaches the age of 60 than ever before, beyond that age life expectancy has improved very little over the past. The Metropolitan Life Insurance Company estimates that the life expectancy of a 60-year-old American male in 1931 was just about the same as it was a century and a half earlier – that is, 14·3 years against the estimated earlier figure of 14·8. For the average American woman the corresponding figures were 15·8 and 16·1. Since 1931, the advent of antibiotics has raised the expectancy at 60 for both sexes by two and a half years. But on the whole, despite all that medicine and science have done, old age overtakes a person at about the same rate and in the same way as it always has. Man has not yet found a way to stave off the gradual weakening and eventual breakdown of the human machine.

As in other forms of machinery, it is the moving parts that go first. The circulatory system – the pulsing heart and arteries – is man's Achilles' heel in the long run. His progress in conquering premature death has raised disorders of this system to the rank of the number one killer. Circulatory diseases are responsible for just over half the deaths in the United States, and of these diseases, a single one, atherosclerosis, accounts for one death out of four.

Atherosclerosis (from Greek words meaning 'mealy hardness') is characterized by grain-like fatty deposits along the inner surface of the arteries, which force the heart to work harder to drive blood through the vessels at a normal pace. The blood pressure rises, and the consequent increase in strain on the small blood vessels may burst them. If this happens in the brain (a particularly vulnerable area) there is a cerebral haemorrhage or 'stroke'. Sometimes the bursting of a vessel is so minor that it occasions only a trifling and temporary discomfort or even goes unnoticed, but a massive collapse of vessels will bring on paralysis or a quick death.

The roughening and narrowing of the arteries introduces another hazard. Because of the increased friction of the blood scraping along the roughened inner surface of the vessels, blood clots are more likely to form, and the narrowing of the vessels heightens the chances that a clot will completely block the blood flow. In the coronary artery, feeding the heart muscle itself, a block ('coronary thrombosis') can produce almost instant death.

Just what causes the formation of deposits on the artery wall is a matter of much debate among medical scientists. Cholesterol certainly seems to be involved, but how it is involved is still far from clear. The plasma of human blood contains 'lipoproteins', which consist of cholesterol and other fatty substances bound to certain proteins. Some of the fractions making up lipoprotein maintain a constant concentration in the blood – in health and in disease, before and after eating, and so on. Others fluctuate, rising after meals. Still others are particularly high in obese individuals. One fraction, rich in cholesterol, is particularly high in overweight people and in those with atherosclerosis.

Atherosclerosis tends to go along with a high blood-fat content, and so does obesity. Overweight people are more prone to atherosclerosis than are thin people. Diabetics also have high blood-fat levels, and they are more prone to atherosclerosis than are normal individuals. And, to round out the picture, the incidence of diabetes among the stout is considerably higher than among the thin.

It is thus no accident that those who live to a great age are so often scrawny, little fellows. Large, fat men may be jolly, but they do not keep the sexton waiting unduly, as a rule. (Of course, there are always exceptions, and one can point to men such as Winston Churchill and Herbert Hoover, who passed their ninetieth birthdays, although they were never noted for leanness.)

The key question, at the moment, is whether atherosclerosis can be fostered or prevented by the diet. Animal fats, such as those in milk, eggs, and butter, are particularly high in cholesterol; plant fats are particularly low in it. Moreover, the fatty acids of plant fats are mainly of the unsaturated type, which has been reported to counter the deposition of cholesterol. Despite the fact that investigations of these matters have yielded no conclusive results one way or the other, people have been flocking to 'low-cholesterol diets', in the hope of staving off thickening of the artery walls. No doubt this will do no harm.

Of course, the cholesterol in the blood is not derived from the cholesterol of the diet. The body can and does make its own cholesterol with great ease, and even though you live on a diet that is completely free of cholesterol, you will still have a generous supply of cholesterol in your blood lipoproteins. It therefore seems reasonable to suppose that what matters is not the mere presence of cholesterol but the individual's tendency to deposit it where it will do the most harm. It may be that there is a hereditary tendency to manufacture excessive amounts of cholesterol. Biochemists are seeking drugs that will inhibit cholesterol formation, in the hope that such drugs may forestall the development of atherosclerosis in those who are prone to the disease.

But even those who escape atherosclerosis grow old. Old age is a disease of universal incidence. Nothing can stop the creeping enfeeblement, the increasing brittleness of the bones, the weakening of the muscles, the stiffening of the joints, the slowing of reflexes, the dimming of sight, the declining agility of the mind. The rate at which this happens is somewhat slower in some than in others, but, fast or slow, the process is inexorable.

Perhaps mankind ought not to complain too loudly about this. If old age and death must come, they arrive unusually slowly. In general, the life-span of mammals correlates with size. The smallest mammal, the shrew, may live $1\frac{1}{2}$ years and a rat may live 4 or 5. A rabbit may live up to 15 years, a dog up to 18, a pig up to 20, a horse up to 40, and an elephant up to 70. To be sure, the smaller the animal the more rapidly it lives – the faster its heartbeat, for instance. A shrew with a heartbeat of 1,000 per minute can be matched against an elephant with a heartbeat of 20 per minute.

In fact, mammals in general seem to live, at best, as long as it takes their hearts to count a thousand million. To this general rule, man himself is the most astonishing exception. Man is considerably smaller than a horse and far smaller than an elephant, yet no mammal can live as long as he can. Even if we discount tales of vast ages from various backwoods where accurate records have not been kept, there are reasonably convincing data for life-spans of up to 115 years. The only vertebrates to outdo this, without question, are certain large, slow-moving tortoises.

Man's heartbeat of about seventy-two per minute is just what is to be expected of a mammal of his size. In seventy years, which is the average life expectancy of man in the technologically advanced areas of the world, the human heart has beaten 2,500 million times; at 115 years it has beaten about 4,000 million times. Even man's nearest relatives, the great apes, cannot match this, even closely. The gorilla, considerably larger than a man, is in extreme old age at fifty.

There is no question but that man's heart out-performs all other hearts in existence. (The tortoise's heart may last longer but it lives nowhere near as intensely.) Why man should be so long-lived is not known, but man, being what he is, is far more interested in asking why he does not live still longer.

What is old age, anyway? So far, there are only speculations. Some have suggested that the body's resistance to infection slowly decreases with age (at a rate depending on heredity). Others speculate that 'clinkers' of one kind or another accumulate in the cells (again at a rate that varies from individual to individual). These supposed side products of normal cellular reactions, which the cell can neither destroy nor get rid of, slowly build up in the cell as the years pass, until they eventually interfere with the cell's metabolism so seriously that it ceases to function. When enough cells are put out of action, so the theory goes, the body dies. A variation of this notion holds that the protein molecules themselves become clinkers, because cross-links develop between them so that they become stiff and brittle and finally bring the cell machinery grinding to a halt.

If this is so, then 'failure' is built into the cell machinery. Carrel's ability to keep a piece of embryonic tissue alive for decades (p. 230) had made it seem that cells themselves might be immortal: it was only the organization into combinations of billions of individual cells that brought death. The organization failed, not the cells.

Not so, apparently. It is now thought that Carrel may (unwittingly) have introduced fresh cells into his preparation in the process of feeding the tissue. Attempts to work with isolated cells or groups of cells in which the introduction of fresh cells was rigorously excluded seems to show that the cells inevitably age – presumably through irreversible changes in the key cell components.

And yet there is man's extraordinarily long life-span. Can it be that human tissue has developed methods of reversing or inhibiting cellular aging effects, that are more efficient than those in any other mammal? Again, birds tend to live markedly longer than mammals of the same size despite the fact that bird-metabolism is even more rapid than mammal-metabolism – again, superior ability of old-age reversal or inhibition.

If old age can be staved off more by some organisms than by others, there seems no reason to suppose that man cannot learn the method and improve upon it. Might not old age then be curable, and might not mankind develop the ability to enjoy an enormously extended life-span – or even immortality?

General optimism in this respect is to be found among some people. Medical miracles in the past would seem to herald unlimited miracles in the future. And if that is so, what a shame to live in a generation that will just miss a cure for cancer, or for arthritis, or for old age!

In the late 1960s, therefore, a movement grew to freeze human bodies at the moment of death, in order that the cellular machinery might remain as intact as possible, until the happy day when whatever it was that marked the death of the frozen individual could be cured. He would then be revived and made healthy, young, and happy.

To be sure, there is no sign at the present moment that any dead body can be restored to life, or that any frozen body – even if alive at the moment of freezing – can be thawed to life. Nor do the proponents of this procedure ('cryonics') give much attention to the complications that might arise in the flood of dead bodies returned to life – the personal hankering for immortality governs all.

Actually, it makes little sense to freeze intact bodies, even if all possible revival could be done. It is wasteful. Biologists have so far had much more luck with the developing of whole organisms from groups of specialized cells. Skin cells or liver cells, after all, have the same genetic equipment as other cells have, and as the original fertilized ovum had in the first place. The cells are specialized because the various genes are inhibited or activated to varying extents; but might not the genes be de-inhibited or de-activated, and might they not then make their cell into the equivalent of a fertilized ovum and develop an organism all over again – the same organism, genetically speaking, as the one of which they had formed part? Surely, this procedure (called 'cloning') offers more hope for a kind of preservation of the personality (if not the memory). Instead of freezing an entire body, chop off the little toe and freeze that.

But do we really want immortality – either through cryonics, through cloning, or through simple reversal of the aging phenomenon in each individual? There are few human beings who wouldn't eagerly accept an immortality reasonably free of aches, pains, and the effects of age – but suppose we were all immortal?

Clearly, if there were few or no deaths on earth, there would have to be few or no births. It would mean a society without babies. Presumably that is not fatal; a society self-centred enough to cling to immortality would not stop at eliminating babies altogether.

But will that do? It would be a society composed of the same brains, thinking the same thoughts, circling the same ruts in the same way, endlessly. It must be remembered that babies possess not only young brains but *new* brains. Each baby (barring identical multiple births) has genetic equipment unlike that of any human individual who ever lived. Thanks to babies, there are constantly fresh genetic combinations injected into mankind, so that the way is open towards improvement and development.

It would be wise to lower the level of the birth-rate, but ought we to wipe it out entirely? It would be pleasant to eliminate pains and discomforts of old age, but ought we to create a species consisting of the old, the tired, the bored, the same, and never allow for the new and the better?

Perhaps the prospect of immortality is worse than the prospect of death.

6 The Species

Varieties of Life

Man's knowledge of his own body is incomplete without a knowledge of his relationship to the rest of life on the earth.

In primitive cultures, the relationship was often considered to be close indeed. Many tribes regarded certain animals as their ancestors or blood brothers, and made it a crime to kill or eat them, except under certain ritualistic circumstances. This veneration of animals as gods or near-gods is called 'totemism' (from an American Indian word), and there are signs of it in cultures that are not so primitive. The animal-headed gods of Egypt were a hangover of totemism, and so, perhaps, is the modern Hindu veneration of cows and monkeys.

On the other hand, Western culture, as exemplified in Greek and Hebrew ideas, very early made a sharp distinction between man and the 'lower animals'. Thus, the Bible emphasizes that man was produced by a special act of creation in the image of God, 'after our likeness' (Genesis 1:26). Yet the Bible attests, nevertheless, to man's remarkably keen interest in the lower animals. Genesis mentions that Adam, in his idyllic early days in the Garden of Eden, was given the task of naming 'every beast of the field, and every fowl of the air'.

Offhand, that seems not too difficult a task – something that one could do in perhaps an hour or two. The scriptural chroniclers put 'two of every sort' of animal in Noah's Ark, whose dimensions were 450 by 75 by 45 feet (if we take the cubit to be 18 inches). The Greek natural philosophers thought of the living world in similarly limited terms: Aristotle could list only about 500 kinds of animals, and his pupil Theophrastus, the most eminent botanist of ancient Greece, listed only about 500 different plants.

Such a list might make some sense if one thought of an elephant as always an elephant, a camel as just a camel, or a flea as simply a flea. Things began to get a little more complicated when naturalists realized that animals had to be differentiated on the basis of whether they could breed with each other. The Indian elephant could not interbreed with the African elephant; therefore, they had to be considered different 'species' of elephant. The Arabian camel (one hump) and the Bactrian camel (two humps) also are separate species. As for the flea, the small biting insects (all resembling the common flea) are divided into 500 different species!

Through the centuries, as naturalists counted new varieties of creatures in the field, in the air, and in the sea, and as new areas of the world came into view through exploration, the number of identified species of animals and plants grew astronomically. By 1800 it had reached 70,000. Today more than 1·25 million different species, two thirds animal and one third plant, are known and no biologist supposes that the count is complete.

The living world would be exceedingly confusing if we were unable to classify this enormous variety of creatures according to some scheme of relationships. One can begin by grouping together the cat, the tiger, the lion, the panther, the leopard, the jaguar, and other cat-like animals in the 'cat family'; likewise, the dog, the wolf, the fox, the jackal, and the coyote form a 'dog family', and so on. On the basis of obvious general criteria one can go on to classify some animals as meat-eaters and others as plant-eaters. The ancients also set up general classifications based on habitat, and so they considered all animals that lived in the sea to be fishes and all that flew in the air to be birds. But this made the whale a fish and the bat a bird. Actually, in a fundamental sense the whale and the bat are more like each other than the one is like a fish or the other like a bird. Both bear live young. Moreover, the whale has air-breathing lungs, rather than the gills of a fish, and the bat has hair instead of the feathers of a bird. Both are classed with the mammals, which give birth to living babies (instead of laying eggs) and feed them on mother's milk.

One of the earliest attempts to make a systematic classification was that of an Englishman named John Ray (or Wray), who in the seventeenth century classified all the known species of plants (about 18,600), and later the species of animals, according to systems which seemed to him logical. For instance, he divided flowering plants into two main groups, on the basis of whether the seed contained one embryonic leaf or two. The tiny embryonic leaf or pair of leaves had the name 'cotyledon', from the Greek word for a kind of cup ('kotyle'), because it lay in a cup-like hollow in the seed. Ray therefore named the two types respectively 'monocotyledonous' and 'dicotyledonous'. The classification (similar, by the way, to one set up 2,000 years earlier by Theophrastus) proved so useful that it is still in effect today. The difference between one embryonic leaf and two in itself is unimportant, but there are a number of important ways in which all monocotyledonous plants differ from all dicotyledonous ones. The difference in the embryonic leaves is just a handy tag which is symptomatic of many general differences. (In the same way, the distinction between feathers and hair is minor in itself but is a handy marker for the vast array of differences that separates birds from mammals.)

Although Ray and others contributed some useful ideas, the real founder of the science of classification, or 'taxonomy' (from a Greek word meaning 'arrangement'), was a Swedish botanist best known by his Latinized name of Carolus Linnaeus, who did the job so well that the main features of his scheme still stand today. Linnaeus set forth his system in 1737 in a book entitled *Systema Naturae*. He grouped species resembling one another into a 'genus' (from a Greek word meaning 'race' or 'sort'), put related genera in turn into an 'order', and grouped similar orders in a 'class'. Each species was given a double name, made up of the name of the genus and of the species itself. (This is much like the system in the telephone book, which lists Smith, John; Smith, William, and so on.) Thus the members of the genus of cats are *Felis domesticus* (the pussycat), *Felis leo* (the lion), *Felis tigris* (the tiger), *Felis pardus* (the leopard), and so on. The genus to which the dog belongs includes *Canis familiaris* (the dog), *Canis lupus* (the European grey wolf), *Canis occidentalis* (the American timber wolf), and so on. The two species of camel are *Camelus bactrianus* (the Bactrian camel) and *Camelus dromedarius* (the Arabian camel).

Around 1800, the French naturalist Georges Leopold Cuvier went beyond 'classes' and added a more general category called the 'phylum' (from a Greek word for 'tribe'). A phylum includes all animals with the same general body plane (a concept that was emphasized and made clear by none other than the great German poet Johann Wolfgang von Goethe). For instance, the mammals, birds, reptiles, amphibia, and fishes are placed in one phylum because all have backbones, a maximum of four limbs, and red blood containing haemoglobin. Insects, spiders, lobsters, and centipedes are placed in another phylum; clams, oysters, and mussels in still another, and so on. In the 1820s, the Swiss botanist Augustin Pyramus de Candolle similarly improved Linnaeus's classification of plants. Instead of grouping species together according to external appearance, he laid more weight on internal structure and functioning.

The tree of life now is arranged as I shall describe in the following paragraphs, going from the most general divisions to the more specific.

We start with the 'kingdoms' – plant, animal, and in-between (that is, those micro-organisms, such as the bacteria, that cannot be classed definitely as plant or animal in nature). The German biologist Ernst Heinrich Haeckel suggested, in 1866, that this in-between group be called 'Protista', a name which is coming into increasing use among biologists even though the world of life is still divided exclusively, in popular parlance, into 'animal and vegetable'.

The plant kingdom, according to one system of classification, is divided into two sub-kingdoms. In the first sub-kingdom, called

Thallophyta, are placed all the plants that do not have roots, stems, or leaves – that is, the algae (one-celled green plants and various seaweeds), which contain chlorophyll, and the fungi (the one-celled moulds plus such organisms as mushrooms), which do not. The members of the second sub-kingdom, the Embryophyta, are divided into two main phyla – the Bryophyta (the various mosses) and the Tracheophyta (plants with systems of tubes for the circulation of sap), which includes all the species that we ordinarily think of as plants.

This last great phylum is made up of three main classes: the Filicineae, the Gymnospermae, and the Angiospermae. In the first class are the ferns, which reproduce by means of spores. The gymnosperms, forming seeds on the surface of the seed-bearing organs, include the various evergreen cone-bearing trees. The angiosperms, with the seeds enclosed in ovules, make up the vast majority of the familiar plants.

As for the animal kingdom, I shall list only the more important phyla.

The Protozoa ('first animals') are, of course, the one-celled animals. Next there are the Porifera, animals consisting of colonies of cells within a pore-bearing skeleton; these are the sponges. The individual cells show signs of specialization but retain a certain independence, for after all are separated by straining through a silk cloth, they may aggregate to form a new sponge.

(In general, as the animal phyla grow more specialized, individual cells and tissues grow less 'independent'. Simple creatures can re-grow to entire organisms even though badly mutilated, a process called 'regeneration'. More complex ones can re-grow limbs. By the time we reach man, however, the capacity for regeneration has sunk quite low. We can re-grow a lost fingernail, but not a lost finger.)

The first phylum whose members can be considered truly multi-celled animals is the Coelenterata (meaning 'hollow gut'). These animals have the basic shape of a cup and consist of two layers of cells – the ectoderm ('outer skin') and the endoderm ('inner skin'). The most common examples of this phylum are the jellyfish and the sea anemones.

All the rest of the animal phyla have a third layer of cells – the mesoderm ('middle skin'). From these three layers, first recognized in 1845 by the German physiologists Johannes Peter Müller and Robert Remak, are formed the many organs of even the most complex animals, including man.

The mesoderm arises during the development of the embryo, and the manner in which it arises divides the animals involved into two 'super-phyla'. Those in which the mesoderm forms at the junction of the ectoderm and the endoderm make up the Annelid super-phylum; those in which the mesoderm arises in the endoderm alone are the Echinoderm super-phylum.

Let us consider the Annelid super-phylum first. Its simplest phylum is Platyhelminthes (Greek for 'flat worms'). This includes not only the parasitic tapeworm but also free-living forms. The flatworms have contractile fibres that can be considered primitive muscles, and they also possess a head, a tail, special reproductive organs, and the beginnings of excretory organs. In addition, the flatworms display bilateral symmetry: that is, they have left and right sides that are mirror images of each other. They move head-first, and their sense organs and rudimentary nerves are concentrated in the head area, so that the flatworm can be said to possess the first step towards a brain.

Next comes the phylum Nematoda (Greek for 'thread worm'), whose most familiar member is the hookworm. These creatures possess a primitive bloodstream – a fluid within the mesoderm that bathes all the cells and conveys food and oxygen to them. This allows the nematodes, in contrast to animals such as the flat tapeworm, to have bulk, for the fluid can bring nourishment to interior cells. The nematodes also possess a gut with two openings, one for the entry of food, the other (the anus) for ejection of wastes.

The next two phyla in this super-phylum have hard external 'skeletons' – that is, shells (which are found in some of the simpler phyla, too). These two groups are the Brachiopoda, which have calcium carbonate shells on top and bottom and are popularly called 'lampshells', and the Mollusca (Latin for 'soft'), whose soft bodies are enclosed in shells originating from the right and left sides instead of the top and bottom. The most familiar molluscs are the clams, oysters, and snails.

A particularly important phylum in the Annelid super-phylum is Annelida. These are worms, but with a difference: they are composed of segments, each of which can be looked upon as a kind of organism in itself. Each segment has its own nerves branching off the main nerve stem, its own blood vessels, its own tubules for carrying off wastes, its own muscles, and so on. In the most familiar annelid, the earthworm, the segments are marked off by little constrictions of flesh which look like little rings around the animal; in fact, Annelida is from a Latin word meaning 'little ring'.

Segmentation apparently endows an animal with superior efficiency, for all the most successful species of the animal kingdom, including man, are segmented. (Of the non-segmented animals, the most complex and successful is the squid.) If you wonder how the human body is segmented, think of the vertebrae and the ribs; each vertebra of the backbone and each rib represents a separate segment of the body, with its own nerves, muscles, and blood vessels.

The annelids, lacking a skeleton, are soft and relatively defenceless. The phylum Arthropoda ('jointed feet'), however, combines segmentation with a skeleton, the skeleton being as segmented as the rest of the body. The skeleton is not only more manoeuvrable for being jointed; it is also light and tough, being made of a polysaccharide called 'chitin' rather than of heavy, inflexible limestone or calcium carbonate. On the whole, the Arthropoda, which includes the lobsters, spiders, centipedes, and insects, is the most successful phylum in existence. At least the phylum contains more species than all the other phyla put together.

This accounts for the main phyla in the Annelid super-phylum. The other super-phylum, the Echinoderm, contains only two important phyla. One is Echinodermata ('spiny skin'), which includes such creatures as the starfish and the sea urchin. The echinoderms differ from other mesoderm-containing phyla in possessing radial symmetry and having no clearly defined head and tail (though in early life echinoderms do show bilateral symmetry, which they lose as they mature).

The second important phylum of the Echinoderm superphylum is important indeed, for it is the one to which man himself belongs.

The general characteristic that distinguishes the members of this phylum (which embraces man, ostrich, snake, frog, mackerel, and a varied host of other animals) is an internal skeleton. No animal outside this phylum possesses one. The particular mark of such a skeleton is the backbone. In fact, the backbone is so important a feature that in common parlance all animals are loosely divided into vertebrates and invertebrates. Actually, there is an in-between group which has a rod of cartilage called a 'notochord' ('backcord') in the place of the backbone. The notochord, first discovered by Von Baer, who had also discovered the mammalian ovum, seems to represent a rudimentary backbone; in fact, it makes its appearance even in mammals during the development of the embryo. So the animals with notochords (various worm-like, slug-like, and mollusc-like creatures) are classed with the vertebrates. The whole phylum was named Chordata in 1880, by the English zoologist Francis Maitland Balfour; it is divided into four sub-phyla, three of which have only a notochord. The fourth, with a true backbone and general internal skeleton, is Vertebrata.

The vertebrates in existence today form two super-classes: the Pisces ('fishes') and the Tetrapoda ('four-footed' animals).

The Pisces group is made up of three classes: (1) the Agnatha ('jawless') fishes, which have true skeletons but no limbs or jaws – the best-known representative, the lamprey, possessing a rasping set of files in a round sucker-like mouth; (2) the Chondrichthyes ('cartilage fish'), with a skeleton of cartilage instead of bone, sharks being the most familiar example; and (3) the Osteichthyes or 'bony fishes'.

The tetrapods, or four-footed animals, all of which breathe by means of lungs, make up four classes. The simplest are the Amphibia ('double life') – for example, the frogs and toads. The double life means that in their immature youth (e.g., as tadpoles) they have no limbs and breathe by means of gills; then as adults they develop four feet and lungs. The amphibians, like fishes, lay their eggs in the water.

The second class are the Reptilia (from a Latin word meaning 'creeping'). They include the snakes, lizards, alligators, and turtles. They breathe with lungs from birth, and hatch their eggs (enclosed in a hard shell) on land. The most advanced reptiles have essentially four-chambered hearts, whereas the amphibian's heart has three chambers and the fish's heart only two.

The final two groups of tetrapods are the Aves (birds) and the Mammalia (mammals). All are warm-blooded: that is, their bodies possess devices which maintain an even internal temperature regardless of the temperature outside (within reasonable limits). Since the internal temperature is usually higher than the external, these animals require insulation. As aids to this end, the birds are equipped with feathers and the mammals with hair, both serving to trap a layer of insulating air next to the skin. The birds lay eggs like those of reptiles. The mammals, of course, bring forth their young already 'hatched' and supply them with milk produced by mammary glands (*mammae* in Latin).

In the nineteenth century zoologists heard reports of a great curiosity so amazing that they refused to believe it. The Australians had found a creature that had hair and produced milk (through mammary glands that lacked nipples), yet laid eggs! Even when the zoologists were shown specimens of the animal (not alive, unfortunately, because it is not easy to keep it alive away from its natural habitat), they were inclined to brand it a clumsy fraud. The beast was a land-and-water animal that looked a good deal like a duck: it had a bill and webbed feet. Eventually

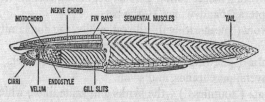

Amphioxus, a primitive, fish-like chordate with a notochord.

the 'duck-billed platypus' had to be recognized as a genuine phenomenon and a new kind of mammal. Another egg-laying mammal, the echidna, has since been found in Australia and New Guinea. Nor is it only in the laying of eggs that these mammals show themselves to be still close to the reptile. They are only imperfectly warm-blooded; on cold days their internal temperature may drop as much as 10°C.

The mammals are now divided into three sub-classes. The egg-laying mammals form the first class, Prototheria (Greek for 'first beasts'). The embryo in the egg is actually well developed by the time the egg is laid, and it hatches out not long afterwards. The second sub-class of mammals, Metatheria ('mid-beasts'), includes the opossums and kangaroos. Their young, though born alive, are in a very undeveloped form and will die in short order unless they manage to reach the mother's protective pouch and stay at the mammary nipples until they are strong enough to move about. These animals are called 'marsupials' (from *marsupium*, Latin for pouch).

Finally at the top of the mammalian hierarchy, we come to the sub-class Eutheria ('true beasts'). Their distinguishing feature is the placenta, a blood-suffused tissue that enables the mother to supply the embryo with food and oxygen and carry off its wastes, so that she can develop the offspring for a long period inside her body (nine months in the case of the human being, two years in the case of elephants and whales). The eutherians are usually referred to as 'placental mammals'.

The placental mammals are divided into well over a dozen orders, of which the following are examples:

Insectivora ('insect-eating') – shrews, moles, and others.

Chiroptera ('hand-wings') – the bats.

Carnivora ('meat-eating') – the cat family, the dog family, weasels, bears, seals, and so on, but not including man.

Rodentia ('gnawing') – mice, rats, rabbits, squirrels, guinea pigs, beavers, porcupines, and so on.

Edentata ('toothless') – the sloths and armadillos, which have teeth, and ant-eaters, which do not.

Artiodactyla ('even toes') – hoofed animals with an even number of toes on each foot, such as cattle, sheep, goats, swine, deer, antelopes, camels, giraffes, and so on.

Perissodactyla ('odd toes') – horses, donkeys, zebras, rhinoceroses, and tapirs.

Proboscidea ('long nose') – the elephants, of course.

Odontoceti ('toothed whales') – the sperm whale and others with teeth.

Mysticeti ('moustached whales') – the right whale, the blue whale, and others that filter their small seafood through fringes of whalebone that look like a colossal moustache inside the mouth.

Primates ('first') – man, apes, monkeys, and some other creatures with which man may be surprised to find himself associated.

The primates are characterized by hands and sometimes feet that are equipped for grasping, with opposable thumbs and big toes. The digits are topped with flattened nails rather than with sharp claws or enclosing hoofs. The brain is enlarged, and the sense of vision is more important than the sense of smell. There are many other, less obvious, anatomical criteria.

The primates are divided into nine families. Some have so few primate characteristics that it is hard to think of them as primates, but so they must be classed. One is the family Tupaiidae, which includes the insect-eating tree-shrews! Then there are the lemurs – nocturnal, tree-living creatures with fox-like muzzles and a rather squirrely appearance, found particularly in Madagascar.

The families closest to man are, of course, the monkeys and apes. There are three families of monkeys (a word possibly derived from the Latin *homunculus*, meaning 'little man').

The two monkey families in the Americas, known as the 'New-World monkeys', are the Cebidae (e.g., the organ-grinder's monkey) and the Callithricidae (e.g., the marmoset). The third, the 'Old World' family, are the Cercopithecidae; they include the various baboons.

The apes all belong to one family, called Pongidae. They are native to the Eastern Hemisphere. Their most noticeable outward differences from the monkeys are, of course, their larger size and their lack of tails. The apes fall into four types: the gibbon, smallest, hairiest, longest-armed, and most primitive of the family; the orangutan, larger, but also a tree-liver like the gibbon; the gorilla, rather larger than a man, mainly ground-dwelling, and a native of Africa; and the chimpanzee, also a dweller in Africa, rather smaller than a man and the most intelligent primate next to man himself.

As for our own family, Hominidae, it consists today of only one genus and, as a matter of fact, only one species. Linnaeus named the species *Homo sapiens* ('man the wise'), and no one has dared change the name, despite provocation.

Evolution

It is almost impossible to run down the roster of living things, as we have just done, without ending with a strong impression that there has been a slow development of life from the very simple to the complex. The phyla can be arranged so that each seems to add something to the one before. Within each phylum, the various classes can be arranged likewise, and within each class the orders.

Furthermore, the species often seem to melt together, as if they were still evolving along their slightly separate roads from common ancestors not very far in the past. Some species are so close together that under special circumstances they will interbreed, as in the case of the horse and donkey, which, by appropriate co-operation, can produce the mule. Cattle can interbreed with buffaloes, and lions with tigers. There are also intermediate species, so to speak – creatures that link together two larger groups of animals. The cheetah is a cat with a smattering of doggish characteristics, and the hyena is a dog with some cattish characteristics. The platypus is a mammal only halfway removed from a reptile. There is a creature called 'peripatus', which seems half worm, half centipede. The dividing lines become particularly thin when we look at certain animals in their youthful stages. The infant frog seems to be a fish, and there is a primitive chordate called 'balanoglossus', discovered in 1825, which as a youngster is so like a young echinoderm that at first it was so classified.

We can trace practically a re-enactment of the passage through the phyla, even in the development of a human being from the fertilized egg. The study of this development ('embryology') began in the modern sense with Harvey, the discoverer of the circulation of the blood. In 1759, the German physiologist Kaspar Friedrich Wolff demonstrated that the change in the egg was really a development; that is, specialized tissues grew out of un-specialized precursors by progressive alteration rather than (as many had previously thought) through the mere growth of tiny, already specialized structures existing in the egg to begin with.

In the course of this development, the egg starts as a single cell (a kind of protozoon), then becomes a small colony of cells (as in a sponge), each of which at first is capable of separating and starting life on its own, as happens when identical twins develop. The developing embryo passes through a two-layered stage (like a coelenterate), then adds a third layer (like an echinoderm), and so it continues to add complexities in roughly the order that the higher and higher species do. The human embryo has at some stage in its development the notochord of a primitive chordate, later gill pouches reminiscent of a fish, and still later the tail and body hair of a lower mammal.

From Aristotle on, many men speculated on the possibility that organisms had evolved from one another. But as Christianity grew in power, such speculations were discouraged. The first chapter of Genesis in the Bible stated flatly that each living thing was created 'after his kind', and, taken literally, this had to mean that the species were 'immutable' and had had the same form from the very beginning. Even Linnaeus, who must have been struck by the apparent kinships among living things, insisted firmly on the immutability of species.

The literal story of Creation, strong as its hold was on the minds of men, eventually had to yield to the evidence of the 'fossils' (from the Latin word meaning 'to dig'). As long ago as 1669, the Danish scientist Nicolaus Steno had pointed out that lower layers ('strata') of rock had to be older than the upper strata. At any reasonable rate of rock formation, it became more and more evident that lower strata had to be *much* older than upper strata. Petrified remnants of once-living things were often found buried so deep under layers of rock that they had to be immensely older than the few thousand years that had elapsed since the creation as described in the Bible. The fossil evidence also pointed to vast changes in the structure of the earth. As long ago as the sixth century B.C., the Greek philosopher Xenophanes of Colophon had noted fossil sea shells in the mountains and had surmised that those mountains had been under water long ages before.

Believers in the literal words of the Bible could and did maintain that the fossils resembled once-living organisms only through accident, or that they had been created deceitfully by the Devil. Such views were most unconvincing, and a more plausible suggestion was made that the fossils were remnants of creatures drowned in the Flood. Sea shells on mountain tops would certainly be evidence for that, since the Biblical account of the Deluge states that water covered all the mountains.

But on close inspection, many of the fossil organisms proved to be different from any living species. John Ray, the early classifier, wondered if they might represent extinct species. A Swiss naturalist named Charles Bonnet went further. In 1770, he suggested that fossils were indeed remnants of extinct species which had been destroyed in ancient geological catastrophes going back to long before the Flood.

It was an English land surveyor named William Smith, however, who laid a scientific foundation for the study of fossils ('palaeontology'). While working on excavations for a canal in 1791, he was impressed by the fact that the rock through which the canal was being cut was divided into strata, and that each stratum contained its own characteristic fossils. It now became possible to put fossils in a chronological order, depending on their place in the series of successive layers, and to associate each fossil with a particular type of rock stratum which would represent a certain period in geological history.

About 1800, Cuvier (the man who invented the notion of the phylum) classified fossils according to the Linnaean system and extended comparative anatomy into the distant past. Although many fossils represented species and genera not found among living creatures, all fitted neatly into one or another of the known phyla and so made up an integral part of the scheme of life. In 1801, for instance, Cuvier studied a long-fingered fossil of a type first discovered twenty years earlier, and demonstrated it to be the remains of a leathery-winged flying creature like nothing now existing – at least like nothing now existing *exactly*. He was able to show from the bone structure that these 'pterodactyls' ('wing-fingers'), as he called them, were nevertheless reptiles, clearly related to the snakes, lizards, alligators, and turtles of today.

Furthermore, the deeper the stratum in which the fossil was to be found, and therefore the older the fossil, the simpler and less highly developed it seemed. Not only that, but fossils sometimes represented intermediate forms connecting two groups of creatures which, as far as living forms were concerned, seemed entirely separate. A particularly startling example, discovered after Cuvier's time, is a very primitive bird called Archaeopteryx (Greek for 'ancient wing'). This now-extinct creature had wings and feathers, but it also had a lizard-like, feather-fringed tail and a beak that contained reptilian teeth! In these and other respects it was clearly midway between a reptile and a bird.

Cuvier still supposed that terrestrial catastrophes, rather than

Archaeopteryx.

evolution, had been responsible for the disappearance of the extinct forms of life, but in the 1830s Charles Lyell's new view of fossils and geological history in his history-making work *Principles of Geology* killed 'catastrophism' deader than a doornail (see Vol. 1, p. 45). Some reasonable theory of evolution became a necessity, if any sense was to be made of the palaeontological evidence.

If animals had evolved from one form to another, what had caused them to do so? This was the main stumbling block in the efforts to explain the varieties of life. The first to attempt an explanation was the French naturalist Jean Baptiste de Lamarck. In 1809, he published a book entitled *Zoological Philosophy*, in which he suggested that the environment caused organisms to acquire small changes which were then passed on to their descendants. Lamarck illustrated his idea with the giraffe (a newly discovered sensation of the time). Suppose that a primitive, antelope-like creature that fed on tree leaves ran out of food within easy reach, and had to stretch its neck as far as it could to get more food. By habitual stretching of its neck, tongue, and legs, it would gradually lengthen those appendages. It would then pass on these developed characteristics to its offspring, which in turn would stretch further and pass on a still longer neck to their descendants, and so on. Little by little, by generation after generation of stretching, the primitive antelope would evolve into a giraffe.

Lamarck's notion of the 'inheritance of acquired characteristics' quickly ran afoul of difficulties. How had the giraffe developed its blotched coat, for instance? Surely no action on its part, deliberate or otherwise, could have effected this change. Furthermore, a sceptical experimenter, the German biologist August Friedrich Leopold Weismann, cut off the tails of mice for generation after generation and reported that the last generation grew tails not one whit shorter than the first. (He might have saved himself the trouble by considering the case of the circumcision of Jewish males, which after more than a hundred generations had produced no shrivelling of the foreskin.)

By 1883, Weismann had observed that the germ cells, which were eventually to produce sperm or ova, separated from the remainder of the embryo at an early stage and remained relatively unspecialized. From this, and from his experiments with rat tails, Weismann deduced the notion of the 'continuity of the germ plasm'. The germ plasm (that is, the protoplasm making up the germ cells) had, he felt, an independent existence, continuous across the generations, with the remainder of the organism but a temporary housing, so to speak, built up and destroyed in each generation. The germ plasm guided the characteristics of the body and was not itself affected by the body. In all this, he was at the extreme opposite to Lamarck and was also wrong, although, on the whole, the actual situation seemed closer to the Weismann view than to that of Lamarck.

Despite its rejection by most biologists, Lamarckism lingered on into the twentieth century and even had a strong but apparently temporary revival in the form of Lysenkoism (hereditary modification of plants by certain treatments) in the Soviet Union. (Trofim Denisovich Lysenko, the exponent of this belief, was powerful under Stalin, retained much influence under Khrushchev, but underwent an eclipse when Khrushchev fell from power in 1964.) Modern geneticists do not exclude the possibility that the action of the environment may bring about certain transmittable changes in simple organisms, but the Lamarckian idea as such was demolished by the discovery of genes and the laws of heredity.

In 1831, a young Englishman named Charles Darwin, a dilettante and sportsman who had spent a more or less idle youth and was restlessly looking for something to do to overcome his boredom, was persuaded by a ship captain and a Cambridge professor to sign on as naturalist on a ship setting off on a five-year voyage around the world. The expedition was to study continental coastlines and make observations of flora and fauna along the way. Darwin, aged twenty-two, made the voyage of the *Beagle* the most important sea voyage in the history of science.

As the ship sailed slowly down the east coast of South America and then up its west coast, Darwin painstakingly collected information on the various forms of plant and animal life. His most striking discovery came in a group of islands in the Pacific, about

650 miles west of Ecuador, called the Galapagos Islands because of giant tortoises living on them (Galapagos coming from the Spanish word for tortoise). What most attracted Darwin's attention during his five-week stay was the variety of finches on the islands; they are known as 'Darwin's finches' to this day. He found the birds divided into at least fourteen different species, distinguished from one another mainly by differences in the size and shape of their bills. These particular species did not exist anywhere else in the world, but they resembled an apparently close relative on the South American mainland.

What accounted for the special character of the finches on these islands? Why did they differ from ordinary finches, and why were they themselves divided into no fewer than fourteen species? Darwin decided that the most reasonable theory was that all of them were descended from the mainland type of finch and had differentiated during long isolation on the islands. The differentiation had resulted from varying methods of obtaining food. Three of the Galapagos species still fed on seeds, as the mainland finch did, but each ate a different kind of seed and varied correspondingly in size, one species being rather large, one medium, and one small. Two other species fed on cacti; most of the others fed on insects.

The problem of the changes in the finches' eating habits and physical characteristics preyed on Darwin's mind for many years. In 1838 he began to get a glimmering of the answer from reading a book that had been published forty years before by an English clergyman named Thomas Robert Malthus. It was entitled *An Essay on the Principle of Population*; in it Malthus maintained that a population always outgrew its food supply, so that eventually starvation, disease or war cut it back. It was in this book that Darwin came across the phrase 'the struggle for existence', which his theories later made famous. Thinking of his finches, Darwin at once realized that competition for food would act as a mechanism favouring the more efficient individuals. When the finches that had colonized the Galapagos multiplied to the point of outrunning the seed supply, only the stronger birds or those particularly adept at obtaining seeds or those able to get new kinds of food would survive. A bird that happened to be equipped with slight variations of the finch characteristics, which enabled it to eat bigger seeds or tougher seeds or, better still, insects, would find an untapped food supply. A bird with a slightly thinner and longer bill could reach food that others could not, or one with an unusually massive bill could use otherwise unusable food. Such birds, and their descendants, would gain in numbers at the expense of the original variety of finch. Each of the adaptive types would find and fill a new, unoccupied niche in the environment. On the Galapagos Islands, virtually empty of bird life to begin with, all sorts of niches were there for the taking, with no established competitors to bar the way. On the South American mainland, with all the niches occupied, the ancestral finch did well merely to hold its own. It proliferated into no further species.

Darwin suggested that every generation of animals was composed of an array of individuals varying randomly from the average. Some would be slightly larger; some would possess organs of slightly altered shape; some abilities would be a trifle above or below normal. The differences might be minute, but those whose make-up was even slightly better suited to the environment would tend to live slightly longer and have more offspring. Eventually, an accumulation of favourable characteristics might be coupled with an inability to breed with the original type or other variations of it, and thus a new species would be born.

Darwin called this process 'natural selection'. According to his view, the giraffe got its long neck not by stretching but because some giraffes were born with longer necks than their fellows, and the longer the neck, the more chance a giraffe had of reaching food. By natural selection, the long-necked species won out. Natural selection explained the giraffe's blotched coat just as easily: an animal with blotches on its skin would blend against the sun-spotted vegetation and thus have more chance of escaping the attention of a prowling lion.

Darwin's view of the way in which species were formed also made clear why it was often so difficult to make clear-cut distinctions between species or between genera. The evolution of species is a continuous process, and, of course, takes a very long time.

There must be any number of species with members which are even now slowly drifting apart into separate species.

Darwin spent many years collecting evidence and working out his theory. He realized that it would shake the foundations of biology and man's thinking about his own place in the scheme of things, and he wanted to be sure of his ground in every possible respect. Darwin started collecting notes on the subject and thinking about it in 1834, even before he read Malthus, and in 1858 he was still working on a book dealing with the subject. His friends (including Lyell, the geologist) knew what he was working on; several had read his preliminary drafts. They urged him to hurry, lest he be anticipated. Darwin would not (or could not) hurry, and he *was* anticipated.

The man who anticipated him was Alfred Russel Wallace, fourteen years younger than Darwin. Wallace's life paralleled that of Darwin. He, too, went on a round-the-world scientific expedition as a young man. In the East Indies, he noticed that the plants and animals in the eastern islands were completely different from those in the western islands. A sharp line could be drawn between the two types of life forms; it ran between Borneo and Celebes, for instance, and between the small islands of Bali and Lombok farther to the south. The line is still called 'Wallace's Line'. (Wallace went on, later in his life, to divide the earth into six large regions, characterized by differing varieties of animals, a division that, with minor modifications, is still considered valid today.)

Now the mammals in the eastern islands and in Australia were distinctly more primitive than those in the western islands and Asia, or indeed in the rest of the world. It looked as if Australia and the eastern islands had split off from Asia at some early time when only primitive mammals existed, and the placental mammals had developed later only in Asia. New Zealand must have been isolated even longer, for it lacked mammals altogether and was inhabited by primitive flightless birds, of which the best-known survivor today is the kiwi.

How had the higher mammals in Asia arisen? Wallace first began puzzling over this in 1855, and in 1858 he, too, came across

Malthus's book and from it, he, too, drew the conclusions Darwin had drawn. But Wallace did not spend fourteen years writing his conclusions. Once the idea was clear in his mind, he sat down and wrote a paper on it in two days. Wallace decided to send his manuscripts to some well-known competent biologist for criticism and review, and he chose Charles Darwin.

When Darwin received the manuscript, he was thunderstruck. It expressed his own thoughts in almost his own terms. At once he passed Wallace's paper to other important scientists and offered to collaborate with Wallace on reports summarizing their joint conclusions. Their reports appeared in the *Journal of the Linnaean Society* in 1858.

The next year Darwin's book was finally published. Its full title is *On the Origin of Species by Means of Natural Selection, or the Preservation of Favoured Races in the Struggle for Life*. We know it simply as *The Origin of Species*.

The theory of evolution has been modified and sharpened since Darwin's time, through knowledge of the mechanism of inheritance, of genes, and of mutations (see Chapter 13). It was not until 1930, indeed, that the English statistician and geneticist Ronald Aylmer Fisher succeeded in showing that Mendelian genetics provided the necessary mechanism for evolution by natural selection. Only then did evolutionary theory gain its modern guise. Nevertheless, Darwin's basic conception of evolution by natural selection has stood firm, and indeed the evolutionary idea has been extended to every field of science – physical, biological, and social.

The announcement of the Darwinian theory naturally blew up a storm. At first, a number of scientists held out against the notion. The most important of these was the English zoologist Richard Owen, who was the successor of Cuvier as an expert on fossils and their classification. Owen stooped to rather unmanly depths in his fight against Darwinism. He not only urged others into the fray while remaining hidden himself, but even wrote anonymously against the theory and quoted himself as an authority.

The English naturalist Philip Henry Gosse tried to wriggle out of the dilemma by suggesting that the earth had been created by God complete with fossils to test man's faith. To most people, however, the suggestion that God would play juvenile tricks on mankind seemed more blasphemous than anything Darwin had suggested.

Its counter-attack blunted, opposition within the scientific world gradually subsided, and, within the generation, nearly disappeared. The opponents outside science, however, carried on the fight much longer and much more intensively. The Fundamentalists (literal interpreters of the Bible) were outraged by the implication that man might be a mere descendant from an ape-like ancestor. Benjamin Disraeli (later to be Prime Minister of Great Britain) created an immortal phrase by remarking acidly: 'The question now placed before society is this, "Is man an ape or an angel?" I am on the side of the angels.' Churchmen, rallying to the angels' defence, carried the attack to Darwin.

Darwin himself was not equipped by temperament to enter violently into the controversy, but he had an able champion in the eminent biologist Thomas Henry Huxley. As 'Darwin's bull-dog', Huxley fought the battle tirelessly in the lecture halls of England. He won his most telling victory almost at the very beginning of the struggle, in the famous debate with Samuel Wilberforce, a bishop of the Anglican Church, a mathematician, and so accomplished and glib a speaker that he was familiarly known as 'Soapy Sam'.

Bishop Wilberforce, after apparently having won the audience, turned at last to his solemn, humourless adversary. As the report of the debate quotes him, Wilberforce 'begged to know whether it was through his grandfather or his grandmother that [Huxley] claimed his descent from a monkey'.

While the audience roared with glee, Huxley rose slowly to his feet and answered: 'If, then, the question is put to me, would I rather have a miserable ape for a grandfather, or a man highly endowed by nature and possessing great means and influence, and yet who employs those faculties and that influence for the mere purpose of introducing ridicule into a grave scientific discussion – I unhesitatingly affirm my preference for the ape.'

Huxley's smashing return apparently not only crushed Wilberforce but also put the Fundamentalists on the defensive. In fact, so clear was the victory of the Darwinian viewpoint that, when Darwin died in 1882, he was buried, with widespread veneration, in Westminster Abbey, where lie England's greats. In addition, the town of Darwin in northern Australia was named in his honour.

Another powerful proponent of evolutionary ideas was the English philosopher Herbert Spencer, who popularized the phrase 'survival of the fittest' and the word 'evolution' – a word Darwin himself rarely used. Spencer tried to apply the theory of evolution to the development of human societies (he is considered the founder of the science of sociology). His arguments, contrary to his intention, were later misused to support war and racism.

The last open battle against evolution took place in 1925; it ended with the anti-evolutionists winning the battle and losing the war.

The Tennessee legislature had passed a law forbidding teachers in publicly supported schools of the state from teaching that man had evolved from lower forms of life. To challenge the law's constitutionality, scientists and educators persuaded a young high-school biology teacher named John T. Scopes to tell his class about Darwinism. Scopes was thereupon charged with violating the law and brought to trial in Dayton, Tennessee, where he taught. The world gave fascinated attention to his trial.

The local population and the judge were solidly on the side of anti-evolution. William Jennings Bryan, the famous orator, three times unsuccessful candidate for the Presidency, and an outstanding Fundamentalist, served as one of the prosecuting attorneys. Scopes had as his defenders the noted criminal lawyer Clarence Darrow and associated attorneys.

The trial was for the most part disappointing, for the judge refused to allow the defence to place scientists on the stand to testify to the evidence behind the Darwinian theory, and restricted testimony to the question whether Scopes had or had not discussed evolution. But the issues nevertheless emerged in the courtroom when Bryan, over the protests of his fellow prosecutors, volunteered to submit to cross-examination on the Fundamentalist position. Darrow promptly showed that Bryan was ignorant of modern developments in science and had only a stereotyped Sunday-school acquaintance with religion and the Bible.

Scopes was found guilty and fined $100. (The conviction was later reversed on technical grounds by the Tennessee Supreme Court.) But the Fundamentalist position (and the state of Tennessee) had stood in so ridiculous a light in the eyes of the educated world that the anti-evolutionists have not made any serious stand since then – at least not in broad daylight.

In fact, if any confirmation of Darwinism were needed, it has turned up in examples of natural selection that have taken place before the eyes of mankind (now that mankind knows what to watch for). A notable example occurred in Darwin's native land.

In England, it seems, the peppered moth exists in two varieties, a light and a dark. In Darwin's time, the white variety was predominant because it was less prominently visible against the light lichen-covered bark of the trees it frequented. It was saved by this 'protective colouration' more often than the clearly visible, dark variety from those animals who would feed on it. In modern, industrialized England, however, soot has killed the lichen cover and blackened the tree bark. Now it is the dark variety that is less visible against the bark and therefore protected. It is the dark variety that is now predominant – through the action of natural selection.

A study of the fossil record has enabled palaeontologists to divide the history of the earth into a series of 'eras'. These were roughed out and named by various nineteenth-century British geologists, including Lyell himself, Adam Sedgwick, and Roderick Impey Murchison. The eras start, as is now known, some 500 or 600 million years ago with the first fossils (when all the phyla except Chordata were already established). The first fossils do not, of course, represent the first life. For the most part, it is only the hard portions of a creature that fossilize, so the clear fossil record contains only animals that possessed shells or bones. Even the simplest and oldest of these creatures are already far advanced and must have a long evolutionary background. One evidence of that is that in 1965 fossil remains of small clam-like creatures were discovered that seem to be about 720 million years old.

Palaeontologists can now do far better than that. It stands to reason that simple one-celled life must extend much farther back in time than anything with a shell, and indeed, signs of blue-green algae and of bacteria have been found in rocks that were a thousand million years old and more. In 1965, the American palaeontologist Elso Sterrenberg Barghoorn discovered minute bacterium-like objects ('micro-fossils') in rocks over 3,000 million years old. They are so small, their structure must be studied by electron microscope.

It would seem then that chemical evolution, moving towards the origin of life, began almost as soon as the earth took on its present shape some 4,600 million years ago. Within 1,500 million years, chemical evolution had reached the stage where systems complicated enough to be called living had formed. About 2,500 million years ago, blue-green algae may have been in existence, and the process of photosynthesis began the slow change from a nitrogen–carbon dioxide atmosphere into a nitrogen–oxygen atmosphere. About a thousand million years ago, the one-celled life of the seas must have been quite diversified and included distinctly animal protozoa, who would then have been the most complicated forms of life in existence – monarchs of the world.

For 2,000 million years after blue-green algae came into existence, the oxygen content must have been very slowly increasing. As the most recent thousand million years of earth's history began to unfold, the oxygen concentration may have been 1 or 2 per cent of the atmosphere, enough to supply a rich source of energy for animal cells beyond anything that had earlier existed. Evolutionary change spurted in the direction of increased complication, and by 600 million years ago there could begin the rich fossil record of elaborate organisms.

The earliest rocks with elaborate fossils are said to belong to the Cambrian age, and the entire 4,000-million-year history of our planet that preceded it has been, until recently, dismissed as the 'pre-Cambrian age'. Now that the traces of life have unmistakably been found in it, the more appropriate name of 'Cryptozoic eon' (Greek for 'hidden life') is used, while the last 600 million years make up the 'Phanerozoic eon' ('visible life').

The Cryptozoic eon is even divided into two sections: the earlier 'Archaeozoic era' ('ancient life'), to which the first traces of unicellular life belong, and the later 'Proterozoic era' ('early life').

The division between the Cryptozoic eon and the Phanerozoic eon is extraordinarily sharp. At one moment in time, so to speak, there are no fossils at all above the microscopic level, and at the next there are elaborate organisms of a dozen different basic types. Such a sharp division is called an 'unconformity', and an unconformity leads invariably to speculations about possible catastrophes. It seems there should have been a more gradual appearance of fossils, and what may have happened is that geological events of some extremely harsh variety wiped out the earlier record.

One fascinating (if highly speculative) suggestion was put forward in 1967 by Walter S. Olson. He suggested that about a thousand million years ago or somewhat less, our moon was captured by the earth – that prior to that the earth was without a satellite. (This possibility is seriously considered by astronomers as a way of explaining certain anomalies of the earth–moon system.) When first captured, the moon was considerably closer than it is now. The enormous tides suddenly set in motion would have broken up the superficial layers of rock on earth and, so to speak, erased any fossil record, which began again – apparently full-born – only after the tides had receded in power to the point where the rocky surface was left relatively untouched.

The broad divisions of the Phanerozoic eon are the Palaeozoic era ('ancient life'), the Mesozoic ('middle life'), and the Cenozoic ('new life'). According to modern methods of geological dating, the Palaeozoic era covered a span of perhaps 350 million years, the Mesozoic 150 million years, and the Cenozoic the last 50 million years of the earth's history.

Each era is in turn subdivided into ages. The Palaeozoic begins, as stated above, with the Cambrian age (named after a location in Wales – actually an ancient tribe that occupied it – where these strata were first uncovered). During the Cambrian period shellfish were the most elaborate form of life. This was the era of the 'trilobites', primitive arthropods of which the modern king crab is the closest living relative. The king crab, because it has survived with few evolutionary changes over long ages, is an example of what is sometimes rather dramatically called a 'living fossil'.

The next age is the Ordovician (named after another Welsh tribe). This was the age, between 450 and 500 million years ago, when the chordates made their first appearance in the form of 'graptolites', small animals living in colonies and now extinct. They are possibly related to the balanoglossus, which, like the graptolites, belongs to the 'hemichordata', the most primitive sub-phylum of the chordate phylum.

Then came the Silurian (named after still another Welsh tribe) and the Devonian (from Devonshire). The Devonian age, between 350 and 400 million years ago, witnessed the rise of fish to dominance in the ocean, a position they still hold. In that age, however, came also the colonization of the dry land by life forms. It is hard to realize, but true, that during perhaps three quarters or more of its history, life was confined to the waters and the land remained dead and barren. Considering the difficulties represented by the lack of water, by extremes of temperature, by the full force of gravity unmitigated by the buoyancy of water, it must be understood that the spread to land of life forms that evolved to meet the conditions of the ocean represented the greatest single victory won by life over the inanimate environment.

The move towards the land probably began when competition for food in the crowded sea drove some organisms into shallow tidal waters, until then unoccupied because the bottom was exposed for hours at a time at low tide. As more and more species crowded into the tide-waters, relief from competition could be attained only by moving farther and farther up the shore, until eventually some mutant organisms were able to establish themselves on dry land.

The first life forms to manage the transition were plants. This took place about 400 million years ago. The pioneers belonged to a now extinct plant group called 'psilopsids' – the first multicellular plants. (The name comes from the Greek word for 'bare' because the stems were bare of leaves, a sign of the primitive nature of these plants.) More complex plants developed and, by 350 million years ago, the land was covered with forest. Once plant life had begun to grow on dry land, animal life could follow suit. Within a few million years the land was occupied by arthropods, molluscs, and worms. All these first land animals were small, because heavier animals, without an internal skeleton, would have collapsed under the force of gravity. (In the ocean, of course, buoyancy largely negated gravity, which was not therefore a factor. Even today the largest animals live in the sea.) The first land creatures to gain much mobility were the insects; thanks to their development of wings, they were able to counteract the force of gravity, which held other animals to a slow crawl.

Finally, 100 million years after the first invasion of the land, there came a new invasion by creatures that could afford to be bulky despite gravity because they had a bracing of bone within. The new colonizers from the sea were bony fishes belonging to the sub-class Crossopterygii ('fringed fins'). Some of their fellow members had migrated to the uncrowded sea deeps, including the coelacanth, which biologists discovered in 1939 to be still in existence (much to their astonishment).

The fishy invasion of land began as a result of competition for oxygen in brackish stretches of fresh water. With oxygen available in unlimited quantities in the atmosphere, those fish best survived who could most effectively gulp air when the oxygen content of water fell below the survival point. Devices for storing such gulped air had survival value, and fish developed pouches in their alimentary canals in which swallowed air could be kept. These pouches developed into simple lungs in some cases. Descendants of these early fish include the 'lungfishes', a few species of which still exist in Africa and Australia. These live in

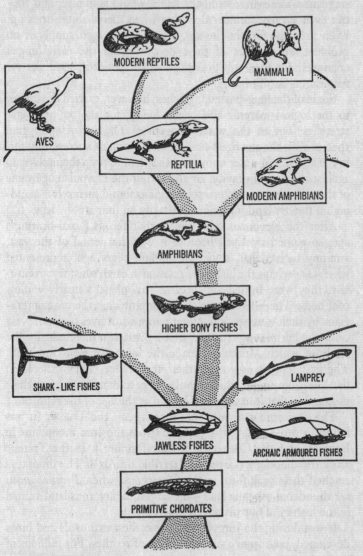

A philogenetic tree, showing evolutionary lines of the vertebrates.

stagnant water where ordinary fishes would suffocate, and they can even survive summer droughts when their habitat dries up. Even fish who live in the sea, where the oxygen supply is no problem, show signs of their descent from the early-lunged creatures, for they still possess air-filled pouches, used not for respiration but for buoyancy.

Some of the lung-possessing fishes, however, carried the matter to the logical extreme and began living, for shorter or longer stretches, out of the water altogether. These crossopterygian species with the strongest fins could do so most successfully for, in the absence of water buoyancy, they had to prop themselves up against the pull of gravity. By the end of the Devonian age some of the primitive-lunged crossopterygians found themselves standing on the dry land, propped up shakily on four stubby legs.

After the Devonian came the Carboniferous ('coal-bearing') age, so named by Lyell because it was the period of the vast, swampy forests that, some 300 million years ago, represented what was perhaps the lushest vegetation in earth's history; eventually, they were buried and became this planet's nearly endless coal beds. This was the age of the amphibians; the crossopterygians by then were spending their entire adult lives on land. Next came the Permian age (named after a district in the Urals, for the study of which Murchison made the long trip from England). The first reptiles now made their appearance. They ushered in the Mesozoic era, in which reptiles were to dominate the earth so thoroughly that it has become known as the age of the reptiles.

The Mesozoic is divided into three ages – the Triassic (it was found in three strata), the Jurassic (from the Jura mountains in France), and the Cretaceous ('chalk-forming'). In the Triassic arose the dinosaurs (Greek for 'terrible lizards'). The dinosaurs reached their peak form in the Cretaceous, when *Tyrannosaurus rex* thundered over the land – the largest carnivorous land animal in the history of our planet.

It was during the Jurassic that the earliest mammals and birds developed, each from a separate group of reptiles. For millions of years these creatures remained inconspicuous and unsuccessful. With the end of the Cretaceous, however, the gigantic reptiles began to disappear (for some unknown reason, so that the cause of the 'great dying' remains the most tantalizing problem in palaeontology), and the mammals and birds came into their own. The Cenozoic era that followed became the age of mammals; it brought in placental mammals and the world we know.

The unity of present life is demonstrated in part by the fact that all organisms are composed of proteins built from the same amino acids. The same kind of evidence has recently established our unity with the past as well. The new science of 'palaeo-biochemistry' (the biochemistry of ancient forms of life) was opened in the late 1950s, when it was shown that certain 300-million-year-old fossils contained remnants of proteins consisting of precisely the same amino acids that make up proteins today – glycine, alanine, valine, leucine, glutamic acid, and aspartic acid. Not one of the ancient amino acids differed from present ones. In addition, traces of carbohydrates, cellulose, fats, and porphyrins were located, with (again) nothing that would be unknown or unexpected today.

From our knowledge of biochemistry we can deduce some of the biochemical changes that may have played a part in the evolution of animals.

Let us take the excretion of nitrogenous wastes. Apparently, the simplest way to get rid of nitrogen is to excrete it in the form of the small ammonia molecule (NH_3), which can easily pass through cell membranes into the blood. Ammonia happens to be extremely poisonous; if its concentration in the blood exceeds one part in a million, the organism will die. For a sea animal, this is no great problem; it can discharge the ammonia into the ocean continuously through its gills. But for a land animal, however, ammonia excretion is out of the question. To discharge ammonia as quickly as it is formed would require such an excretion of urine that the animal would quickly be dehydrated and die. Therefore a land organism must produce its nitrogenous wastes in a less toxic form than ammonia. The answer is urea. This substance can be carried in the blood in concentrations up to one part in a thousand without serious danger.

Now fish eliminate nitrogenous wastes as ammonia, and so do tadpoles. But when a tadpole matures to a frog, it begins to eliminate nitrogenous wastes as urea. This change in the chemistry of the organism is every bit as crucial for the change-over from life in the water to life on land as is the visible change from gills to lungs.

Such a biochemical change must have taken place when the crossopterygians invaded the land and became amphibians. Thus, there is every reason to believe that biochemical evolution played as great a part in the development of organisms as 'morphological' evolution (that is, changes in form and structure).

Another biochemical change was necessary before the great step from amphibian to reptile could be taken. If the embryo in a reptile's egg excreted urea, it would build up to toxic concentrations in the limited quantity of water within the egg. The change that took care of this problem was the formation of uric acid instead of urea. Uric acid (a purine molecule resembling the adenine and guanine that occur in nucleic acids) is insoluble in water; it is therefore precipitated in the form of small granules and thus cannot enter the cells. This change-over from urea excretion to uric-acid excretion was as essential in the development of reptiles as was the change-over, for instance, from a three-chambered heart to an essentially four-chambered one.

In adult life, reptiles continue eliminating nitrogenous wastes as uric acid. They have no urine in the liquid sense. Instead, the uric acid is eliminated as a semi-solid mass through the same body opening that serves for the elimination of faeces. This single body opening is called the 'cloaca' (Latin for 'sewer').

Birds and egg-laying mammals, which lay eggs of the reptilian type, preserve the uric-acid mechanism and the cloaca. In fact, the egg-laying mammals are often called 'monotremes' (from Greek words meaning 'one hole').

Placental mammals, on the other hand, can easily wash away the embryo's nitrogenous wastes, for the embryo is connected, indirectly, to the mother's circulatory system. Mammalian embryos, therefore, manage well with urea. It is transferred to the mother's bloodstream and passes out through the mother's kidneys.

An adult mammal has to excrete substantial amounts of urine to get rid of its urea. This calls for two separate openings: an anus to eliminate the indigestible solid residues of food and a urethral opening for the liquid urine.

The account just given of nitrogen excretion demonstrates that although life is basically a unity there are systematic minor variations from species to species. Furthermore, these variations seem to be greater as the species considered are farther removed from each other in the evolutionary sense.

Consider, for instance, that antibodies can be built up in animal blood to some foreign protein or proteins as, for example, those in human blood. Such 'anti-sera', if isolated, will react strongly with human blood, coagulating it, but will not react in this fashion with the blood of other species. (This is the basis of the tests indicating that bloodstains are of human origin – or not – which sometimes lend drama to murder investigations.) Interestingly, anti-sera that will react with human blood will respond weakly with chimpanzee blood, while anti-sera that will react strongly with chicken blood will react weakly with duck blood, and so on. Antibody specificity thus can be used to indicate close relationships among life forms.

Such tests indicate, not surprisingly, the presence of minor differences in the complex protein molecule – differences small enough in closely related species to allow some overlapping in anti-serum reactions.

When biochemists developed techniques for determining the precise amino-acid structure of proteins, in the 1950s, this method of arranging species according to protein structure was vastly sharpened.

In 1965, even more detailed studies were reported on the haemoglobin molecules of various types of primates, including man. Of the two kinds of peptide chains in haemoglobin, one, the 'alpha chain', varied little from primate to primate. The other, the 'beta chain', varied considerably. Between a particular primate and man there were only six differences in the amino acids and the alpha chain, but twenty-three in those of the beta chains. Judging by differences in the haemoglobin molecules, it is believed man diverged from the other apes about 75 million years ago, or just about the time the ancestral horses and donkeys diverged.

Still broader distinctions can be made by comparing molecules of 'cytochrome c', an iron-containing protein molecule made up of about 105 amino acids and found in the cells of every oxygen-breathing species – plant, animal, or bacterial. By analysing the cytochrome-c molecules from different species it was found that the molecules in man differed from those of the rhesus monkey in only one amino acid in the entire chain. Between the cytochrome c of man and that of a kangaroo, there were 10 differences in amino acid; between those of man and a tuna fish, 21 differences; between those of man and a yeast cell, some 40 differences.

With the aid of computer analysis, biochemists have decided it takes on the average some 7 million years for a change in one amino-acid residue to establish itself, and estimates can be made of the time in the past when one type of organism diverged from another. It was about 2,500 million years ago, judging from cytochrome-c analysis, that higher organisms diverged from bacteria (that is, it was about that long ago that a living creature was last alive who might be considered a common ancestor). Similarly, it was about 1,500 million years ago that plants and animals had a common ancestor, and 1,000 million years ago that insects and vertebrates had a common ancestor.

If mutations in the DNA chain, leading to changes in amino-acid pattern, were established by random factors only, it might be supposed that the rate of evolution would continue at an approximately constant rate. Yet there are occasions when evolution seems to progress more rapidly than at others – when there is a sudden flowering of new species or a sudden spate of deaths of old ones. The period at the end of the Cretaceous, when all the dinosaurs then living – together with other groups of organisms – utterly died out over a relatively short period of time (while other groups of organisms lived on undisturbed), is a case in point. It may be that the rate of mutations is greater at some periods in earth's history than at others, and these more frequent mutations may establish an extraordinary number of new species or render an extraordinary number of old ones unviable. (Or else some of the new species may prove more efficient than the old and compete them to death.)

One environmental factor which encourages the production of mutation is energetic radiation, and earth is constantly bombarded by energetic radiation from all directions at all times. The atmosphere absorbs most of it, but even the atmosphere is helpless to ward off cosmic radiation. Can it be that cosmic radiation is greater at some periods than at others?

A difference can be postulated in each of two different ways. Cosmic radiation is diverted to some extent by earth's magnetic field. However, the magnetic field varies in intensity, and there are periods, at varying intervals, when it sinks to zero intensity. Bruce Heezen suggested in 1966 that these periods when the magnetic field, in the process of reversal, goes through a time of zero intensity may also be periods when unusual amounts of cosmic radiation reach the surface of the earth, bringing about a jump in mutation-rate. This is a sobering thought in view of the fact that the earth seems to be heading towards such a period of zero intensity.

Then, too, what about the occurrence of supernovas in earth's vicinity – close enough to the solar system, that is, to produce a distinct increase in the intensity of bombardment by cosmic rays of the earth's surface? Two American astronomers, K. D. Terry and Wallace H. Tucker, have speculated on that possibility. Could a combination of these two effects, taking place with fortuitous simultaneity – a nearby supernova just when earth's magnetic field had temporarily faded – account for the sudden dying of the dinosaurs? Well, perhaps, but as yet there is no firm evidence.

The Descent of Man

James Ussher, the seventeenth-century Irish archbishop, dated the creation of man precisely in the year 4004 B.C.

Before Darwin, few men dared to question the Biblical interpretation of man's early history. The earliest reasonably definite date to which the events recorded in the Bible can be referred is the reign of Saul, the first king of Israel, who is believed to have become king about 1025 B.C. Bishop Ussher and other Biblical scholars who worked back through the chronology of the Bible came to the conclusion that man and the universe could not be more than a few thousand years old.

The documentary history of man, as recorded by Greek historians, began only about 700 B.C. Beyond this hard core of history, dim oral traditions went back to the Trojan War, about 1200 B.C., and more dimly still to a pre-Greek civilization on the island of Crete under a King Minos. Nothing beyond documentation – the writings of historians in known languages, with all the partiality and distortion that might involve – was known to moderns concerning the everyday life of ancient times prior to the eighteenth century. Then, in 1738, the cities of Pompeii and Herculaneum, buried in an eruption of Vesuvius in A.D. 79, began to be excavated. For the first time, historians grew aware of what could be done by digging, and the science of archaeology got its start.

At the beginning of the nineteenth century, archaeologists began to get their first real glimpses of human civilizations that came before the periods described by the Greek and Hebrew historians. In 1799, during General Bonaparte's invasion of Egypt, an officer in his army, named Boussard, discovered an inscribed stone in the town of Rosetta, on one of the mouths of the Nile. The slab of black basalt had three inscriptions, one in Greek, one in an ancient form of Egyptian picture writing called 'hieroglyphic' ('sacred writing'), and one in a simplified form of Egyptian writing called 'demotic' ('of the people').

The inscription in Greek was a routine decree of the time of Ptolemy V, dated the equivalent of 27 March 196 B.C. Plainly, it must be a translation of the same decree given in the other two languages on the slab (compare the no-smoking signs and other official notices that often appear in three languages in public places, especially airports, today). Archaeologists were overjoyed: at last they had a 'pony' with which to decipher the previously undecipherable Egyptian scripts. Important work was done in 'cracking the code' by Thomas Young, the man who had earlier established the wave theory of light (see Vol. 1, p. 360), but it fell to the lot of a French student of antiquities, Jean François Champollion, to solve the 'Rosetta stone' completely. He ventured the guess that Coptic, a still-remembered language of certain Christian sects in Egypt, could be used as a guide to the ancient Egyptian language. By 1821, he had cracked the hieroglyphs and the demotic script and opened the way to reading all the inscriptions found in the ruins of ancient Egypt.

An almost identical find later broke the undeciphered writing of ancient Mesopotamia. On a high cliff near the ruined village of Behistun in western Iran, scholars found an inscription that had been carved about 520 B.C. at the order of the Persian emperor Darius I. It announced the manner in which he had come to the throne after defeating a usurper; to make sure that everyone could read it, Darius had had it carved in three languages – Persian, Sumerian, and Babylonian. The Sumerian and Babylonian writings were based on pictographs formed as long ago as 3100 B.C. by indenting clay with a stylus; these had developed into a 'cuneiform' ('wedge-shaped') script, which remained in use until the first century A.D.

An English army officer, Henry Creswicke Rawlinson, climbed the cliff, transcribed the entire inscription, and, by 1846, after ten years of work, managed to work out a complete translation, using local dialects as his guide where necessary. The deciphering of the cuneiform scripts made it possible to read the history of the ancient civilizations between the Tigris and the Euphrates.

Expedition after expedition was sent to Egypt and Mesopotamia to look for tablets and the remains of the ancient civilizations. In 1854 a Turkish scholar, Hurmuzd Rassam, discovered the remnants of a library of clay tablets in the ruins of Nineveh, the capital of ancient Assyria – a library that had been collected by the last great Assyrian king, Ashurbanipal, about 650 B.C. In 1873 the English Assyriologist George Smith discovered clay tablets giving legendary accounts of a flood so like the story of Noah that it became clear that much of the first part of the book of Genesis was based on Babylonian legend. Presumably, the

Jews picked up the legends during their Babylonian captivity in the time of Nebuchadrezzar, a century after the time of Ashurbanipal. In 1877, a French expedition to Iraq uncovered the remains of the culture preceding the Babylonian – that of the aforementioned Sumerians. This carried the history of the region back to earliest Egyptian times. And in 1921, remains of a totally unexpected civilization were discovered along the Indus Valley in what is now Pakistan. It had flourished between 2500 and 2000 B.C.

Yet Egypt and Mesopotamia were not quite in the same league with Greece when it came to dramatic finds on the origins of modern Western culture. Perhaps the most exciting moment in the history of archaeology came in 1873 when a German ex-grocer's boy found the most famous of all legendary cities.

Heinrich Schliemann as a boy developed a mania for Homer. Although most historians regarded the *Iliad* as mythology, Schliemann lived and dreamed of the Trojan War. He decided that he must find Troy, and by nearly superhuman exertions he raised himself from grocer's boy to millionaire so that he could finance the quest. In 1868, at the age of forty-six, he set forth. He persuaded the Turkish government to give him permission to dig in Asia Minor, and, following only the meagre geographical clues afforded by Homer's accounts, he finally settled upon a mound near the village of Hissarlik. He browbeat the local population into helping him dig into the mound. Excavating in a completely amateurish, destructive and unscientific manner, he began to uncover a series of buried ancient cities, each built on the ruins of the other. And then, at last, success: he uncovered Troy – or at least a city he proclaimed as Troy. Actually, the particular ruins he named Troy are now known to be far older than Homer's Troy, but Schliemann had proved that Homer's tales were not mere legends.

Inexpressibly excited by his triumph, Schliemann went on to mainland Greece and began to dig at the site of Mycenae, a ruined village which Homer had described as the once powerful city of Agamemnon, leader of the Greeks in the Trojan War. Again he uncovered an astounding find – the ruins of a city with gigantic walls which we now know to date back to 1500 B.C.

Schliemann's successes prompted the British archaeologist Arthur John Evans to start digging on the island of Crete, described in Greek legends as the site of a powerful early civilization under a King Minos. Evans, exploring the island in the 1890s, laid bare a brilliant, lavishly ornamented 'Minoan' civilization that stretched back many centuries before the time of Homer's Greece. Here, too, written tablets were found. They were in two different scripts, one of which, called 'Linear B', was finally deciphered in the 1950s and shown to be a form of Greek through a remarkable feat of cryptography and linguistic analysis by a young English architect named Michael Vestris.

As other early civilizations were uncovered – the Hittites and the Mittanni in Asia Minor, the Indus civilization in India, and so on – it became obvious that the history recorded by Greece's Herodotus and the Hebrews' Old Testament represented comparatively advanced stages of human civilization. Man's earliest cities were at least several thousand years old, and his prehistoric existence in less civilized modes of life must stretch many thousands of years farther into the past.

Anthropologists find it convenient to divide cultural history into three major periods: the Stone Age, the Bronze Age, and the Iron Age (a division first suggested by the Roman poet and philosopher, Lucretius, and introduced to modern science by the Danish palaeontologist Christian Jurgenson Thomsen in 1834). Before the Stone Age, there may have been a 'Bone Age', when pointed horns, chisel-like teeth, and club-like thigh bones served men at a time when the working of relatively intractable rock had not yet been perfected.

The Bronze and Iron Ages are, of course, very recent; as soon as we delve into the time before written history, we are back in the Stone Age. What we call civilization (from the Latin word for 'city') began perhaps around 6000 B.C., when man first turned from hunting to agriculture, learned to domesticate animals, invented pottery and new types of tools, and started to develop permanent communities and a settled way of life. Because the archaeological remains from this period of transition are marked by advanced stone tools formed in new ways, it is called the New Stone Age, or the 'Neolithic' period.

This Neolithic Revolution seems to have started in the Near East, at the crossroads of Europe, Asia, and Africa (where later the Bronze and Iron Ages also were to originate). From there, it appears, the revolution slowly spread in widening waves to the rest of the world. It did not reach western Europe and India until 3000 B.C., northern Europe and eastern Asia until 2000 B.C., and central Africa and Japan until perhaps 1000 B.C. or later. Southern Africa and Australia remained in the Old Stone Age until the eighteenth and nineteenth centuries. Most of America also was still in the hunting phase when the Europeans arrived in the sixteenth century, although a well-developed civilization, possibly originated by the Mayas, had developed in Central America and Peru as early as the first centuries of the Christian era.

Evidences of man's pre-Neolithic cultures began to come to light in Europe at the end of the eighteenth century. In 1797 an Englishman named John Frere dug up in Suffolk some crudely fashioned flint tools too primitive to have been made by Neolithic man. They were found thirteen feet underground, which, allowing for the normal rate of sedimentation, testified to great age. In the same stratum with the tools were bones of extinct animals. More and more signs of the great antiquity of tool-making man were discovered, notably by two nineteenth-century French archaeologists, Jacques Boucher de Perthes and Édouard Armand Lartet. Lartet, for instance, found a mammoth tooth on which some early man had scratched an excellent drawing of the mammoth, obviously from living models. The mammoth was a hairy species of elephant that disappeared from the earth well before the beginning of the New Stone Age.

Archaeologists launched upon an active search for early stone tools. They found that these could be assigned to a relatively short Middle Stone Age ('Mesolithic') and a long Old Stone Age ('Palaeolithic'). The Palaeolithic was divided into Lower, Middle, and Upper periods. The earliest objects that could be considered tools ('eoliths', or 'dawn stones') seemed to date back nearly a million years!

What sort of creature had made the Old Stone Age tools? It turned out that Palaeolithic man, at least in his late stages, was far more than a hunting animal. In 1879, a Spanish nobleman, the Marquis de Sautuola, explored some caves that had been discovered a few years earlier – after having been blocked off by rock slides since prehistoric times – at Altamira in northern Spain near the city of Santander. While he dug into the floor of a cave, his five-year-old daughter, who had come along to watch papa dig, suddenly cried: 'Toros! Toros!' ('Bulls! Bulls!'). The father looked up, and there on the walls of the cave were drawings of various animals, in vivid colour and vigorous detail.

Anthropologists found it hard to believe that these sophisticated drawings could have been made by primitive man. But some of the pictured animals were plainly extinct types. The French archaeologist Henri Édouard Prosper Breuil found similar art in caves in southern France. All the evidence finally forced archaeologists to agree with Breuil's firmly expressed views and to conclude that the artists must have lived in the late Palaeolithic, say about 10,000 B.C.

Something was already known about the physical appearance of these Palaeolithic men. In 1868, workmen excavating a roadbed for a railway had uncovered the skeletons of five human beings in the so-called Cro-Magnon caves in south-west France. The skeletons were unquestionably *Homo sapiens*, yet some of them, and similar skeletons soon found elsewhere, seemed to be up to 35,000 or 40,000 years old, according to the geological evidence. They were given the name 'Cro-Magnon man'. Taller than the average modern man and equipped with a large braincase, Cro-Magnon man is pictured by artists as a handsome, stalwart fellow, modern enough, it would certainly appear, to be able to interbreed with present-day human beings.

Mankind, traced thus far back, was not a planet-wide species as it is now. Prior to 20,000 B.C. or so, he was confined to the great 'world island' of Africa, Asia, and Europe. It was only later that hunting bands began to migrate across narrow ocean passages into the Americas, Indonesia, and Australia. It was not until 400 B.C., and later, that daring Polynesian navigators crossed wide stretches of the Pacific, without compasses, and in what were little more than canoes, to colonize the islands of the Pacific.

Finally, it was not until the twentieth century that the foot of man rested on Antarctica.

But if we are to trace the fortunes of prehistoric man at a time when he was confined to only part of the earth's land area, there must be some manner of dating events, at least roughly. A variety of ingenious methods have been used.

Archaeologists have, for instance, used tree rings for the purpose, a technique ('dendrochronology') introduced in 1914 by the American astronomer Andrew Ellicott Douglass. Tree rings are widely separated in wet summers when much new wood is laid down, and closely spaced in dry summers. The pattern over the centuries is quite distinctive. A piece of wood forming part of a primitive abode can have its ring pattern matched with the one place of the scheme where it will fit, and, in this way, its date can be obtained.

A similar system can be applied to layers of sediment or 'varves' laid down summer after summer by melting glaciers in such places as Scandinavia. Warm summers will leave thick layers, cool summers thin ones, and again there is a distinctive pattern. In Sweden, events can be traced back 18,000 years in this way.

An even more startling technique is that developed in 1946 by the American chemist Willard Frank Libby. Libby's work had its origin in the discovery in 1939 by the American physicist Serge Korff that cosmic-ray bombardment of the atmosphere produced neutrons. Nitrogen reacts with these neutrons, producing radioactive carbon 14 in nine reactions out of every ten, and radioactive hydrogen 3 in the tenth reaction.

As a result, the atmosphere would always contain small traces of carbon 14 (and even smaller traces of hydrogen 3). Libby reasoned that radioactive carbon 14 created in the atmosphere by cosmic rays would enter all living tissue via carbon dioxide, first absorbed by plants and then passed on to animals. As long as a plant or animal lived, it would continue to receive radio-carbon and maintain it at a constant level in its tissues. But when the organism died and ceased to take in carbon, the radio-carbon in its tissues would begin to diminish by radioactive breakdown, at a rate determined by its 5,600-year half-life. Therefore, any piece of preserved bone, any bit of charcoal from an ancient camp-fire, or organic remains of any kind could be dated by measuring the amount of radio-carbon left. The method is reasonably accurate for objects up to 30,000 years old, and this covers archaeological history from the ancient civilizations back to the beginnings of Cro-Magnon man. For developing this technique of 'archaeometry', Libby was awarded the Nobel Prize for chemistry in 1960.

Cro-Magnon was not the first early man dug up by the archaeologists. In 1857, in the Neanderthal valley of the German Rhineland, a digger discovered part of a skull and some long bones that looked human in the main but only crudely human. The skull had a sharply sloping forehead and very heavy brow ridges. Some archaeologists maintained that they were the re-mains of a human being whose bones had been deformed by dis-ease, but as the years passed other such skeletons were found, and a detailed and consistent picture of Neanderthal man was de-veloped. Neanderthal was a short, squat, stooping biped, the men averaging a little taller than five feet, the women somewhat shorter. The skull was roomy enough for a brain nearly as large as modern man's. Anthropological artists picture the creature as barrel-chested, hairy, beetle-browed, chinless, and brutish in ex-pression, a picture originated by the French palaeontologist Marcellin Boule, who was the first to describe a nearly complete Neanderthal skeleton in 1911. Actually, he was probably not as sub-human as pictured. Modern examination of the skeleton described by Boule shows it to have belonged to a badly arthritic creature. A normal skeleton gives rise to a far more human image. In fact, give a Neanderthal man a shave and a haircut, dress him in well-fitted clothes, and he could probably walk down New York's Fifth Avenue without getting much notice.

Traces of Neanderthal man were eventually found not only in Europe but also in northern Africa, in Russia and Siberia, in Palestine, and in Iraq. About a hundred different skeletons have now been located at some forty different sites, and men of this sort may still have been alive as recently as 30,000 years ago.

Skeletal remains somewhat resembling Neanderthal man were discovered in still more widely separated places; these were Rhodesian man, dug up in northern Rhodesia in southern Africa in 1921, and Solo man, found on the banks of the Solo River in Java in 1931. They were considered separate species of the genus *Homo*, and so the three types were named *Homo neanderthalensis*, *Homo rhodesiensis*, and *Homo solensis*. But some anthropologists and evolutionists maintain that all three should be placed in the same species as *Homo sapiens*, as 'varieties' or 'sub-species' of man. There were men that we call *sapiens* living at the same time as Neanderthal, and intermediate forms have been found which suggest that there may have been interbreeding between them. If Neanderthal and his cousins can be classed as *sapiens*, then our species is perhaps 200,000 years old. Another apparent relative of Neanderthal, 'Heidelberg man' (of which only a jawbone was discovered in 1907 and nothing additional since then), is much older, and if we include him in our species, the history of *Homo sapiens* can be pushed even farther back. Indeed, at Swanscombe in England archaeologists have found skull fragments which seem to be definitely *sapiens* and even older than Neanderthal. One *Homo sapiens* remnant, discovered just west of Budapest in 1966, may be 500,000 years old.

Darwin's *Origin of Species* launched a great hunt for man's distinctly sub-human ancestors – what the popular press came to call the 'missing link' between man and his presumably ape-like forerunners. This hunt, in the very nature of things, could not be an easy one. Primates are quite intelligent, and few allow themselves to be trapped in situations that lead to fossilization. It has been estimated that the chance of finding a primate skeleton by random search is only one in a thousand billion.

In the 1880s, a Dutch palaeontologist, Marie Eugène François Thomas Dubois, got it into his head that the ancestors of man might be found in the East Indies (modern Indonesia), where great apes still flourished (and where he could work conveniently because those islands then belonged to the Netherlands). Surprisingly enough, Dubois, working in Java, the most populous of the Indonesian islands, did turn up a creature somewhere between an ape and a man! After three years of hunting, he found the top of a skull which was larger than an ape's but smaller than any recognized as human. The next year he found a similarly intermediate thighbone. Dubois named his 'Java man' *Pithecanthropus erectus* ('erect apeman'). Half a century later, in the 1930s, another Dutchman, Gustav H. R. von Koenigswald, discovered more bones of *Pithecanthropus*, and they composed a clear picture of a small-brained, very beetling-browed creature with a distant resemblance to Neanderthal.

Meanwhile other diggers had found, in a cave near Peking, skull, jaws, and teeth of a primitive man they called 'Peking man'. Once this discovery was made, it came to be realized that such teeth had been located earlier – in a Peking chemist's shop, where they were kept for medicinal purposes. The first intact skull was located in December 1929, and Peking man was eventually recognized as markedly similar to 'Java man'. It lived perhaps half a million years ago, used fire, and had tools of bone and stone. Eventually, fragments from forty-five individuals were accumulated, but they disappeared in 1941 during an attempted evacuation of the fossils in the face of the advancing Japanese.

Peking man was named *Sinanthropus pekinensis* ('China man of Peking'), but closer examination of more and more of these comparatively small-brained 'hominids' ('man-like' creatures) made it seem that it was poor practice to place Peking man and Java man in separate genera. The German-American biologist Ernst Walter Mayr felt it wrong to place them in a separate genus from modern man, so that Peking man and Java man are now considered two varieties of the species *Homo erectus*.

It is unlikely that mankind originated in Java, despite the existence there of a small-brained hominid. For a while the vast continent of Asia, early inhabited by 'Peking man', was suspected of being the birthplace of man, but as the twentieth century progressed, attention focused more and more firmly on Africa, which, after all, is the continent richest in primate life generally, and of the higher primates particularly.

The first significant African finds were made by two English scientists, Raymond Dart and Robert Broom. One spring day in

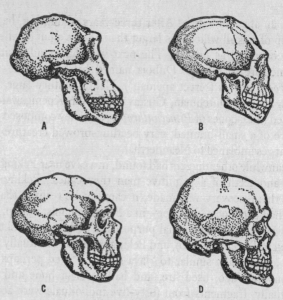

Reconstructed skulls of (A) Zinjanthropus, (B) Pithecanthropus, (C) Neanderthal, and (D) Cro-Magnon.

1924, workers blasting in a limestone quarry near Taungs in South Africa picked up a small skull that looked nearly human. They sent it to Dart, an anatomist working in Johannesburg. Dart immediately identified it as a being between an ape and a man, and he called it *Australopithecus africanus* ('southern ape of Africa'). When his paper announcing the find was published in London, anthropologists thought he had blundered, mistaking a chimpanzee for an apeman. But Broom, an ardent fossil-hunter who had long been convinced that man originated in Africa, rushed to Johannesburg and proclaimed *Australopithecus* the closest thing to a missing link that had yet been discovered.

Through the following decades Dart, Broom, and several anthropologists searched for and found many more bones and teeth of the South African apeman, as well as clubs that he used to kill game, the bones of animals that he killed, and caves in which he lived. *Australopithecus* was a short, small-brained creature with a snout-like face, in many ways less human than Java man. But he had more human brows and more human teeth than *Pithecanthropus*. He walked erect, used tools, and probably had a primitive form of speech. In short, he was an African variety of hominid living at least half a million years ago and definitely more primitive than *Homo erectus*.

There were no clear grounds for suspecting priority between the African and Asian varieties of *Homo erectus* at first, but the balance swung definitely and massively towards Africa with the work of the Kenya-born Englishman Louis Seymour Bazett Leakey and his wife Mary. With patience and persistence, the Leakeys combed likely areas in eastern Africa for early fossil hominids. The most promising was Olduvai Gorge, in what is now Tanzania, and there, on 17 July 1959, Mary Leakey crowned a more than quarter-century search by discovering fragments of a skull that, when pieced together, proved to encase the smallest brain of any hominid yet discovered. Other features showed this hominid, however, to be closer to man than to ape, for he walked upright and the remains were surrounded by small tools formed out of pebbles. The Leakeys named their find *Zinjanthropus* ('East African man', using the Arabic word for East Africa).

Zinjanthropus does not seem to be in the direct line of ancestry of modern man. Still older fossils, some 2 million years old, may qualify. These, given the name of *Homo habilis* ('nimble man'), were $4\frac{1}{2}$-foot-tall creatures who already had hands with opposable thumbs which were nimble enough (hence the name) to make them utterly man-like in this respect.

Prior to *Homo habilis*, we approach fossils that are too primitive to be called hominids and begin to come nearer to the common ancestor of man and the other apes. There is *Ramapithecus*, of which an upper jaw was located in northern India in the early 1930s by G. Edward Lewis. The upper jaw was distinctly closer to the human than is that of any living primate other than man himself; it was perhaps 3 million years old. In 1962, Leakey discovered an allied species which isotope studies showed to be 14 million years old.

In 1948, Leakey had discovered a still older fossil (perhaps 25 million years old), which was named 'Proconsul'. (This name, meaning 'before Consul', honoured Consul, a chimpanzee in the London Zoo.) Proconsul seems to be the common ancestor of the larger great apes, the gorilla, chimpanzee, and orangutan. Farther back, then, there must be a common ancestor of Proconsul and *Ramapithecus* (and of the primitive ape that was ancestral to the smallest modern ape, the gibbon). Such a creature, the first of all the ape-like creatures, would date back perhaps 40 million years.

For many years anthropologists were greatly puzzled by a fossil that did look like a missing link, but of a curious and incredible kind. In 1911, near a place called Piltdown Common in Sussex, England, workmen building a road found an ancient, broken skull in a gravel bed. The skull came to the attention of a lawyer named Charles Dawson, and he took it to a palaeontologist, Arthur Smith Woodward, at the British Museum. The skull was high-browed, with only slight brow ridges; it looked more modern than Neanderthal. Dawson and Woodward went searching in the gravel pit for other parts of the skeleton. One day Dawson, in Woodward's presence, came across a jawbone in about the place where the skull fragments had been found. It had the same reddish-brown hue as the other fragments, and therefore appeared to have come from the same head. But the jawbone, in contrast to the human upper skull, was like that of an ape! Equally strange, the teeth in the jaw, though ape-like, were ground down as human teeth are by chewing.

Woodward decided that this half-ape, half-man might be an early creature with a well-developed brain and a backward jaw. He presented the find to the world as the 'Piltdown man', or *Eoanthropus dawsoni* ('Dawson's dawn man').

Piltdown man became more and more of an anomaly as anthropologists found that in all other fossil finds that included the jaw, jawbone development did keep pace with skull development. Finally, in the early 1950s three British scientists, Kenneth Oakley, Wilfrid Le Gros Clark, and Joseph Sidney Weiner, decided to investigate the possibility of fraud. It *was* a fraud. The jawbone, that of a modern ape, had been planted.

Another odd story of primate relics had a happier ending. In 1935, von Koenigswald had come across a huge but man-like fossil-tooth for sale in a Hong Kong pharmacy. The Chinese pharmacist considered it a 'dragon tooth' of valuable medicinal properties. Von Koenigswald ransacked other Chinese pharmacies and had four such molars before the Second World War temporarily ended his activities.

The man-like nature of the teeth made it seem that gigantic human beings, possibly nine feet high, once roamed the earth. There was a tendency to accept this, perhaps, because the Bible says (Genesis 6:4), 'There were giants on the earth in those days.'

Between 1956 and 1968, however, four jawbones were discovered into which such teeth would fit. The creature 'Gigantopithecus' was seen to be the largest primate ever known to exist, but was distinctly an ape and not a hominid, for all its human-appearing teeth. Very likely it was a gorilla-like creature, standing nine feet tall when upright and weighing six hundred pounds. It may have existed contemporaneously with *Homo erectus* and possessed the same feeding habits (hence the similarity in teeth). It has, of course, been extinct for at least a million years and could not possibly have been responsible for that Biblical verse.

It is important to emphasize that the net result of human evolution has been the production today of a single species. That is, while there may have been a number of species of hominids, one only has survived. All men today, regardless of differences in appearances, are *Homo sapiens*, and the difference between blacks and whites is approximately that between horses of different colouring.

Still, ever since the dawn of civilization, men have been more or less acutely conscious of racial differences, and usually they have viewed other races with the emotions generally evoked by strangers, ranging from curiosity to contempt to hatred. But seldom has racism had such tragic and long-persisting results as the modern conflict between white men and Negroes. (White men are often referred to as 'Caucasians', a term first used, in 1775, by the German anthropologist Johann Friedrich Blumenbach, who was under the mistaken impression that the Caucasus contained the most perfect representatives of the group. Blumenbach also classified Negroes as 'Ethiopians' and east Asians as 'Mongolians', terms that are still sometimes used.)

The racist conflict between white and Negro, between Caucasian and Ethiopian, so to speak, entered its worst phase in the fifteenth century, when Portuguese expeditions down the west coast of Africa began a profitable business of carrying off Negroes into slavery. As the trade grew and nations built their economies on slave labour, rationalizations to justify the Negroes' enslavement were invoked in the name of the Scriptures, of social morality, and even of science.

According to the slave-holders' interpretation of the Bible – an interpretation believed by many to this day – Negroes were descendants of Ham and, as such, an inferior tribe subject to Noah's curse . . . 'a servant of servants shall he be unto his brethren' (Genesis 9:25). Actually, the curse was laid upon Ham's son, Canaan, and on his descendants the 'Canaanites', who were reduced to servitude by the Israelites when the latter conquered the land of Canaan. No doubt the words in Genesis 9:25 represent a comment after the fact, written by the Hebrew writers of the Bible to justify the enslavement of the Canaanites. In any case, the point of the matter is that the reference is to the Canaanites only, and the Canaanites were certainly white men. It was a twisted interpretation of the Bible that the slave-holders used, with telling effect in centuries past, to defend their subjugation of the Negro.

The 'scientific' racists of more recent times took their stand on even shakier ground. They argued that the Negro was inferior to the white man because he obviously represented a lower stage of evolution. Were not his dark skin and wide nose, for instance, reminiscent of the apes? Unfortunately for their case, this line of reasoning actually leads to the opposite conclusion. The Negro is the least hairy of all the groups of mankind; in this respect and in the fact that his hair is crisp and woolly, rather than long and straight, the Negro is farther from the ape than the white man is!

The same can be said of the Negro's thick lips; they resemble those of an ape less than do the white man's thin lips.

The fact of the matter is that to attempt to rank the various groups of *Homo sapiens* on the evolutionary ladder is to try to do fine work with blunt tools. Humanity consists of but one species, and so far the variations that have developed in response to natural selection are quite superficial.

The dark skin of dwellers in the earth's tropical and subtropical regions has obvious value in preventing sunburn. The fair skin of northern Europeans is useful to absorb as much ultraviolet radiation as possible from the comparatively feeble sunlight in order that enough vitamin D be formed from the sterols in the skin. The narrowed eyes of the Eskimo and the Mongol have survival value in lands where the glare from snow or desert sands is intense. The high-bridged nose and narrow nasal passages of the European serve to warm the cold air of the northern winter. And so on.

Since the tendency of *Homo sapiens* has been to make our planet one world, no basic differences in the human constitution have developed in the past, and they are even less likely to develop in the future. Interbreeding is steadily evening out man's inheritance. The American Negro is one of the best cases in point. Despite social barriers against intermarriage, nearly four fifths of the Negroes in the United States, it is estimated, have some white ancestry. By the end of the twentieth century probably there will be no 'pure-blooded' Negroes in North America.

Anthropologists nevertheless are keenly interested in race, primarily as a guide to the migrations of early man. It is not easy to identify specific races. Skin colour, for instance, is a poor guide; the Australian aborigine and the African Negro are both dark in colour but are no more closely related to each other than either is to the European. Nor is the shape of the head – 'dolichocephalic' (long) versus 'brachycephalic' (wide), terms introduced in 1840 by the Swedish anatomist Anders Adolf Retzius – much better despite the classifications of Europeans into subgroups on this basis. The ratio of head length to head width multiplied by a hundred ('cephalic index', or, if skull measurements were substituted, 'cranial index') served to divide Europeans into 'Nordics', 'Alpines', and 'Mediterraneans'. The differences, however, from one group to another are small, and the spread within a group is wide. In addition, the shape of the skull is affected by environmental factors such as vitamin deficiencies, the type of cradle in which the infant slept, and so on.

But the anthropologists have found an excellent marker for race in blood groups. The Boston University biochemist William Clouser Boyd was prominent in this connection. He pointed out that blood groups are inherited in a simple and known fashion, are unaltered by the environment, and show up in distinctly different distributions in the various races.

The American Indian is a particularly good example. Some tribes are almost entirely O; others are O but with a heavy admixture of A; virtually no Indians have B or AB blood. An American Indian testing as a B or AB is almost certain to possess some European ancestry. The Australian aborigines are likewise high in O and A, with B virtually non-existent. But they are distinguished from the American Indian in being high in the more recently discovered blood group M and low in blood group N, while the American Indian is high in N and low in M.

In Europe and Asia, where the population is more mixed, the differences between peoples are smaller, yet still distinct. For instance, in London 70 per cent of the population has O blood, 26 per cent A, and 5 per cent B. In the city of Kharkov, Russia, on the other hand, the corresponding distribution is 60, 25, and 15. In general, the percentage of B increases as one travels eastwards in Europe, reaching a peak of 40 per cent in central Asia.

Now the blood-type genes show the not-yet-entirely-erased marks of past migrations. The infiltration of the B gene into Europe may be a dim mark of the invasion by the Huns in the fifth century and by the Mongols in the thirteenth. Similar blood studies in the Far East seem to indicate a comparatively recent infiltration of the A gene into Japan from the south-west and of the B gene into Australia from the north.

A particularly interesting, and unexpected, echo of early human migrations in Europe showed up in Spain. It came out in a study of Rh blood distribution. (The Rh blood groups are so named from the reaction of the blood to anti-sera developed against the red cells of a rhesus monkey. There are at least eight alleles of the responsible gene; seven are called 'Rh positive', and the eighth, recessive to all the others, is called 'Rh negative' because it shows its effect only when a person has received the allele from both parents.) In the United States about 85 per cent of the population is Rh positive, 15 per cent Rh negative. The same proportion holds in most of the European peoples. But curiously, the Basques of northern Spain stand apart, with something like 60 per cent Rh negative to 40 per cent Rh positive. And the Basques are also notable in having a language unrelated to any other European language.

The conclusion that can be drawn from this is that the Basques are a remnant of a prehistoric invasion of Europe by a Rh-negative people. Presumably a later wave of invasions by Rh-positive tribes penned them up in their mountainous refuge in the western corner of the continent, where they remain the only sizable group of survivors of the 'early Europeans'. The small residue of Rh-negative genes in the rest of Europe and in the American descendants of the European colonizers may represent a legacy from those early Europeans.

The peoples of Asia, the African Negroes, the American Indians, and the Australian aborigines are almost entirely Rh positive.

Man's Future

Attempting to foretell the future of the human race is a risky proposition that had better be left to mystics and science-fiction writers (though, to be sure, I am a science-fiction writer myself, among other things). But of one thing we can be fairly sure. Provided there are no worldwide catastrophes, such as a full-scale nuclear war, or a massive attack from outer space, or a pandemic of a deadly new disease, the human population will increase rapidly. It is now nearly three times as large as it was only a century and a half ago. Some estimates are that the total number of human beings who have lived over a period of 600,000 years comes to 77,000 million. If so, then 4 per cent of all the human beings who have ever lived are living at this moment. And the world population is still growing at a tremendous rate – indeed, at a faster rate than ever before.

Since we have no censuses of ancient populations, we must estimate them roughly on the basis of what we know about the conditions of human life. Ecologists have estimated that the pre-agricultural food supply – obtainable by hunting, fishing, collecting wild fruit and nuts, and so on – could not have supported a world population of more than 20 million, and in all likelihood the actual population during the Palaeolithic era was only a third or half of this at most. This means that as late as 6000 B.C. it could not have numbered more than 6 to 10 million people – roughly the population of a single present city such as Tokyo or New York. (When America was discovered, the food-gathering Indians occupying what is now the United States probably numbered not much more than 250,000, which is like imagining the population of Dayton, Ohio, spread out across the continent.)

The first big jump in world population came with the Neolithic Revolution and agriculture. The British biologist Julian Sorrell Huxley (grandson of the Huxley who was 'Darwin's bulldog') estimates that the population began to increase at a rate which doubled its numbers every 1,700 years or so. By the opening of the Bronze Age, the world population may have been about 25 million; by the beginning of the Iron Age, 70 million; by the start of the Christian era, 150 million, with one third crowded into the Roman Empire, another third into the Chinese Empire, and the rest scattered. By 1600, the earth's population totalled perhaps 500 million, considerably less than the present population of China alone.

At that point the smooth rate of growth ended and the population began to explode. World explorers opened up some 18 million square miles of almost empty land on new continents to colonization by the Europeans. The eighteenth-century Industrial Revolution accelerated the production of food and of people. Even backward China and India shared in the population explosion. The doubling of the world's population now took place not in a period of nearly two millennia but in less than two centuries. The population expanded from 500 million in 1600 to 900 million in 1800. Since then it has grown at an even faster rate. By 1900, it had reached 1,600 million. In the first seventy years of the twentieth century, it had climbed to 3,600 million, despite two world wars.

Currently, the world population is increasing at the rate of 220,000 each day, or 70 million each year. This is an increase at the rate of 2 per cent each year (as compared with an estimated increase of only 0·3 per cent per year in 1650). At this rate, the population of the earth will double in about thirty-five years and in some regions, such as Latin America, the doubling will take place in a shorter time. There is every reason to fear that by the year 2000 the earth's population will be over 6,000 million.

At the moment students of the population explosion are leaning strongly towards the Malthusian view, which has been unpopular ever since it was advanced in 1798. As I said earlier, Thomas Robert Malthus maintained in *An Essay on the Principle of Population* that population always tends to grow faster than the food supply, with the inevitable result of periodic famines and wars. Despite his predictions, the world population has grown apace without any serious setbacks in the past century and a half. But for this postponement of catastrophe we can be grateful, in large measure, that large areas of the earth were still open for the expansion of food production. Now we are running out of tillable new lands. A majority of the world's population is underfed, and mankind must make mighty efforts to wipe out this chronic undernourishment. To be sure, the sea can be more rationally exploited and its food yield multiplied. The use of chemical fertilizers must yet be introduced to wide areas. Proper use of pesticides will reduce the loss of food to insect depredation in areas where such loss has not yet been countered. There are also ways of encouraging growth directly. Plant hormones such as 'gibberellin' (studied by Japanese biochemists before the Second World War and coming to Western attention in the 1950s) could accelerate plant growth, while small quantities of antibiotics added to animal feed will accelerate animal growth (perhaps by suppressing intestinal bacteria that otherwise compete for the food supply passing through the intestines and by suppressing mild but debilitating infections). Nevertheless, with new mouths to feed multiplying as fast as they are, it will take Herculean efforts merely to keep the world's population up to the present none-too-good mark in which some 300 million children under five, the world over, are undernourished to the point of suffering permanent brain damage.

Even so common (and, till recently, disregarded) a resource as fresh water is beginning to feel the pinch. Fresh water is now being used at the rate of nearly 2 billion gallons a day the world over; although total rainfall, which at the moment is the main source of fresh water, is fifty times this quantity, only a fraction of the rainfall is easily recoverable. And in the United States where fresh water is used at a total rate of 350,000 million gallons a day at a larger per capita rate than in the world generally, some 10 per cent of the total rainfall is being consumed one way or another.

The result is that the world's lakes and rivers are being quarrelled over more intensely than ever. (The quarrel of Syria and Israel over the Jordan, and of Arizona and California over the Colorado River, are cases in point.) Wells are being dug ever deeper, and in many parts of the world the ground-water level is sinking dangerously. Attempts to conserve fresh water have included the use of cetyl alcohol as a cover for lakes and reservoirs in such regions as Australia, Israel, and East Africa. Cetyl alcohol spreads out into a film one molecule thick, cutting down on water evaporation without polluting the water. (Of course, water pollution by sewage and by industrial wastes is an added strain on the diminishing fresh-water surplus.)

Eventually, it seems, it will be necessary to obtain fresh water from the oceans, which, for the foreseeable future, offer an unlimited supply. The most promising methods of de-salting sea water include distillation and freezing. In addition, experiments are proceeding with membranes that will selectively permit water molecules to pass, but not the various ions. Such is the importance of this problem that the Soviet Union and the United States are discussing a joint attack on it, at a time when cooperation between these two competing nations is, in other respects, exceedingly difficult to arrange.

But let us be as optimistic as we can and admit no reasonable limits to human ingenuity. Let us suppose that by miracles of technology we raise the productivity of the earth tenfold; suppose that we mine the metals of the ocean, bring up gushers of oil in the Sahara, find coal in Antarctica, harness the energy of sunlight, develop fusion power. Then what? If the rate of increase of the human population continues unchecked at its present rate, all our science and technical invention will still leave us struggling uphill like Sisyphus.

If you are not certain whether to accept this pessimistic appraisal, let us consider the powers of a geometric progression. It has been estimated that the total quantity of living matter on earth is now equal to 2×10^{19} grams. If so, the total mass of humanity is about $1/100,000$ of the mass of all life.

If the earth's population continues to double every thirty-five years (as it is now doing) then by A.D. 2570 it will have increased 100,000-fold. It may prove extremely difficult to increase the mass of life as a whole which the earth can support (though one species can always multiply at the expense of others). In that case, by A.D. 2570 the mass of humanity would comprise all of life and we would be reduced to cannibalism if some were to survive.

Even if we could imagine artificial production of foodstuffs out of the inorganic world via yeast culture, hydroponics (the growth of plants in solutions of chemicals), and so on, no conceivable advance could match the inexorable number increase involved in this doubling-every-thirty-five years. At that rate, by A.D. 2600, it would reach 630 billion! Our planet would have standing room only, for there would be only two and a half square feet per person on the entire land surface, including Greenland and Antarctica. In fact, if the human species could be imagined as continuing to multiply further at the same rate, by A.D. 3550 the total mass of human tissue would be equal to the mass of the earth.

If there are those who see a way out in emigration to other planets, they may find food for thought in the fact that, assuming there were a billion other inhabitable planets in the universe and people could be transported to any of them at will, at the present rate of increase of human numbers every one of those planets would be crowded literally to standing room only by A.D. 5000. By A.D. 7000 the mass of humanity would be equal to the mass of the known universe!

Obviously, the human race cannot increase at the present rate for very long, regardless of what is done with respect to the supply of food, water, minerals, and energy. I do not say 'will not' or 'dare not' or 'should not'; I say quite flatly 'cannot'.

Indeed, it is not mere numbers that will set limit to our growth if it continues at our present rate. It is not only that there are more men, women, and children each minute, but that each individual uses (on the average) more of earth's unrenewable resources, expends more energy, and produces more waste and pollution each minute. Where population doubles every thirty-five years at the moment, energy-utilization is increasing at such a rate that in thirty-five years it will have increased not twice but sevenfold.

The blind urge to waste and poison faster and faster each year is driving us to destruction even more rapidly, then, than mere multiplication alone. For instance, smoke from burning coal and oil is freely dumped into the air by home and factory, as is the gaseous chemical refuse from industrial plants. Cars by the hundreds of millions discharge fumes of petrol and of its breakdown and oxidation products, to say nothing of carbon monoxide and lead compounds. Oxides of sulphur and nitrogen (produced either directly or through later oxidation by ultra-violet light from the sun), together with other substances, can corrode metals, weather construction materials, embrittle rubber, damage crops, cause and exacerbate respiratory diseases, and even serve as one of the causes of lung cancer.

When atmospheric conditions are such that the air over a city remains stagnant for a period of time, the pollutants collect, seriously contaminating the air and encouraging the formation of a smoky fog ('smog') that was first publicized in Los Angeles but had long existed in many cities and now exists in more. At its worst, it can take thousands of lives among those who, out of age or illness, cannot tolerate the added stress placed on their lungs. Such disasters took place in Donora, Pennsylvania, in 1948, and in London in 1952.

The fresh waters of the earth are polluted by chemical wastes, and occasionally one of them will come to dramatic notice. Thus, in 1970, it was found that mercury compounds heedlessly dumped into the world's waters were finding their way into sea organisms in sometimes dangerous quantities. At this rate, far from finding the ocean a richer source of food, we may make a good beginning at poisoning it altogether.

Indiscriminate use of long-lingering pesticides results in their incorporation first into plants, then into animals. Because of the poisoning, some birds find it increasingly difficult to form normal eggshells, so that in attacking insects we are bringing perilously close to extinction the peregrine falcon.

Almost every new so-called technological advance, hastened into without due caution by the eagerness to overreach one's competitors and multiply one's profits, can bring about difficulties. Since the Second World War, synthetic detergents have replaced soaps. Important ingredients of those detergents are various phosphates, which washed into the water supply and greatly accelerated the growth of micro-organisms that, however, used up the oxygen supply of the waters – thus leading to the death of other sea-organisms. These deleterious changes in water habitats ('eutrophication') are rapidly aging the Great Lakes, for instance – the shallow Lake Erie in particular – and are shortening their natural lives by millions of years. Thus Lake Erie may become Swamp Erie, while the swampy Everglades may dry up altogether.

Living species are utterly interdependent. There are obvious cases like the interconnection of plants and bees, where the plants are pollinated by the bees and the bees are fed by the plants, and a million other cases less obvious. Every time life is made easier or more difficult for one particular species, dozens of others are affected – sometimes in hard-to-predict ways. The study of this interconnectability of life, ecology, is only now attracting attention, for in many cases mankind, in an effort to achieve some short-term benefit for himself, has so altered the ecological structure as to bring about some long-term difficulty. Clearly man must learn to look far more carefully before he leaps.

Even so apparently other-worldly an affair as rocketry must be carefully considered. A single large rocket may inject over a hundred tons of exhaust gases into the atmosphere at levels above sixty miles. Such quantities of material could appreciably change the properties of the thin upper atmosphere and lead to hard-to-predict climatic changes. There is the prospect of mass use of gigantic supersonic transport planes (SSTs) to travel through the stratosphere at higher-than-sound velocities. Those who object to their use cite not only the noise factor involved in sonic booms but also the chance of climate-affecting pollution.

Another factor that makes the increase in numbers even worse is the uneven distribution of mankind over the face of the earth. Everywhere there is a trend towards accumulation within metropolitan areas. In the United States, even while the population goes up and up, certain farming states not only do not share in the explosion but are actually decreasing in population. It is estimated that the urban population of the earth is doubling not every thirty-five years but every eleven years. By A.D. 2005, when the earth's total population will have doubled, the metropolitan population will, at this rate, have increased over ninefold.

This is serious. We are already witnessing a breakdown in the social structure – a breakdown that is most strongly concentrated in just those advanced nations where urbanization is most apparent. Within those nations, it is most concentrated in the cities, especially in their most crowded portions. There is no question but that when living beings are crowded beyond a certain point, many forms of pathological behaviour became manifest. This has been found to be true in laboratory experiments on rats, and the newspaper and our own experience should convince us that this is also true for human beings.

It would seem obvious, then, that *if present trends continue unchanged*, the world's social and technological structure will have broken down well within the next half-century, with incalculable consequences. Mankind, in sheer madness, may even resort to the ultimate catastrophe of thermonuclear warfare.

But *will* present trends continue?

Clearly, changing them will require a massive effort and will mean that man must change long-cherished beliefs. For most of man's history, he has lived in a world in which life was brief and many children died while still infants. If the tribal population were not to die out, women had to bear as many babies as they could. For this reason, motherhood was deified and every trend that might lower the birthrate was stamped out. The status of women was lowered so that they might be nothing but baby-making and baby-rearing machines. Sexual mores were so controlled that only those actions were approved of that led to conception; everything else was considered perverted and sinful.

But now we live in a crowded world. If we are to avoid catastrophe, motherhood must become a privilege sparingly doled out. Our views on sex and on its connection with childbirth must be changed.

Again, the problems of the world – the really serious problems – are global in nature. The dangers posed by over-population, over-pollution, the disappearance of resources, the risk of nuclear war, affect every nation, and there can be no real solutions unless all nations cooperate. What this means is that a nation can no longer go its own way, heedless of the others; nations can no longer act on the assumption that there is such a thing as a 'national security' whereby something good can happen to them if something bad happens to someone else. In short, an effective world government is necessary.

Mankind is moving in this direction (very much against its will in many cases) but the question is whether the movement will be quick enough.

I do not wish to make it appear as though there is no hope and as though mankind is in a blind alley from which there is no escape at all. While things look dark and difficult, there may yet be ways out.

One possible source of optimism is an impending revolution in communications. The proliferation of communications satellites may make it possible in the near future for every person to be within reach of every other person. Underdeveloped nations can leapfrog over the earliest communications networks' necessity of involving large capital investments and move directly into a world in which every man has his own television station, so to speak, for receipt and emission of messages.

The world will become so much smaller as to resemble in social structure a kind of neighbourhood village. (Indeed, the phrase 'global village' has come into use to describe the new situation.) Education can penetrate every corner of the global village with the ubiquity of television. The new generation of every underdeveloped nation may grow up learning about modern agricultural methods, about the proper use of fertilizers and pesticides, and about the techniques of birth control.

There may even be, for the first time in earth's history, a tendency towards decentralization. With ubiquitous television making all parts of the world equally accessible to business conferences and libraries and cultural programmes, there will be less need to conglomerate everything into a large, decaying mass.

Who knows, then? Catastrophe seems to have the edge, but the race for salvation is perhaps not quite over.

Assuming that the race for salvation is won; that the population levels off and a slow and humane decrease begins to take place; that an effective and sensible world government is instituted, allowing local diversity but not local murder; that the ecological structure is cared for and the earth systematically preserved – what then?

For one thing, man will probably continue to extend his range. Beginning as a primitive hominid in east Africa – at first perhaps no more widespread or successful than the modern gorilla – he slowly moved outwards until by 15,000 years ago he had colonized the entire 'world island' (Asia, Africa and Europe). He then made the leap into the Americas, Australia, and even through the Pacific islands. By the twentieth century, the population remained thin in particularly undesirable areas – such as the Sahara, the Arabian Desert, and Greenland – but no sizable area was utterly un-inhabited by man except for Antarctica. Now scientific stations, at least, are permanently established even on that least habitable of continents.

Where next?

One possible answer is the sea. It was in the sea that life originated and where it still flourishes best in terms of sheer quantity. Every kind of land animal, except for the insects, has tried the experiment of returning to the sea for the sake of its relatively unfailing food supply and for the relative equability of the environment. Among mammals, such examples as the otter, the seal, or the whale, indicate progressive stages of re-adaptation to a watery environment.

Can man return to the sea, not by the excessively slow altera-tion of his body through evolutionary change, but by the rapid help of technological advance? Encased in the metal walls of sub-marines and bathyscaphes, he has penetrated the ocean to its very deepest floor.

For bare submergence, much less is required. In 1943, the French oceanographer Jacques-Ives Cousteau invented the aqua-lung. This device brings oxygen to a man's lungs from a cylinder of compressed air worn on his back and makes possible the modern sport of scuba diving ('scuba' is an acronym for 'self-contained underwater-breathing apparatus'). This makes it possible for a man to stay under water for considerable periods in his skin, so to speak, without being encased in ships or even in enclosed suits.

Cousteau also pioneered in the construction of under-water living quarters in which men could remain submerged for even longer periods. In 1964, for instance, two men lived two days in an air-filled tent 432 feet below sea level. (One was Jon Lindbergh, son of the aviator.) At shallower depths, men have remained under water for many weeks.

Even more dramatic is the fact that beginning in 1961, the biologist Johannes A. Kylstra, at the University of Leyden, began to experiment with actual water-breathing in mammals. The lung and the gill act similarly, after all, except that the gill is adapted to work on lower levels of oxygenation. Kylstra made use of a water solution sufficiently like mammalian blood to avoid damaging lung tissue and then oxygenated it heavily. He found that both mice and dogs could breathe such liquid for extended periods without apparent ill effect.

Hamsters have been kept alive under ordinary water when they were enclosed in a sheet of thin silicone rubber through which oxygen could pass from water to hamster and carbon dioxide from hamster to water. The membrane was virtually an artificial gill. With such advances and still others to be expected, can man look forward to a future in which he can remain under water for indefinite periods and make all the planet's surface – land and sea – his home?

And what of outer space? Need man remain on his home planet, or can he venture to other worlds?

Once the first satellites were launched into orbit in 1957, the thought naturally arose that the dream of space travel, till then celebrated only in science-fiction stories, might become an actuality. It took only three and a half years after the launching of Sputnik I for the first step to be taken.

On 12 April 1961, the Soviet cosmonaut Yuri Alexeyevich Gagarin was launched into orbit and returned safely. Three months later, on 6 August, another Soviet cosmonaut, Gherman Stepanovich Titov, flew seventeen orbits before landing, spending 24 hours in free flight. On 20 February 1962, the United States put its first man in orbit when the astronaut John Herschel Glenn circled the earth three times. Since then dozens of men have left the earth and, in some cases, remained in space for weeks. Included is one woman – the Soviet cosmonaut Valentina V. Tereshkova – who was launched on 16 June 1963 and remained in free flight for 71 hours, making seventeen orbits altogether.

Rockets have left the earth carrying two and three men at a time. The first such launching was that of the Soviet cosmonauts Vladimir M. Komarov, Konstantin P. Feokstistov, and Boris G.

Yegorov, on 12 October 1964. The Americans launched Virgil I. Grissom and John W. Young in the first multi-manned U.S. rocket on 23 March 1965.

The first man to leave his rocket ship in space was the Soviet cosmonaut Aleksei A. Leonov, who did so on 18 March 1965. This 'space walk' was duplicated by the American astronaut Edward H. White on 3 June 1965.

Although most of the space 'firsts' through 1965 had been made by the Soviets, the Americans thereafter went into the lead. Manned vehicles manoeuvred in space, rendezvoused with each other, docked, and began to move farther and farther out.

The space programme, however, did not continue without tragedy. In January 1967, three American astronauts – Grissom, White, and Roger Chaffee – died on the ground in a fire that broke out in their space capsule during routine tests. Then, on 23 April 1967, Komarov died when his parachute fouled during re-entry. He was the first man to die in the course of a space flight.

The American plans to reach the moon by means of three-man vessels (the 'Apollo' programme) were delayed by the tragedy while the space capsules were redesigned for greater safety, but the plans were not abandoned. The first manned Apollo vehicle, Apollo 7, was launched on 11 October 1968, with its three-man crew under the command of Walter M. Schirra. Apollo 8, launched on 21 December 1968 under the command of Frank Borman, approached the moon, circling it at close quarters. Apollo 10, launched on 18 May 1969, also approached the moon, detached the lunar module, and sent it down to within nine miles of the lunar surface.

Finally, on 16 July 1969, Apollo 11 was launched under the command of Neil A. Armstrong. On 20 July, Armstrong was the first human being to stand on the soil of another world.

Since then three other Apollo vehicles have been launched. Three of them, Apollo 12, 14, and 15, completed their missions with outstanding success. Apollo 13 had trouble in space and was forced to return without landing on the moon, but *did* return safely without loss of life.

The Soviet space programme has not as yet included manned

flights to the moon. However, on 12 September 1970, an unmanned vessel was fired to the moon. It soft-landed safely, gathered up specimens of soil and rock, then safely brought these back to earth. Still later, an automatic Soviet vehicle landed on the moon and moved about under remote control for months, sending back data.

And the future?

The space programme has been expensive and has met with growing resistance from scientists who think that too much of it has been public-relations-minded and too little scientific, or who think it obscures other programmes of greater scientific importance. It has also met with growing resistance from the general public, which considers it too expensive, particularly in the light of urgent sociological problems on earth.

Nevertheless, the space programme will probably continue, if only at a reduced pace; and if mankind can figure out how to spend less of its energies and resources on the suicidal folly of war, the programme may even accelerate. There are plans for the establishment of space stations – in effect, large vehicles in more or less permanent orbit about the earth and capable of housing sizable numbers of men and women for extended periods – so that observations and experiments can be conducted that will presumably be of great value. Shuttle vessels, perhaps reusable, will be devised.

It is to be hoped that further trips to the moon will eventually result in the establishment of more or less permanent colonies there that, we may further hope, can exploit lunar resources and become independent of earth's day-to-day help.

But can man penetrate beyond the moon?

In theory, there is no reason why he cannot, but flights to the next nearest world on which he can land, Mars (Venus, though closer, is too hot for manned landing), will require flights not of days, as in the case of the moon, but of months. And for those months he will have to take a livable environment along with him.

Man has already had some experience along these lines in descending into the ocean depths in submarines and vessels such as the bathyscaphe. As on those voyages, he will go into space in a bubble of air enclosed in a strong metal shell, carrying a full supply of the food, water, and other necessities he will require for the journey. But the take-off into space is complicated enormously by the problem of overcoming gravity. In the space ship, a large proportion of the weight and volume must be devoted to the engine and fuel, and the possible 'payload' of crew and supplies will at first be small indeed.

The food supply will have to be extremely compact: there will be no room for any indigestible constituents. The condensed, artificial food might consist of lactose, a bland vegetable oil, an appropriate mixture of amino acids, vitamins, minerals, and a dash of flavouring, the whole enclosed in a tiny carton made of edible carbohydrate. A carton containing 180 grams of solid food would suffice for one meal. Three such cartons would supply 3,000 calories. To this a gram of water per calorie ($2\frac{1}{2}$ to 3 litres per day per person) would have to be added; some of it might be mixed in the food to make it more palatable, increasing the size of the carton. In addition, the ship would have to carry oxygen for breathing in the amount of about one litre (1,150 grams) of oxygen in liquid form per day per person.

Thus the daily requirement for each person would be 540 grams of dry food, 2,700 grams of water, and 1,150 grams of oxygen. Total, 4,390 grams, or roughly $9\frac{1}{2}$ pounds. Imagine a trip to the moon, then, taking one week each way and allowing two days on the moon's surface for exploration. Each man on the ship would require about 150 pounds of food, water, and oxygen. This can probably be managed at present levels of technology.

For an expedition to Mars and back, the requirements are vastly greater. Such an expedition might well take $2\frac{1}{2}$ years, allowing for a wait on Mars for a favourable phase of the planetary orbital positions to start the return trip. On the basis I have just described, such a trip would call for about 5 tons of food, water, and oxygen per man. To transport such a supply in a space ship is, under present technological conditions, unthinkable.

The only reasonable solution for a long trip is to make the space ship self-sufficient, in the same sense that the earth, itself a massive 'ship' travelling through space, is self-sufficient. The food, water, and air taken along to start with would have to be endlessly reused by recycling the wastes.

Such 'closed systems' have already been constructed in theory. The recycling of wastes sounds unpleasant, but this is, after all, the process that maintains life on the earth. Chemical filters on the ship could collect the carbon dioxide and water vapour exhaled by the crew members; urea, salt, and water could be recovered by distillation and other processes from the urine and faeces; the dry faecal residue could be sterilized of bacteria by ultra-violet light; and along with the carbon dioxide and water could then be fed to algae growing in tanks. By photosynthesis the algae would convert the carbon dioxide and nitrogenous compounds of the faeces to organic food, plus oxygen, for the crew. The only thing that would be required from outside the system is energy for the various processes, including photosynthesis, and this could be supplied by the sun.

It has been estimated that as little as 250 pounds of algae per man could take care of the crew's food and oxygen needs for an indefinite period. Adding the weight of the necessary processing equipment, the total weight of supplies per man would be perhaps 350 pounds, certainly no more than 1,000 pounds. Studies have also been made with systems in which hydrogen-using bacteria are employed. These do not require light, merely hydrogen which can be obtained through the electrolysis of water. The efficiency of such systems is much higher, according to the report, than that of photosynthesizing organisms.

Aside from supply problems, there is that of prolonged weightlessness. Men have survived weeks of continuous weightlessness without permanent harm, but there have been enough minor disturbances to make prolonged weightlessness a disturbing factor. Fortunately, there are ways to counteract it. A slow rotation of the space vehicle, for instance, could produce the sensation of weight by virtue of the centrifugal force, acting like the force of gravity.

More serious and less easily countered are the hazards of high acceleration and sudden deceleration, which space travellers will inevitably encounter in taking off and landing on rocket flights.

The normal force of gravity at the earth's surface is called 1 g. Weightlessness is 0 g. An acceleration (or deceleration) that doubles the body's weight is 2 g, a force tripling the weight is 3 g, and so on.

The body's position during acceleration makes a big difference. If you are accelerated head first (or decelerated feet first), the blood rushes away from your head. At a high enough acceleration (say 6 g for 5 seconds), this means 'blackout'. On the other hand, if you are accelerated feet first (called 'negative acceleration', as opposed to the 'positive' head-first acceleration), the blood rushes to your head. This is more dangerous, because the heightened pressure may burst blood vessels in the eyes or brain. The investigators of acceleration call it 'redout'. An acceleration of $2\frac{1}{2}$ g for 10 seconds is enough to damage some of the vessels.

By far the easiest to tolerate is 'transverse' acceleration – i.e., with the force applied at right angles to the long axis of the body, as in a sitting position. Men have withstood transverse accelerations as high as 10 g for more than 2 minutes in a centrifuge without losing consciousness.

For shorter periods the tolerances are much higher. Astounding records in sustaining high g decelerations were made by Colonel John Paul Stapp and other volunteers on the sled track of the Holloman Air Force Base in New Mexico. On his famous ride of 10 December 1954, Stapp took a deceleration of 25 g for about a second. His sled was brought to a full stop from a speed of more than 600 miles per hour in just 1·4 seconds. This, it was estimated, amounted to driving a car into a brick wall at 120 miles per hour! Of course, Stapp was strapped in the sled in a manner to minimize injury. He suffered only bruises, blisters, and painful eye shocks that produced two black eyes.

An astronaut, on take-off, must absorb (for a short while) as much as $6\frac{1}{2}$ g and, at re-entry, up to 11 g.

Devices such as contour couches, harnesses, and perhaps even immersion in water in a water-filled capsule or space suit will give a sufficient margin of safety against high g forces.

Similar studies and experiments are being made on the radiation hazards, the boredom of long isolation, the strange experience of being in soundless space where night never falls, and other eerie conditions that space fliers will have to endure. All in all, those preparing for man's first venture away from his home planet see no insurmountable obstacles ahead.

7 The Mind

The Nervous System

Physically speaking, man is a rather unimpressive specimen, as organisms go. He cannot compete in strength with most other animals his size. He walks awkwardly, compared with, say, the cat; he cannot run with the dog or the deer; in vision, hearing, and the sense of smell he is inferior to a number of other animals. His skeleton is ill-suited to his erect posture: man is probably the only animal that develops 'low back pain' from his normal posture and activities. When we think of the evolutionary perfection of other organisms – the beautiful efficiency of the fish for swimming or the bird for flying, the great fecundity and adaptability of the insects, the perfect simplicity and efficiency of the virus – man seems a clumsy and poorly designed creature indeed. As sheer organism, he could scarcely compete with the creatures occupying any specific environmental niche on earth. He has come to dominate the earth only by grace of one rather important specialization – his brain.

A cell is sensitive to a change in its surroundings ('stimulus') and will react appropriately ('response'). Thus, a protozoon will swim towards a drop of sugar solution deposited in the water near it, or away from a drop of acid. Now this direct, automatic sort of response is fine for a single cell, but it would mean chaos for a collection of cells. Any organism made up of a number of cells must have a system that coordinates their responses. Without such a system, it would be like a city of men completely out of communication with one another and acting at cross-purposes. So even the coelenterates, the most primitive multicelled animals, have the beginnings of a nervous system. We can see in them the first nerve cells ('neurons') – special cells with fibres that extend from the main cell body and put out extremely delicate branches.

The functioning of nerve cells is so subtle and complex that even at this simple level we are already a little beyond our depth when it comes to explaining just what happens. In some way not yet understood, a change in the environment acts upon the nerve cell. It may be a change in the concentration of some substance, or in the temperature, or in the amount of light, or in the movement of the water, or it may be an actual touch by some object. Whatever the stimulus, the impulse jumps a tiny gap ('synapse') to the next nerve cell; and so it is transmitted from cell to cell. (In well-developed nervous systems, a nerve cell may make thousands of synapses with its neighbours.) In the case of a coelenterate, such as a jellyfish, the impulse is communicated throughout the organism. The jellyfish responds by contracting some part or all of its body. If the stimulus is a contact with a food particle, the organism engulfs the particle by contraction of its tentacles.

All this is strictly automatic, of course, but since it helps the jellyfish, we like to read purpose into the organism's behaviour. Indeed, man, as a creature who behaves in a purposeful, motivated way, naturally tends to attribute purpose even to inanimate nature. Scientists call this attitude 'teleological', and they try to avoid such a way of thinking and speaking as much as they can. But in describing the results of evolution it is so convenient to speak in terms of development towards more efficient ends that even among scientists all but the most fanatical purists occasionally lapse into teleology. (Readers of this book have noticed, of course, that I have sinned often.) Let us, however, try to avoid teleology in considering the development of the nervous system and the brain. Nature did not design the brain; it came about as the result of a long series of evolutionary accidents, so to speak,

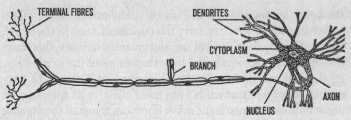

TERMINAL FIBRES — DENDRITES — CYTOPLASM — BRANCH — AXON — NUCLEUS

A nerve cell.

which happened to produce helpful features that at each stage
gave an advantage to organisms possessing them. In the fight for
survival, an animal that was more sensitive to changes in the environment than its competitors, and could respond to them faster,
would be favoured by natural selection. If, for instance, an animal
happened to possess some spot on its body that was exceptionally
sensitive to light, the advantage would be so great that evolution
of eye spots, and eventually of eyes, would follow almost inevitably.

Specialized groups of cells that amount to rudimentary 'sense
organs' begin to appear in the Platyhelminthes, or flatworms.
Furthermore, the flatworms also show the beginnings of a nervous
system that avoids sending nerve impulses indiscriminately
throughout the body, but instead speeds them to the critical
points of response. The development that accomplishes this is a
central nerve cord. The flatworms are the first to develop a
'central nervous system'.

This is not all. The flatworm's sense organs are localized in its
head end, the first part of its body that encounters the environment as it moves along, and so naturally the nerve cord is particularly well developed in the head region. That knob of development is the beginning of a brain.

Gradually the more complex phyla add new features. The sense
organs increase in number and sensitivity. The nerve cord and its
branches grow more elaborate, developing a widespread system of
afferent ('carrying to') nerve cells that bring messages to the cord
and efferent ('carrying away') fibres that transmit messages to the
organs of response. The knot of nerve cells at the crossroads in

the head becomes more and more complicated. Nerve fibres evolve into forms that can carry the impulses faster. In the squid, the most highly developed of the unsegmented animals, this faster transmission is accomplished by a thickening of the nerve fibre. In the segmented animals, the fibre develops a sheath of fatty material ('myelin') which is even more effective in speeding the nerve impulse. In man some nerve fibres can transmit the impulse at 100 metres per second (about 225 miles per hour), compared to only about 1/10 of a mile per hour in some of the invertebrates.

The chordates introduce a radical change in the location of the nerve cord. In them this main nerve trunk (better known as the spinal chord) runs along the back instead of along the belly, as in all lower animals. This may seem a step backwards – putting the cord in a more exposed position. But the vertebrates have the cord well protected within the bony spinal column. The backbone, though its first function was protecting the nerve cord, produced amazing dividends, for it served as a girder upon which chordates could hang bulk and weight. From the backbone they can extend ribs that enclose the chest, jawbones that carry teeth for chewing, and long bones that form limbs.

The chordate brain develops from three structures which are already present in simple form in the most primitive vertebrates. These structures, at first mere swellings of nerve tissue, are the 'fore-brain', 'mid-brain', and 'hind-brain', a division first noted by the Greek anatomist Erasistratus of Chios about 280 B.C. At the head end of the spinal cord, the cord widens smoothly into the hind-brain section known as the 'medulla oblongata'. On the front side of this section in all but the most primitive chordates is a bulge called the 'cerebellum' ('little brain'). Forward of this is the mid-brain. In the lower vertebrates the mid-brain is concerned chiefly with vision and has a pair of 'optic lobes', while the fore-brain is concerned with smell and taste and contains 'olfactory bulbs'. The fore-brain, reading from front to rear, is divided into the olfactory-bulb section, the 'cerebrum', and the 'thalamus', the lower portion of which is the 'hypothalamus'. (Cerebrum is Latin for 'brain'; in man, at least, the cerebrum is the largest and most important part of the organ.) By removing the cerebrum from animals and observing the results, the French anatomist Marie Jean Pierre Flourens was able to demonstrate in 1824 that it was indeed the cerebrum that was responsible for acts of thought and will.

It is the roof of the cerebrum, moreover, the cap called the cerebral cortex, that is the star of the whole show. In fishes and amphibians this is merely a smooth covering (called the 'pallium', or cloak). In reptiles a patch of new nerve tissue, called the 'neopallium' ('new cloak') appears. This is the real forerunner of things to come. It will eventually take over the supervision of vision and other sensations. In the reptiles the clearing house for visual messages has already moved from the mid-brain to the fore-brain in part; in birds this move is completed. With the first mammals, the neopallium begins to take charge. It spreads virtually over the entire surface of the cerebrum. At first it remains a smooth coat, but as it goes on growing in the higher mammals, it becomes so much larger in area than the surface of the cerebrum that it is bent into folds, or 'convolutions'. This folding is

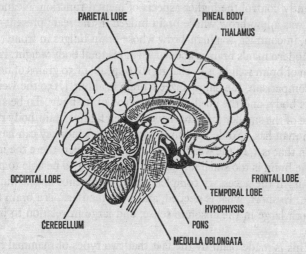

The human brain.

responsible for the complexity and capacity of the brain of a higher mammal, notably that of man.

More and more, as one follows this line of species development, the cerebrum comes to dominate the brain. The mid-brain fades to almost nothing. In the case of the primates, which gain in the sense of sight at the expense of the sense of smell, the olfactory lobes of the fore-brain shrink to mere blobs. By this time the cerebrum has expanded over the thalamus and the cerebellum.

Even the early man-like fossils had considerably larger brains than the most advanced apes. Whereas the brain of the chimpanzee or of the orangutan weighs less than 400 grams (under 14 ounces), and the gorilla, though far larger than a man, has a brain that averages about 540 grams (19 ounces), *Pithecanthropus*'s brain apparently weighed about 850 to 1,000 grams (30 to 35 ounces). And these were the 'small-brained' hominids. Rhodesian man's brain weighed about 1,300 grams (46 ounces); the brain of Neanderthal and of modern *Homo sapiens* comes to about 1,500 grams (53 ounces or 3.3 pounds). Modern man's mental gain over Neanderthal apparently lies in the fact that a larger proportion of his brain is concentrated in the fore-regions, which apparently control the higher aspects of mental function. Neanderthal was a low-brow whose brain bulged in the rear; present-day man, in contrast, is a high-brow whose brain bulges in front.

Modern man's brain is about 1/50 of his total body weight. Each gram of brain weight is in charge, so to speak, of 50 grams of body. In comparison, the chimpanzee's brain is about 1/150 the weight of its body, and the gorilla's about 1/500 of its body. To be sure, some of the smaller primates have an even higher brain/body ratio than man has. (So do the humming birds.) A monkey can have a brain that is 1/18 the weight of its body. However, there the mass of the brain is too small in absolute terms for it to be able to pack into itself the necessary complexity for intelligence on the human scale. In short, what is needed, and what man has, is a brain that is both large in the absolute sense, and large in relation to body size.

This is made plain by the fact that two types of mammal have brains that are distinctly larger than the human brain and yet that do not lend those mammals super-intelligence. The largest elephants can have brains as massive as 6,000 grams (about thirteen pounds) and the largest whales can have brains that reach a mark of 9,000 grams (or nearly nineteen pounds). The size of the bodies those brains must deal with is, however, enormous. The elephant's brain, despite its size, is only 1/1,000 the weight of its body, and the brain of a large whale may be only 1/10,000 the weight of its body.

In only one direction, however, does man have a possible rival. The dolphins and porpoises, small members of the whale family, show possibilities. Some of these are no heavier than man and yet have brains that are larger (with weights up to 1,700 grams, or 60 ounces) and more extensively convoluted.

It is not safe to conclude from this evidence alone that the dolphin is more intelligent than man, because there is the question of the internal organization of the brain. The dolphin's brain (like that of Neanderthal man) may be oriented more in the direction of what we might consider 'lower functions'.

The only safe way to tell is to attempt to gauge the intelligence of the dolphin by actual experiment. Some investigators, notably John C. Lilly, seem convinced that dolphin intelligence is indeed comparable to our own, that dolphins and porpoises have a speech pattern as complicated as ours, and that possibly a form of interspecies communication may yet be established.

Even if this is so, there can be no question but that dolphins, however intelligent, lost their opportunity to translate that intelligence into control of the environment when they re-adapted to sea life. It is impossible to make use of fire under water, and it was the discovery of the use of fire that first marked off mankind from all other organisms. More fundamentally still, rapid locomotion through a medium as viscous as water requires a thoroughly streamlined shape. This has made impossible in the dolphin the development of anything equivalent to the human arm and hand with which the environment could be delicately investigated and manipulated.

In effective intelligence, at least, *Homo sapiens* stands without a peer on earth at present or, so far as we know, in the past.

While considering the difficulty in determining the precise intelligence level of a species such as the dolphin, it might be well to say that no completely satisfactory method exists for measuring the precise intelligence level of individual members of our own species.

In 1904, the French psychologists Alfred Binet and Théodore Simon devised means of testing intelligence by answers given to judiciously chosen questions. Such 'intelligence tests' gave rise to the expression 'intelligence quotient' (or 'IQ'), representing the ratio of the mental age, as measured by the test, to the chronological age; this ratio being multiplied by one hundred to remove decimals. The public was made aware of the significance of IQ chiefly through the work of the American psychologist Lewis Madison Terman.

The trouble is that no test has been devised that is not culturally centred. Simple questions about ploughs might stump an intelligent city boy, and simple questions about escalators might stump an equally intelligent farm boy. Both would puzzle an equally intelligent Australian aborigine, who might nevertheless dispose of questions about boomerangs that would leave us gasping.

Another familiar test is aimed at an aspect of the mind even more subtle and elusive than intelligence. This consists of ink-blot patterns first prepared by a Swiss doctor, Hermann Rorschach, between 1911 and 1921. Subjects are asked to convert these ink blots into images; from the type of image a person builds into such a 'Rorschach test', conclusions concerning his personality are drawn. Even at best, however, such conclusions are not likely to be truly conclusive.

Oddly enough, many of the ancient philosophers almost completely missed the significance of the organ under man's skull. Aristotle considered the brain merely an air-conditioning device, so to speak, designed to cool the overheated blood. In the generation after Aristotle, Herophilus of Chalcedon, working at Alexandria, correctly recognized the brain as the seat of intelligence, but, as usual, Aristotle's errors carried more weight than did the correctness of others.

The ancient and medieval thinkers therefore often tended to place the seat of emotions and personality in organs such as the heart, the liver, and the spleen (*vide* the expressions 'broken-hearted', 'lily-livered', 'vents his spleen').

The first modern investigator of the brain was a seventeenth-century English physician and anatomist named Thomas Willis; he traced the nerves that led to the brain. Later, a French anatomist named Félix Vicq d'Azyr and others roughed out the anatomy of the brain itself. But it was the eighteenth-century Swiss physiologist Albrecht von Haller who made the first crucial discovery about the functioning of the nervous system.

Von Haller found that he could make a muscle contract much more easily by stimulating a nerve than by stimulating the muscle itself. Furthermore, this contraction was involuntary; he could even produce it by stimulating a nerve after the organism had died. Von Haller went on to show that the nerves carried sensations. When he cut the nerves attached to specific tissues, these tissues could no longer react. The physiologist concluded that the brain received sensations by way of nerves and then sent out, again by way of nerves, messages that led to such responses as muscle contraction. He supposed that the nerves all came to a junction at the centre of the brain.

In 1811 the Austrian physician Franz Joseph Gall focused attention on the 'grey matter' on the surface of the cerebrum (which is distinguished from the 'white matter' in that the latter consists merely of the fibres emerging from the nerve-cell bodies, these fibres being white because of their fatty sheaths). Gall suggested that the nerves did not collect at the centre of the brain, as von Haller had thought, but that each ran to some definite portion of the grey matter, which he considered the coordinating region of the brain. Gall reasoned that different parts of the cerebral cortex were in charge of collecting sensations from different parts of the body and sending out the messages for responses to specific parts as well.

If a specific part of the cortex was responsible for a specific property of the mind, what was more natural than to suppose that the degree of development of that part would reflect a person's character or mentality? By feeling for bumps on a person's skull one might find out whether this or that portion of the brain was enlarged and so judge whether he was particularly generous or particularly depraved or particularly something else. With this reasoning, some of Gall's followers founded the pseudo-science of 'phrenology', which had quite a vogue in the nineteenth century and is not exactly dead even today. (Oddly enough, although Gall and his followers emphasized the high forehead and domed head as a sign of intelligence – a view that still influences people today – Gall himself had an unusually small brain, about 15 per cent smaller than the average.)

But the fact that phrenology, as developed by charlatans, is nonsense does not mean that Gall's original notion of the specialization of functions in particular parts of the cerebral cortex was wrong. Even before specific explorations of the brain were attempted, it was noted that damage to a particular portion of the brain might result in a particular disability. In 1861, the French surgeon Pierre Paul Broca, by assiduous post-mortem study of the brain, was able to show that patients with 'aphasia' (the inability to speak, or to understand speech) usually possessed physical damage to a particular area of the left cerebrum, an area called 'Broca's convolution' as a result.

Then, in 1870, two German scientists, Gustav Fritsch and Eduard Hitzig, began to map the supervisory functions of the brain by stimulating various parts of it and observing what muscles responded. A half-century later, this technique was greatly refined by the Swiss physiologist Walter Rudolf Hess, who was awarded a share of the 1949 Nobel Prize for medicine and physiology in consequence.

It was discovered by such methods that a specific band of the cortex was particularly involved in the stimulation of the various voluntary muscles into movement. This band is therefore called the 'motor area'. It seems to bear a generally inverted relationship to the body; the uppermost portions of the motor area, towards the top of the cerebrum, stimulate the lowermost portions of the leg; as one progresses downwards in the motor area, the muscles higher in the leg are stimulated, then the muscles of the torso, then those of the arm and hand, and finally those of the neck and hand.

Behind the motor area is another section of the cortex that receives many types of sensation and is therefore called the 'sensory area'. As in the case of the motor area, the regions of the sensory area in the cerebral cortex are divided into sections that seem to bear an inverse relation to the body. Sensations from the foot are at the top of the area, followed successively as we go downwards with sensations from the leg, hip, trunk, neck, arm, hand, fingers, and, lowest of all, the tongue. The sections of the sensory area devoted to the lips, tongue, and hand are (as one might expect) larger in proportion to the actual size of those organs than are the sections devoted to other parts of the body.

If, to the motor area and the sensory area, are added those sections of the cerebral cortex primarily devoted to receiving the impressions from the major sense organs, the eye and ear, it still leaves a major portion of the cortex without any clearly assigned and obvious function.

It is this apparent lack of assignment that has given rise to the statement, often encountered, that the human being 'uses only one fifth of his brain'. That, of course, is not so; the best we can really say is that one fifth of man's brain has an obvious function. We might as well suppose that a construction firm engaged in building a skyscraper is using only one fifth of its employees because that one fifth was actually engaged in raising steel beams, laying down electric cables, transporting equipment, and such. This would ignore the executives, secretaries, filing clerks, supervisors, and others. Analogously, the major portion of the brain is engaged in what we might call white-collar work, in the assembling of sensory data, in its analysis, in deciding what to ignore, what to act upon, and just how to act upon it. The cerebral cortex has distinct 'association areas' – some for sound sensations, some for visual sensations, some for others.

When all these association areas are taken into account, there still remains one area of the cerebrum that has no specific and easily definable function. This is the area just behind the forehead, which is called the 'prefrontal lobe'. Its lack of obvious function is such that it is sometimes called the 'silent area'. Tumours have made it necessary to remove large areas of the prefrontal lobe without any particular significant effect on the individual; yet surely it is not a useless mass of nerve tissue.

One might even suppose it to be the most important portion of the brain if one considers that in the development of the human nervous system there has been a continual piling-up of complication at the forward end. The prefrontal lobe might therefore be the brain area most recently evolved and most significantly human.

In the 1930s, it seemed to a Portuguese surgeon, Antonio Egas Moniz, that where a mentally ill patient was at the end of his tether, it might be possible to help by taking the drastic step of severing the prefrontal lobes from the rest of the brain. The patient might then be cut off from a portion of the associations he had built up, which were, apparently, affecting him adversely, and make a fresh and better start with the brain he had left. This operation, 'prefrontal lobotomy', was first carried out in 1935; in a number of cases it did indeed seem to help. Moniz shared (with W. R. Hess) the Nobel Prize for medicine and physiology in 1949 for his work. Nevertheless, the operation never achieved popularity and is less popular now than ever. Too often, the cure is literally worse than the disease.

The cerebrum is actually divided into two 'cerebral hemispheres' connected by a tough bridge of white matter, the 'corpus callosum'. In effect, the hemispheres are separate organs, unified in action by the nerve fibres that cross the corpus callosum and act to coordinate the two. Despite this, the hemispheres remain potentially independent.

The situation is somewhat analogous to that of our eyes. Our two eyes act as a unit, ordinarily, but if one eye is lost, the other can meet our needs. Similarly, the removal of one of the cerebral hemispheres does not make an experimental animal brainless. The remaining hemisphere learns to carry on.

Ordinarily, each hemisphere is largely responsible for a particular side of the body: the left cerebral hemisphere for the right side, the right cerebral hemisphere for the left side. If both hemispheres are left in place and the corpus callosum is cut, coordination is lost, and the two body halves come under more or less independent control. A literal case of twin brains, so to speak, is set up.

Monkeys can be so treated (with further operation upon the optic nerve to make sure that each eye is connected to only one hemisphere), and when this is done each eye can be separately trained to do particular tasks. A monkey can be trained to select a cross over a circle to indicate, let us say, the presence of food. If only the left eye is kept uncovered during the training period, only the left eye will be useful in this respect. If the right eye is uncovered and the left eye covered, the monkey will have no right-eye memory of his training. He will have to hunt for his food by trial and error. If the two eyes are trained to contradictory tasks and if both are then uncovered, the monkey alternates activities, as the hemispheres politely take their turns.

Naturally, in any such two-in-charge situation, there is always the danger of conflict and confusion. To avoid that, one cerebral hemisphere (almost always the left one in human beings) is dominant, when both are normally connected. Broca's convolution, which controls speech, is in the left hemisphere, for instance. The 'gnostic area', which is an over-all association area, a kind of court of highest appeal, is also in the left hemisphere. Since the left cerebral hemisphere controls the motor activity of the right-hand side of the body, it is not surprising that most people are right-handed (though even left-handed people usually have a dominant left cerebral hemisphere). Where clear-cut dominance is not established between left and right, there may be ambidexterity, rather than a clear right-handedness or left-handedness, along with some speech difficulties and, perhaps, manual clumsiness.

The cerebrum is not the whole of the brain. There are areas of grey matter embedded below the cerebral cortex. These are called the 'basal ganglia'; included is a section called the 'thalamus'. The thalamus acts as a reception centre for various sensations. The more violent of these, such as pain, extreme heat or cold, or rough touch, are filtered out. The milder sensations from the muscles – the gentle touches, the moderate temperatures – are passed on to the sensory area of the cerebral cortex. It is as though mild sensations can be trusted to the cortex, where they can be considered judiciously and where reaction can come after a more or less prolonged interval of consideration. The rough sensations, however, which must be dealt with quickly and for which there is no time for consideration, are handled more or less automatically in the thalamus.

Underneath the thalamus is the 'hypothalamus', centre for a variety of devices for controlling the body. The body's appestat, mentioned on p. 286 as controlling the body's appetite, is located there; so is the control of the body's temperature. It is through the hypothalamus, moreover, that the brain exerts at least some influence over the pituitary gland (see p. 290); this is an indication of the manner in which the nervous controls of the body and the chemical controls (the hormones) can be unified into a master supervisory force.

In 1954, the physiologist James Olds discovered another and rather frightening function of the hypothalamus. It contains a region that, when stimulated, apparently gives rise to a strongly pleasurable sensation. An electrode affixed to the 'pleasure centre' of a rat, so arranged that it can be stimulated by the animal itself, will be stimulated up to 8,000 times an hour for hours or days at a time, to the exclusion of food, sex, and sleep. Evidently, all the desirable things in life are desirable only in so far as they stimulate the pleasure centre. To stimulate it directly makes all else unnecessary.

The hypothalamus also contains an area that has to do with the wake–sleep cycle, since damage to parts of it induces a sleep-like state in animals. The exact mechanism by which the hypothalamus performs its function is uncertain. One theory is that it sends signals to the cortex, which sends signals back in response, in mutually stimulating fashion. With continuing wakefulness, the coordination of the two fails, the oscillations become ragged, and the individual becomes sleepy. A violent stimulus (a loud noise, a persistent shake of the shoulder, or, for that matter, a sudden interpretation of a steady noise) will arouse one. In the absence of such stimuli, coordination will be restored, eventually, between hypothalamus and cortex, and sleep will end spontaneously; or perhaps sleep will become so shallow that a perfectly ordinary stimulus, of which the surroundings are always full, will suffice to wake one.

During sleep, dreams – sensory data more or less divorced from reality – will take place. Dreaming is apparently a universal phenomenon; people who report dreamless sleep are merely failing to remember their dreams. The American physiologist W. Dement, studying sleeping subjects in 1952, noticed periods of rapid eye movements that sometimes persisted for minutes ('REM sleep'). During this period breathing, heartbeat, and blood pressure rose to waking levels. This takes place about a quarter of the sleeping time. If a sleeper was awakened during these periods, he generally reported having had a dream. Furthermore, if a sleeper was continually disturbed during these periods, he began to suffer psychological distress; the periods of distress were multiplied during succeeding nights as though to make up for the lost dreaming.

It would seem, then, that dreaming has an important function in the working of the brain. It is suggested that dreaming is a device whereby the brain runs over the events of the day to remove the trivial and repetitious that might otherwise clutter it and reduce its efficiency. Sleep is the natural time for such an activity, for the brain is then relieved of many of its waking functions. Failure to accomplish this task (because of interruption) may so clog the brain that clearing attempts must be made during waking periods, producing hallucinations (that is, waking dreams, so to speak) and other unpleasant symptoms. One might naturally wonder if this is not a chief function of sleep, since there is very little physical resting in sleep that cannot be duplicated by quiet wakefulness. REM sleep even occurs in infants, who spend half their sleeping time at it and who would seem to lack anything about which to dream. It may be that REM sleep helps the development of the nervous system. (It has been observed in mammals other than man, too.)

*

Below the cerebrum is the smaller cerebellum (also divided into two 'cerebellar hemispheres') and the 'brain stem', which narrows and leads smoothly into the 'spinal cord' that extends about eighteen inches down the hollow centre of the spinal column.

The spinal cord consists of grey matter (at the centre) and white matter (on the periphery); to it are attached a series of nerves that are largely concerned with the internal organs – the heart, lungs, digestive system, and so on – organs that are more or less under involuntary control.

In general, when the spinal cord is severed, through disease or through injury, that part of the body lying below the severed segment is disconnected, so to speak. It loses sensation and is paralysed. If the cord is severed in the neck region, death follows, because the chest is paralysed, and with it the action of the lungs. It is this that makes a 'broken neck' fatal, and hanging a feasible form of quick execution. It is the severed cord, rather than a broken bone, that is fatal.

The entire structure of the 'central nervous system', consisting of the cerebrum, cerebellum, brain stem, and spinal cord, is carefully coordinated. The white matter of the spinal cord is made up of bundles of nerve fibres that run up and down the cord, unifying the whole. Those that conduct impulses downwards from the brain are the 'descending tracts' and those that conduct them upwards to the brain are the 'ascending tracts'.

In 1964, research specialists at Cleveland's Metropolitan General Hospital reported the isolation from rhesus monkeys of brains which were then kept independently alive for as long as eighteen hours. This offers the possibility of detailed specific study of the brain's metabolism through a comparison of the nutrient medium entering the blood vessels of the isolated brain and of the same medium leaving it.

The next year they were transplanting dogs' heads to the necks of other dogs, hooking them up to the host's blood-supply, and keeping the brains in the transplanted heads alive and working for as long as two days. By 1966, dogs' brains were lowered to temperatures near freezing for six hours and then revived to the point of showing clear indications of normal chemical and electrical activity. Brains are clearly tougher than they might seem to be.

Nerve Action

It is not only the various portions of the central nervous system that are hooked together by nerves, but, clearly, all the body that, in this fashion, is placed under the control of that system. The nerves interlace the muscles, the glands, the skin; they even invade the pulp of the teeth (as we learn to our cost at every toothache).

The nerves themselves were observed in ancient times but the structure and function were consistently misunderstood. Until modern times, they were felt to be hollow and to function as carriers of a subtle fluid. Rather complicated theories developed by Galen involved three different fluids carried by the veins, the arteries and the nerves, respectively. The fluid of the nerves, usually referred to as 'animal spirits', was the most rarefied of the three. When Galvani discovered that muscles and nerves could be stimulated by an electric discharge, this laid the foundation for a series of studies that eventually showed nerve action to be associated with electricity, a subtle fluid, indeed, more subtle than Galen could have imagined.

Specific work on nerve action began in the early nineteenth century with the German physiologist Johannes Peter Müller, who, among other things, showed that sensory nerves always produced their own sensations regardless of the nature of the stimulus. Thus, the optic nerve registered a flash of light, whether it was stimulated by light itself, or by the mechanical pressure of a punch in the eye. (In the latter case, you 'see stars'.) This emphasizes that our contact with the world is not a contact with reality at all, but a contact with specialized stimuli that the brain usually interprets in a useful manner, but can interpret in a non-useful manner.

Study of the nerves was advanced greatly in 1873, when an

Italian physiologist, Camillo Golgi, developed a cellular stain involving silver salts that was well adapted to react with nerve cells, making clear their finest details. He was able to show, in this manner, that nerves were composed of separate and distinct cells, and that the processes of one cell might approach very closely to those of another, but that they did not fuse. There remained the tiny gap of the synapse. In this way, Golgi bore out, observationally, the contentions of a German anatomist, Wilhelm von Waldeyer, to the effect that the entire nervous system consisted of individual nerve cells or neurons (this contention being termed the 'neuron theory').

Golgi did not, however, himself support the neuron theory. This proved to be the task of the Spanish neurologist Santiago Ramon y Cajal, who, by 1889, using an improved version of Golgi's stain, worked out the connections of the cells in the grey matter of the brain and spinal cord and fully established the neuron theory. Golgi and Ramon y Cajal, although disputing the fine points of their findings, shared the Nobel Prize for medicine and physiology in 1906.

These nerves form two systems: the 'sympathetic' and the 'parasympathetic'. (The terms date back to the semi-mystical notions of Galen.) Both systems act on almost every internal organ, exerting control by opposing effects. For instance, the sympathetic nerves act to accelerate the heartbeat, the parasympathetic nerves to slow it; the sympathetic nerves slow up secretion of digestive juices, the parasympathetic stimulate such secretions, and so on. Thus, the spinal cord, together with the sub-cerebral portions of the brain, regulates the workings of the organs in an automatic fashion. This set of involuntary controls was investigated in detail by the British physiologist John Newport Langley in the 1890s, and he named it the 'autonomic nervous system'.

In the 1830s, the English physiologist Marshall Hall had studied another type of behaviour which seemed to have voluntary aspects but proved to be really quite involuntary. When you accidentally touch a hot object with your hand, the hand draws away instantly. If the sensation of heat had to go to the brain, be considered and interpreted there, and evoke the appropriate message to the hand, your hand would be pretty badly scorched by the time it got the message. The unthinking spinal cord disposes of the whole business automatically and much faster. It was Hall who gave the process the name 'reflex'.

The reflex is brought about by two or more nerves working in coordination, to form a 'reflex arc'. The simplest possible reflex arc is one consisting of two neurons, a sensory (bringing sensations to a 'reflex centre' in the central nervous system, usually at some point in the spinal cord) and a motor (carrying instructions for movement from the central nervous system). The two neurons may be connected by one or more 'connector neurons'. A particular study of such reflex arcs and of their function in the body was made by the English neurologist Charles Scott Sherrington, who won a share in the 1932 Nobel Prize for medicine and physiology in consequence. It was Sherrington who, in 1897, coined the word 'synapse'.

Reflexes bring about so rapid and certain a response to a particular stimulus that they offer simple methods for checking the general integrity of the nervous system. A familiar example is the 'patellar reflex' or, as it is commonly called, the knee jerk. When the legs are crossed a sudden blow below the knee of the upper leg will cause it to make a quick, kicking motion – a fact first brought into medical prominence in 1875 by the German

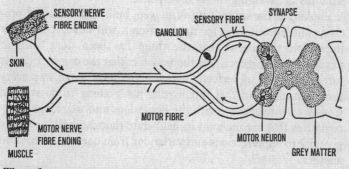

The reflex arc.

neurologist Carl Friedrich Otto Westphal. The patellar reflex is not important in itself, but its non-appearance can mean some serious disorder involving the portion of the nervous system in which that reflex arc is to be found.

Sometimes, damage to a portion of the central nervous system brings about the appearance of an abnormal reflex. If the sole of the foot is scratched the normal reflex brings the toes together and bent downwards. Certain types of damage to the central nervous system will cause the big toe to bend upwards in response to this stimulus, and the little toes to spread apart as they bend down. This is the 'Babinski reflex', named after a French neurologist, Joseph François Félix Babinski, who described it in 1896.

In man, reflexes are decidedly subordinate to the conscious will. The lower phyla of animals, on the other hand, not only are much more strictly controlled by their reflexes but also have them far more highly developed.

One of the best examples is a spider spinning its web. Here the reflexes produce such an elaborate pattern of behaviour that it is difficult to think of it as mere reflex action; instead, it is usually called 'instinctive' behaviour. (Because the word 'instinct' is often misused, biologists prefer the term 'innate' behaviour.) The spider is born with a nerve-wiring system in which the switches have been pre-set, so to speak. A particular stimulus sets it off on weaving a web, and each act in the process in turn acts as a stimulus determining the next response.

Looking at the spider's intricate web, built with beautiful precision and effectiveness for the function it will serve, it is almost impossible to believe that the thing has been done without purposeful intelligence. Yet the very fact that the complex task is carried through so perfectly and in exactly the same way every time is itself proof that intelligence has nothing to do with it. Conscious intelligence, with the hesitations and weighings of alternatives that are inherent in deliberate thought, will inevitably give rise to imperfections and variations from one construction to another.

With increasing intelligence, animals tend more and more to

shed instincts and inborn skills. Thereby they doubtless lose something of value. A spider can build its amazingly complex web perfectly the first time, although it has never seen web-spinning, or even a web, before. Man, on the other hand, is born almost completely unskilled and helpless. A newborn baby can automatically suck on a nipple, wail if it is hungry, and hold on for dear life if about to fall, but it can do very little else. Every parent knows how painfully and with what travail a child comes to learn the simplest forms of suitable behaviour. And yet, a spider or an insect, though born with perfection, cannot deviate from it. The spider builds a beautiful web, but if its preordained web should fail, it cannot learn to build another type of web. A boy, on the other hand, reaps great benefits from being unfettered by inborn perfection. He may learn slowly and attain only imperfection at best, but he can attain a variety of imperfection of his own choosing. What man has lost in convenience and security, he has gained in an almost limitless flexibility.

Recent work, however, emphasizes the fact that there is not always a clear division between instinct and learned behaviour. It would seem, on casual observation, for instance, that chicks or ducklings, fresh out of the shell, follow their mothers out of instinct. Closer observation shows that this is not so.

The instinct, however, is not to follow their mother, but merely to follow something of a characteristic shape or colour or faculty of movement. Whatever object provides this sensation at a certain period of early life is followed by the young creature and is thereafter treated as the mother. This may really be the mother; it almost invariably is, in fact, but it need not be! In other words, following is instinctive, but the 'mother' that is followed is learned. (Much of the credit for this discovery goes to the remarkable Austrian naturalist, Konrad Zacharias Lorenz. Lorenz, during the course of studies now some thirty years old, was followed hither and yon by a gaggle of goslings.)

The establishment of a fixed pattern of behaviour in response to a particular stimulus encountered at a particular time of life is called 'imprinting'. The specific time at which imprinting takes place is a 'critical period'. For chicks the critical period of 'mother-imprinting' lies between 13 and 16 hours after hatching. For a puppy there is a critical period between 3 and 7 weeks, during which the stimulations it is usually likely to encounter imprint various aspects of what we consider normal doggish behaviour.

Imprinting is the most primitive form of learned behaviour, one that is so automatic, takes place in so limited a time, and under so general a set of conditions that it is easily mistaken for instinct.

A logical reason for imprinting is that it allows a certain desirable flexibility. If a chick were born with some instinctive ability of distinguishing its true mother so that it might follow only her, and if the true mother were for any reason absent in the chick's first day of life, the little creature would be helpless. As it is, the question of motherhood is left open for just a few hours and the chick may imprint itself to any hen in the vicinity and thus adopt a foster mother.

As stated earlier, it had been Galvani's experiments just before the opening of the nineteenth century that had first indicated some connection between electricity and the actions of muscle and nerve.

The electrical properties of muscle led to a startling medical application, thanks to the work of the Dutch physiologist Willem Einthoven. In 1903, he developed an extremely delicate galvanometer, one delicate enough to respond to the tiny fluctuations of the electric potential of the beating heart. By 1906, Einthoven was recording the peaks and troughs of this potential (the recording being an 'electrocardiogram') and correlating them with various types of heart disorder.

The more subtle electrical properties of nerve impulses were thought to have been initiated and propagated by chemical changes in the nerve. This was elevated from mere speculation to experimental demonstration by the nineteenth-century German physiologist Emil Du Bois-Reymond, who by means of a delicate galvanometer was able to detect tiny electric currents in stimulated nerves.

With modern electronic instruments, researches into the electrical properties of the nerve have been incredibly refined. By placing tiny electrodes at different spots on a nerve fibre and by detecting electrical changes through an oscilloscope, it is possible to measure a nerve impulse's strength, duration, speed of propagation, and so on. For such work, the American physiologists Joseph Erlanger and Herbert Spencer Gasser were awarded the 1944 Nobel Prize for medicine and physiology.

If you apply small electric pulses of increasing strength to a single nerve cell, up to a certain point there is no response whatever. Then suddenly the cell fires: an impulse is initiated and travels along the fibre. The cell has a threshold; it will not react at all to a stimulus below the threshold, and to any stimulus above the threshold it will respond only with an impulse of a certain fixed intensity. The response, in other words, is 'all or nothing'. And the nature of the impulse elicited by the stimulus seems to be the same in all nerves.

How can such a simple yes–no affair, identical everywhere, lead to the complex sensations of sight, for instance, or to the complex finger responses involved in playing a violin? It seems that a nerve, such as the optic nerve, contains a large number of individual fibres, some of which may be 'firing' and others not, and where the 'firing' may be in rapid succession or slowly, forming a pattern, possibly a complex one, shifting continuously with changes in the over-all stimulus. (For work in this field, the English physiologist Edgar Douglas Adrian shared, with Sherrington, the 1932 Nobel Prize in medicine and physiology.) Such a changing pattern may be continually 'scanned' by the brain and interpreted appropriately. But nothing is known about how the interpretation is made or how the pattern is translated into action such as the contraction of a muscle or secretion by a gland.

The firing of the nerve cell itself apparently depends on the movement of ions across the membrane of the cell. Ordinarily, the inside of the cell has a comparative excess of potassium ions, while outside the cell there is an excess of sodium ions. Somehow the cell holds potassium ions in and keeps sodium ions out so that the concentrations on the two sides of the cell membrane do not equalize. It is now believed that a 'sodium pump' of some kind inside the cell keeps pumping out sodium ions as fast as they come in. In any case, there is an electric potential difference of about one tenth of a volt across the cell membrane, with the inside negatively charged with respect to the outside. When the nerve cell is stimulated, the potential difference across the membrane collapses, and this represents the firing of the cell. It takes a couple of thousandths of a second for the potential difference to be re-established, and during that interval the nerve will not react to another stimulus. This is the 'refractory period'.

Once the cell fires, the nerve impulse travels down the fibre by a series of firings, each successive section of the fibre exciting the next in turn. The impulse can travel only in the forward direction, because the section that has just fired cannot fire again until after a resting pause.

Research that related, in the fashion just described, nerve action and ion permeability led to the award of the 1963 Nobel Prize for medicine and physiology to two British physiologists, Alan Lloyd Hodgkin and Andrew Fielding Huxley, and to an Australian physiologist, John Carew Eccles.

What happens, though, when the impulse travelling along the length of the nerve fibre comes to a synapse – a gap between one nerve cell and the next? Apparently, the nerve impulse also involves the production of a chemical that can drift across the gap and initiate a nerve impulse in the next nerve cell. In this way, the impulse can travel from cell to cell.

One of the chemicals definitely known to affect the nerves is the hormone adrenalin. It acts upon nerves of the sympathetic system, which slows the activity of the digestive system and accelerates the rate of respiration and the heartbeat. When anger or fear excites the adrenal glands to secrete the hormone, its stimulation of the sympathetic nerves sends a faster surge of blood through the body, carrying more oxygen to the tissues, and by slowing down digestion for the duration it saves energy during the emergency.

The American psychologists and police officers John Augustus Larsen and Leonard Keeler took advantage of this finding in 1921 to devise a machine to detect the change in blood pressure, pulse rate, breathing rate, and a perspiration brought on by emotion. This device, the 'polygraph', detected the emotional effort involved in telling a lie, which always carries with it the fear of detection in any reasonably normal individual and therefore brings adrenalin into play. While not infallible, the polygraph has gained great fame as a 'lie detector'.

In the normal course, the nerve endings of the sympathetic nervous system themselves secrete a compound very like adrenalin, called 'noradrenalin'. This chemical serves to carry the nerve impulses across the synapses, transmitting the message by stimulating the nerve endings on the other side of the gap.

In the early 1920s the English physiologist Henry Dale and the German physiologist Otto Loewi (who were to share the Nobel Prize in physiology and medicine in 1930) studied a chemical that performed this function for most of the nerves other than those of the sympathetic system. The chemical is called acetylcholine. It is now believed to be involved not only at the synapses but also in conducting the nerve impulse along the nerve fibre itself. Perhaps acetylcholine acts upon the 'sodium pump'. At any rate, the substance seems to be formed momentarily in the nerve fibre and to be broken down quickly by an enzyme called 'cholinesterase'. Anything that inhibits the action of cholinesterase will interfere with this chemical cycle and will stop the transmission of nerve impulses. The deadly substances now known as 'nerve gases' are cholinesterase inhibitors. By blocking the conduction of nerve impulses they can stop the heartbeat and produce death within minutes. The application to warfare is obvious. They can be used, less immorally, as insecticides.

A less drastic interference with cholinesterase is that of local anaesthetics, which in this way suspend (temporarily) those nerve impulses associated with pain.

Thanks to the electric currents involved in nerve impulses, it is possible to 'read' the brain's activity, in a way, though no one has yet been able to translate fully what the brain waves are saying. In 1929, a German psychiatrist, Hans Berger, reported earlier work in which he applied electrodes to various parts of the head and was able to detect rhythmic waves of electrical activity.

Berger gave the most pronounced rhythm the name of 'alpha wave'. In the alpha wave, the potential varies by about 20 microvolts in a frequency of roughly ten times a second. The alpha wave is clearest and most obvious when the subject is resting with his eyes closed. When the eyes are open, but are viewing featureless illumination, the alpha wave persists. If, however, the ordinary variegated environment is in view, the alpha wave vanishes, or is drowned, by other more prominent rhythms. After a while, if nothing visually new is presented, the alpha wave reappears. Typical names for other types of waves are 'beta waves', 'delta waves', and 'theta waves'.

Electroencephalograms ('electrical wirings of the brain' or, as they are abbreviated, 'EEG') have since been extensively studied, and they show that each individual has his own pattern, varying with excitement and in sleep. Although the electroencephalogram is still far from being a method of 'reading thoughts' or tracing the mechanism of the intellect, it does help in the diagnosis of major upsets of brain function, particularly epilepsy. It can also help locate areas of brain damage or brain tumours.

In the 1960s specially designed computers were called into battle. If a particular small environmental change is applied to a subject, it is to be presumed that there will be some response in the brain that will reflect itself in a small alteration in the EEG pattern at the moment when the change is introduced. The brain will be engaged in many other activities, however, and the small alteration in the EEG will not be noticeable. Notwithstanding, if the process is repeated over and over again, a computer can be programmed to average out the EEG pattern and find the consistent difference.

By 1964, the American psychologist Manfred Clynes reported analyses fine enough to be able to tell, by a study of the EEG pattern alone, what colour a subject was looking at. The English neurophysiologist William Grey Walter similarly reported a brain-signal pattern that seems characteristic of the learning process. It comes when the subject under study has reason to think he is about to be presented with a stimulus that will call for thought or action. Walter calls it the 'expectancy wave' and points out that it is absent in children under three and in certain psychotics. The reverse phenomenon, that of bringing about specific actions through direct electrical stimulation of the brain, was also reported in 1965. José Manuel Rodriguez Delgado of Yale, transmitting electrical stimulation by radio signals, caused animals to walk, climb, yawn, sleep, mate, switch emotions, and so on at command. Most spectacularly, a charging bull was made to stop short and trot peacefully away.

Human Behaviour

Unlike physical phenomena, such as the motions of planets or the behaviour of light, the behaviour of living things has never been reduced to rigorous natural laws and perhaps never will be. There are many who insist that the study of human behaviour cannot become a true science, in the sense of being able to explain or predict behaviour in any given situation on the basis of universal natural laws. Yet life is no exception to the rule of natural law, and it can be argued that living behaviour would be fully explainable if all the factors were known. The catch lies in that last phrase. It is unlikely that all the factors will ever be known; they are too many and too complex. Man need not, however, despair of ever being able to understand himself. There is ample room for better knowledge of his own mental complexities, and even if we never reach the end of the road, we may yet hope to travel along it quite a way.

Not only is the subject particularly complex, but the study of the subject has not been progressing for long. Physics came of age in 1600, and chemistry in 1775, but the much more complex study of 'experimental psychology' dates only from 1879, when the German physiologist Wilhelm Wundt set up the first laboratory devoted to the scientific study of human behaviour. Wundt interested himself primarily in sensation and in the manner in which men perceived the details of the universe about them.

At almost the same time, the study of human behaviour in one particular application – that involving man as an industrial cog – arose. In 1881, the American engineer Frederick Winslow Taylor began measuring the time required to do certain jobs, and to work out methods for so organizing the work as to minimize that time. He was the first 'efficiency expert' and was (like all efficiency experts who are prone to lose sight of values beyond the stop watch) unpopular with the workers.

But as we study human behaviour, step by step, either under controlled conditions in a laboratory or empirically in a factory, it does seem that we are tackling a fine machine with blunt tools.

In the simple organisms we can see direct, automatic responses of the kind called 'tropisms' (from a Greek word meaning to 'turn'). Plants show 'phototropism' (turning towards light), 'hydrotropism' (turning towards water, in this case by the roots), and 'chemotropism' (turning towards particular chemical substances). Chemotropism is also characteristic of many animals, from protozoa to ants. Certain moths are known to fly towards a scent as far as two miles away. That tropisms are completely automatic is shown by the fact that a phototropic moth will even fly into a candle flame.

The reflexes mentioned earlier in this chapter do not seem to progress far beyond tropisms, and imprinting, also mentioned, represents learning, but in so mechanical a fashion as scarcely to deserve the name. Yet neither reflexes nor imprinting can be regarded as characteristic of the lower animals only; man has his share.

The human infant from the moment of birth will grasp a finger tightly if it touches his palm and will suck at a nipple if that is put to his lips. The importance of such instincts to keep the infant secure from falling and from starvation is obvious.

It seems almost inevitable that the infant is subject also to imprinting. This is not a fit subject for experimentation, of course, but knowledge can be gained through incidental observations. Children who, at the babbling stage, are not exposed to the sounds of actual speech may not develop the ability to speak later, or do so to an abnormally limited extent. Children brought up in impersonal institutions where they are efficiently fed and their physical needs are amply taken care of, but where they are not fondled, cuddled, and dandled, become sad little specimens indeed. Their mental and physical development is greatly retarded and many die for no other reason, apparently, than lack of 'mothering' – by which may be meant the lack of adequate stimuli to bring about the imprinting of necessary behaviour patterns. Similarly, children who are unduly deprived of the stimuli involved in the company of other children during critical periods in childhood develop personalities that may be seriously distorted in one fashion or another.

Of course, one can argue that reflexes and imprinting are a matter of concern only for infancy. When a man achieves adulthood, he is then a rational being who responds in more than a mechanical fashion. But does he? To put it another way: Does man possess 'free will' (as he likes to think), or is his behaviour in some respects absolutely determined by the stimulus, as the bull's was in Delgado's experiment described on p. 399.

One can argue for the existence of free will on philosophical or theological grounds, but I know of no one who has ever found a way to demonstrate it experimentally. To demonstrate 'determinism', the reverse of free will, is not exactly easy, either. Attempts in that direction, however, have been made. Most notable were those of the Russian physiologist Ivan Petrovich Pavlov.

Pavlov started with a specific interest in the mechanism of digestion. He showed in the 1880s that gastric juice was secreted in the stomach as soon as food was placed on a dog's tongue; the stomach would secrete this juice even if food never reached it. But if the vagus nerve (which runs from the medulla oblongata to various parts of the alimentary canal) was cut near the stomach, the secretion stopped. For his work on the physiology of digestion, Pavlov received the Nobel Prize in physiology and medicine in 1904. But like some other Nobel laureates (notably, Ehrlich and Einstein) Pavlov went on to other discoveries that dwarfed the accomplishments for which he actually received the prize.

He decided to investigate the automatic, or reflex, nature of secretions, and he chose the secretion of saliva as a convenient, easy-to-observe example. The sight or odour of food causes a dog (and a man, for that matter) to salivate. What Pavlov did was to ring a bell every time he placed food before a dog. Eventually, after twenty to forty associations of this sort, the dog salivated when it heard the bell even though no food was present. An association had been built up. The nerve impulse that carried the sound of the bell to the cerebrum had become equivalent to one representing the sight or odour of food.

In 1903, Pavlov invented the term 'conditioned reflex' for this phenomenon; the salivation was a 'conditioned response'. Willy-nilly, the dog salivated at the sound of the bell just as it would at the sight of food. Of course the conditioned response could be wiped out – for instance, by repeatedly denying food to the dog when the bell was rung and subjecting it to a mild electric shock instead. Eventually, the dog would not salivate but instead would wince at the sound of the bell, even though it received no electric shock.

Furthermore, Pavlov was able to force dogs to make subtle decisions by associating food with a circular patch of light and an electric shock with an elliptical patch. The dog could make the distinction, but as the ellipse was made more and more nearly circular, distinction became more difficult. Eventually, the dog, in an agony of indecision, developed what could only be called a 'nervous breakdown'.

Conditioning experiments have thus become a powerful tool in psychology. Through them, animals sometimes almost 'talk' to the experimenter. The technique has made it possible to investigate the learning abilities of various animals, their instincts, their visual abilities, their ability to distinguish colours, and so on. Of all the investigations, not the least remarkable are those of the Austrian naturalist Karl von Frisch. Von Frisch trained bees to go to dishes placed in certain locations for their food, and he learned that these foragers soon told the other bees in their hive where the food was located. From his experiments von Frisch learned that the bees could distinguish certain colours, including ultra-violet, but excluding red, which they communicated with one another by means of a dance on the honeycombs, that the nature and vigour of the dance told the direction and distance of the food dish from the hive and even how plentiful or scarce the food supply was, and that the bees were able to tell direction from the polarization of light in the sky. Von Frisch's fascinating discoveries about the language of the bees opened up a whole new field of study of animal behaviour.

In theory, all learning can be considered to consist of conditioned responses. In learning to type, for instance, you start by watching the typewriter keyboard and gradually substitute certain automatic movements of the fingers for visual selection of the proper key. Thus the thought 'k' is accompanied by a specific movement of the middle finger of the right hand; the thought 'the' causes the first finger of the left hand, the first finger of the right hand, and the second finger of the left hand to hit certain spots in that order. These responses involve no conscious thought. Eventually a practised typist has to stop and think to recall where the letters are. I am myself a rapid and completely mechanical typist, and if I am asked where the letter 'f', say, is located on the keyboard, the only way I can answer (short of looking at the keyboard) is to move my fingers in the air as if typing and try to catch one of them in the act of typing 'f'. Only my fingers know the keyboard; my conscious mind does not.

The same principle may apply to more complex learning, such as reading or playing a violin. Why, after all, does the design CRAYON in black print on this piece of paper automatically evoke a picture of a pigmented stick of wax and a certain sound that represents a word? You do not need to spell out the letters or search your memory or reason out the possible message contained in the design; from repeated conditioning you automatically associate the symbol with the thing itself.

In the early decades of this century the American psychologist John Broadus Watson built a whole theory of human behaviour, called 'behaviourism', on the basis of conditioning. Watson went so far as to suggest that people had no deliberate control over the way they behaved; it was all determined by conditioning. Although his theory was popular for a time, it never gained wide support among psychologists. In the first place, even if the theory is basically correct – if behaviour is dictated solely by conditioning – behaviourism is not very enlightening on those aspects of human behaviour that are of most interest to us, such as creative intelligence, artistic ability, and the sense of right and wrong. It would be impossible to identify all the conditioning influences and relate them to the pattern of thought and belief in any measurable way; and something that cannot be measured is not subject to any really scientific study.

In the second place, what does conditioning have to do with a process such as intuition? The mind suddenly puts two previously unrelated thoughts or events together, apparently by sheer chance, and creates an entirely new idea or response.

Cats and dogs, in solving tasks (as in finding out how to work a lever in order to open a door), may do so by a process of trial and error. They may move about randomly and wildly until some motion of theirs trips the lever. If they are set to repeating the task a dim memory of the successful movement may lead them to it sooner, and then still sooner at the next attempt, until finally they move to the lever at once. The more intelligent the animal, the fewer attempts will be required to graduate from sheer trial and error to purposive useful action.

By the time we reach man, memory is no longer feeble. The tendency might be to search for a dropped coin by glances randomly directed at the floor, but from past experience he may look in places where he has found the coin before, or look in the direction of the sound, or institute a systematic scanning of the floor. Similarly, if he were in a closed place, he might try to escape by beating and kicking at the walls randomly; but he would also know what a door would look like and would concentrate his efforts on that.

A man can, in short, simplify trial and error by calling on years of experience, and transfer it from thought to action. In seeking a solution, he may do nothing, he may merely act in thought. It is this etherealized trial and error we call reason, and it is not even entirely restricted to the human species.

Apes, whose patterns of behaviour are simpler and more mechanical than man's, show some spontaneous insight, which may be called reason. The German psychologist Wolfgang Köhler, trapped in one of the German colonies in Africa by the advent of the First World War, discovered some striking illustrations of this in his famous experiments with chimpanzees. In one case a chimp, after trying in vain to reach bananas with a stick that was too short, suddenly picked up another bamboo stick that the experimenter had left lying handy, joined the two sticks together, and so brought the fruit within reach. In another instance a chimp piled one box on another to reach bananas hanging overhead. These acts had not been preceded by any training or experience that might have formed the association for the animal; apparently they were sheer flashes of inspiration.

To Köhler, it seemed that learning involved the entire pattern of a process, rather than individual portions of it. He was one of the founders of the 'Gestalt' school of psychology (that being the German word for 'pattern').

The power of conditioning has turned out to be greater than had been expected, in fact. For a long time it had been assumed that certain body functions such as heartbeat, blood pressure, and intestinal contractions were essentially under the control of the autonomic nervous system and therefore beyond conscious control. There were catches, of course. A man adept at yoga can produce effects on his heartbeat by control of chest muscles, but that is no more significant than stopping the blood-flow through a wrist artery by applying thumb-pressure. Again, one can make one's heart beat faster by fantasying a state of anxiety, but that is the conscious manipulation of the autonomic nervous system. Is it possible simply to will the heart to beat faster or the blood pressure to rise without extreme manipulation of either the muscles or the mind?

The American psychologist Neal Elgar Miller and his co-workers have carried out conditioning experiments where rats were rewarded when they happened to increase their blood pressure for any reason, or where their heartbeat was increased or decreased. Eventually, for the sake of the reward, they learned to perform voluntarily a change effected by the autonomic nervous system – just as they might learn to press a lever, and for the same purpose.

At least one experimental programme, using human volunteers

(male) who were rewarded by flashes of light revealing photographs of nude girls, demonstrated the volunteers' ability to produce increases or decreases in blood pressure in response. The volunteers did not know what was expected of them in order to produce the flashing light – and the nude – but just found that, as time went along, they caught the desired glimpses more often.

There is more subtlety to the autonomic body controls than had once been suspected, too. Since living organisms are subjected to natural rhythms – the ebb and flow of the tides, the somewhat slower alternation of day and night, the still slower swing of the seasons – it is not surprising that they themselves respond rhythmically. Trees shed their leaves in the autumn and bud in the spring; men grow sleepy at night and rouse themselves at dawn.

What did not come to be fully appreciated until lately was the complexity and multiplicity of the rhythmic responses, and their automatic nature, which persisted even in the absence of the environmental rhythm.

Thus, the leaves of plants rise and fall in a day-long rhythm to match the coming and going of the sun. This is made apparent by time-lapse photography. Seedlings grown in darkness showed no such cycle, but the potentiality was there. One exposure to light – one only – was enough to convert that potentiality into actuality. The rhythm then began, and it continued even if the light was cut off again. From plant to plant, the exact period of rhythm varied – anywhere from 24 to 26 hours in the absence of light – but it was always about 24 hours, under the regulating effect of the sun. A 20-hour cycle could be established if artificial light were used on a 10-hour-on and 10-hour-off cycle, but as soon as the light was turned off altogether the about-24-hour rhythm re-established itself.

This daily rhythm, a kind of 'biological clock' that works even in the absence of outside hints, permeates all life. Franz Halberg of the University of Minnesota named it 'circadian rhythm', from the Latin 'circa dies', meaning 'about a day'.

Human beings are not immune to such rhythms. Men and

women have voluntarily lived for months at a time in caves where they separated themselves from any time-telling mechanism and had no idea whether it was night or day outside. They soon lost all track of time and ate and slept rather erratically. However, they also noted their temperature, pulse, blood pressure, and brain waves, and sent these and other measurements to the surface, where observers kept track of them in connection with time. It turned out that, however time-confused the cave-dwellers were, the bodily rhythm was not. The rhythm remained stubbornly at a period of about a day, with all measurements rising and falling regularly, through all the stay in the cave.

This is by no means only an abstract matter. In nature, the earth's rotation remains steady, and the alternation of day and night remains constant and beyond human interference – but only if you remain in the same spot on earth, or only shift north or south. If you travel east or west for long distances and quite rapidly, however, you change the time of day. You may land in Japan at lunchtime (for Japanese) when your biological clock tells you it is time to go to bed. The jet-age traveller often has difficulty matching his activity to that of the at-home people surrounding him. If he does so – with his pattern of hormone-secretion, for instance, not matching the pattern of his activity – he will be tired and inefficient, suffering from 'jet fatigue'.

Less dramatically, the ability of an organism to withstand a dose of X-rays or various types of medication often depends on the setting of the biological clock. It may well be that medical treatment ought to vary with the time of day or, for maximum effect and minimum side-effect, be restricted to one particular time of day.

What keeps the biological clock so well regulated? Suspicion in this respect has fallen upon the pineal gland. In some reptiles the pineal gland is particularly well developed and seems to be similar in structure to the eye. In the tuatara, a lizard-like reptile that is the last surviving species of its order and is found only on some small islands off New Zealand, the 'pineal eye' is a skin-covered patch on top of its skull, particularly prominent for about six months after birth and definitely sensitive to light.

The pineal gland does not 'see' in the ordinary sense of the word, but it may produce some chemical that rises and falls in rhythmic response to the coming and going of light. It thus may regulate the biological clock and do so even after light ceases to be periodic (having learned its chemical lesson by a kind of conditioning).

But then how does the pineal gland work in mammals, where it is no longer located just under the skin at the top of the head but is buried deep in the centre of the brain? Can there be something more penetrating than light – something that is rhythmic in the same sense? There are speculations that cosmic rays might be the answer. These have a circadian rhythm of their own, thanks to earth's magnetic field and the solar wind, and perhaps this force is the external regulator.

Even if the external regulator is found, is the internal biological clock something that can be identified? Is there some chemical reaction in the body that rises and falls in a circadian rhythm and that controls all the other rhythms? Is there some 'master-reaction' that we can tab as *the* biological clock? If so, it has not yet been found.

Simple reasoning from the conditioned reflex finds it difficult to encompass the subtleties of intuition and reason. An attack upon human behaviour has been launched by methods that are themselves highly intuitive. These methods can be traced back nearly two centuries to an Austrian physician, Franz Anton Mesmer, who became the sensation of Europe for his experiments with a powerful tool for probing human behaviour. He used magnets at first, and then his hands only, obtaining his effects by what he called 'animal magnetism' (soon renamed 'mesmerism'); he would put a patient into a trance and pronounce the patient cured of his illness. He may well have produced some cures (since some disorders can be treated by suggestion) and gained many ardent followers, including the Marquis de Lafayette, fresh from his American triumph. However, Mesmer, an ardent astrologer and all-round mystic, was investigated sceptically but fairly by a committee, which included Lavoisier and Benjamin Franklin, and was then denounced as a fake and eventually retired in disgrace.

Nevertheless, he had started something. In the 1850s a British surgeon named James Braid revived hypnotism (he was the first to use this term) as a medical device, and other physicians also took it up. Among them was a Viennese doctor named Josef Breuer, who in the 1880s began to use hypnosis specifically for mental and emotional disorders.

Hypnotism (Greek for 'putting to sleep') had been known, of course, since ancient times, and had often been used by mystics. But Breuer and others now began to interpret its effects as evidence of the existence of an 'unconscious' level of the mind. Motivations of which the individual was unaware were buried there, and they could be brought to light by hypnosis. It was tempting to suppose that these motivations were suppressed from the conscious mind because they were associated with shame or guilt, and that they might account for useless, irrational, or even vicious behaviour.

Breuer set out to employ hypnosis to probe the hidden causes of hysteria and other behaviour disorders. Working with him was a pupil named Sigmund Freud. For a number of years they treated patients together, putting the patients under light hypnosis and encouraging them to speak. They found that the patients' venting of experiences or impulses buried in the unconscious often acted as a cathartic, relieving their symptoms after they awoke from the hypnosis.

Freud came to the conclusion that practically all of the suppressed memories and motivations were sexual in origin. Sexual impulses tabooed by society and the child's parents were driven underground, but they still strove for expression and generated intense conflicts which were the more damaging for being unrecognized and unadmitted.

In 1894, after breaking with Breuer because the latter disagreed with his concentration on the sexual factor, Freud went on alone to develop his ideas about the causes and treatment of mental disturbances. He dropped hypnosis and urged his patients to babble in a virtually random manner – to say anything that came into their minds. As the patient came to feel that the physician was listening sympathetically without any moral censure, slowly – sometimes very slowly – the individual began to unburden himself, to remember things long repressed and forgotten. Freud called this slow analysis of the 'psyche' (Greek for soul or mind) 'psychoanalysis'.

Freud's involvement with the sexual symbolism of dreams and his description of infantile wishes to substitute for the parent of the same sex in the marital bed (the 'Oedipus complex' in the case of boys and the 'Electra complex' in girls – named after characters in Greek mythology) horrified some and fascinated others. In the 1920s, after the dislocations of the First World War and amid the further dislocations of Prohibition in America and changing mores in many parts of the world, Freud's views struck a sympathetic note, and psychoanalysis attained the status, almost, of a popular fad.

More than half a century after its beginnings, however, psychoanalysis still remains an art rather than a science. Rigorously controlled experiments, such as those conducted in physics and the other 'hard' sciences, are, of course, exceedingly difficult in psychiatry. The practitioners must base their conclusions largely on intuition or subjective judgement. Psychiatry (of which psychoanalysis is only one of the techniques) has undoubtedly helped many patients, but it has produced no spectacular cures and has not notably reduced the incidence of mental disease. Nor has it developed any all-embracing and generally accepted theory, comparable to the germ theory of infectious disease. In fact, there are almost as many schools of psychiatry as there are psychiatrists.

Serious mental illness takes various forms, ranging from chronic depression to a complete withdrawal from reality into a world in which some, at least, of the details do not correspond to the way most of us see things. This form of psychosis is usually called 'schizophrenia', a term introduced by the Swiss psychiatrist Eugen Bleuler. The word covers such a multitude of disorders that it can no longer be described as a specific disease. About 60 per cent of all the chronic patients in our mental hospitals are diagnosed as schizophrenics.

Until recently, drastic treatments, such as prefrontal lobotomy, or shock therapy using electricity or insulin (the latter technique introduced in 1933 by the Austrian psychiatrist Manfred Sakel), were all that could be offered. Psychiatry and psychoanalysis have been of little avail, except occasionally in the early stages when a physician is still able to communicate with the patient. But some recent discoveries concerning drugs and the chemistry of the brain ('neurochemistry') have introduced an encouraging note.

Even the ancients knew that certain plant juices could induce hallucinations (fantasies of vision, hearing, and so on) and others could bring on happy states. The Delphic priestesses of ancient Greece chewed some plant before they pronounced their cryptic oracles. Indian tribes of the south-western United States have made a religious ritual of chewing peyote or mescal buttons (which produce hallucinations in colour). Perhaps the most dramatic case was that of the Moslem sect in a mountain stronghold in Iran who used 'hashish', the juice of hemp leaves, more familiarly known to us as 'marijuana'. The drug, taken in their religious ceremonies, gave the communicants the illusion that they caught glimpses of the paradise to which their souls would go after death, and they would obey any command of their leader, called 'the Old Man of the Mountains', to receive this key to heaven. His commands took the form of ordering them to kill enemy rulers and hostile Moslem government officials; this gave rise to the word 'assassin', from 'hashishin' (a user of hashish). The sect terrorized the region throughout the twelfth century, until the Mongol invaders in 1256 swarmed into the mountains and killed every last assassin.

The modern counterpart of the euphoric herbs of earlier times (aside from alcohol) is the group of drugs known as the 'tranquillizers'. As a matter of fact, one of the tranquillizers had been known in India as long ago as 1000 B.C. in the form of a plant called *Rauwolfia serpentinum*. It was from the dried roots of this plant that American chemists in 1952 extracted 'reserpine', the first of the currently popular tranquillizing drugs. Several substances with similar effects but simpler chemical structure have since been synthesized.

The tranquillizers are sedatives, but with a difference. They reduce anxiety without appreciably depressing other mental activity. Nevertheless, they do tend to make people sleepy, and they may have other undesirable effects. They were at once found to be immensely helpful in relieving and quieting mental patients, including some schizophrenics. The tranquillizers are not cures for any mental illness, but they suppress certain symptoms that stand in the way of adequate treatment. By reducing the hostilities and rages of patients, and by quieting their fears and anxieties, they reduce the necessity for drastic physical restraints, make it easier for psychiatrists to establish contacts with patients, and increase the patients' chances of release from the hospital.

But where the tranquillizers had their runaway boom was among the public at large, which apparently seized upon them as a panacea to banish all cares.

Reserpine turns out to have a tantalizing resemblance to an important substance in the brain. A portion of its complex molecule is rather similar to the substance called 'serotonin'. Serotonin was discovered in the blood in 1948, and it has greatly intrigued physiologists ever since. It was found to be present in the hypothalamus region of the human brain and proved to be widespread in the brain and nerve tissues of other animals, including invertebrates.

What is more, various other substances that affect the central nervous system have turned out to resemble serotonin closely. One of them is a compound in toad venom called 'bufotenin'. Another is mescaline, the active drug in mescal buttons. Most dramatic of all is a substance named 'lysergic acid diethylamide' (popularly known as LSD). In 1943, a Swiss chemist named Albert Hofmann happened to absorb some of this compound in the laboratory and was overcome by strange sensations. Indeed, what he seemed to perceive by way of his senses in no way matched what we would take to be the objective reality of the environment. He suffered what we call hallucinations, and LSD is an example of what we now call a 'hallucinogen'.

Those who take pleasure in the sensations they experience when under the influence of a hallucinogen refer to this as 'mind-expansion', apparently indicating that they sense – or think they sense – more of the universe than they would under ordinary conditions. But then, so do drunks once they bring themselves to the stage of 'delirium tremens'. The comparison is not as unkind as it may seem, for investigations have shown that a small dose of LSD, in some cases, can produce many of the symptoms of schizophrenia!

What can all this mean? Well, serotonin (which is structurally like the amino acid tryptophan) can be broken down by means of an enzyme called 'amine oxidase', which occurs in brain cells. Suppose that this enzyme is taken out of action by a competitive substance with a structure like serotonin's – lysergic acid, for example. With the breakdown enzyme removed, serotonin will accumulate in the brain cells, and its level may rise too high. This will upset the serotonin balance in the brain and may bring on the schizophrenic state.

Is it possible that schizophrenia arises from some naturally induced upset of this sort? The manner in which a tendency to schizophrenia is inherited certainly makes it appear that some metabolic disorder (one, moreover, that is gene-controlled) is involved. In 1962, it was found that with a certain course of treatment the urine of schizophrenics often contained a substance absent from the urine of non-schizophrenics. The substance eventually turned out to be a chemical called 'dimethoxyphenylethylamine', with a structure that lies somewhere between adrenalin and mescaline. In other words, certain schizophrenics seem, through some metabolic error, to form their own hallucinogens and to be, in effect, on a permanent drug-high.

Not everyone reacts identically to a given dose of one drug or another. Obviously, however, it is dangerous to play with the chemical mechanism of the brain. To become a mental cripple is a price surely too high for any amount of 'mind-expanding' fun. Nevertheless, the reaction of society to drug use – particularly to that of marijuana, which has not yet been definitely shown to be as harmful as other hallucinogens – tends to be over-strenuous. Many of those who inveigh against the use of drugs of one sort or another are themselves thoroughly addicted to the use of alcohol or tobacco, both of which, in the mass, are responsible for much harm both to the individual and to society. Hypocrisy of this sort tends to decrease the credibility of much of the anti-drug movement.

Neurochemistry also offers a hope for understanding that elusive mental property known as 'memory'. There are, it seems, two varieties of memory: short-term and long-term. If you look up a phone number, it is not difficult to remember it until you have dialled; it is then automatically forgotten and, in all probability, will never be recalled again. A telephone number you use frequently, however, enters the long-term memory category. Even after a lapse of months, you can dredge it up.

Yet even of what we would consider long-term memory items, much is lost. We forget a great deal and even, alas, forget much of vital importance (as every student facing an examination is woefully aware). Yet is it forgotten? Has it really vanished, or is it simply so well stored that it is difficult to recall – buried, so to speak, under too many extraneous items?

The tapping of such hidden memories has become an almost literal tap. The American-born surgeon Wilder Graves Penfield at McGill University in Montreal, while operating on a patient's brain, accidentally touched a particular spot that caused the patient to hear music. That happened over and over again. The patient could be made to relive an experience in full, while remaining quite conscious of the present. Proper stimulation can apparently reel off memories with great accuracy. The area involved is called the 'interpretative cortex'. It may be that the accidental tapping of this portion of the cortex gives rise to the phenomenon of *déjà vu* (the feeling that something has happened before) and other manifestations of 'extrasensory perception'.

But if memory is so detailed, how can the brain find room for it all? It is estimated that, in a lifetime, a brain can store 1,000,000,000,000,000 (a thousand billion) units of information. To store so much, the units of storage must be of molecular size. There would be room for nothing more.

Suspicion is currently falling on ribonucleic acid (RNA), in which the nerve cell, surprisingly enough, is richer than almost any other type of cell in the body. This is surprising, because RNA is involved in the synthesis of protein (see Chapter 13) and is therefore usually found in particularly high quantity in those tissues producing large quantities of protein either because they are actively growing or because they are producing copious quantities of protein-rich secretions. The nerve cell falls into neither classification.

A Swedish neurologist, Holger Hyden, developed techniques that could separate single cells from the brain and then analyse them for RNA content. He took to subjecting rats to conditions where they were forced to learn new skills – that of balancing on a wire for long periods of time, for instance. By 1959, he had discovered that the brain cells of rats that were forced to learn increased their RNA content up to 12 per cent higher than that of the brain cells of rats allowed to go their normal way.

The RNA molecule is so very large and complex that if each unit of stored memory is marked off by an RNA molecule of distinctive pattern we need not worry about capacity. So many different RNA patterns are available that even a number such as a thousand billion is insignificant in comparison.

But ought one to consider RNA by itself? RNA molecules are formed according to the pattern of DNA molecules in the chromosomes. Is it that each person carries a vast supply of potential memories – a 'memory bank', so to speak – in the DNA molecules he was born with, called upon and activated by actual events with appropriate modifications?

And is RNA the end? The chief function of RNA is to form specific protein molecules. Is it the protein, rather than the RNA, that is truly related to the memory function?

One way of testing this is to make use of a drug called 'puromycin', which interferes with protein formation by way of RNA. The American man-and-wife team Louis Barkhouse Flexner and Josepha Barbar Flexner conditioned mice to solve a maze, then immediately injected puromycin. The mice forgot what they had learned. The RNA molecule was still there, but the key protein molecule could not be formed. Using puromycin, the Flexners showed that while short-term memory could be erased in this way in rats, long-term memory could not. The proteins for the latter had, presumably, already been formed.

And yet it may be that memory is more subtle and is not to be fully explained on the simple molecular level. There are indications that patterns of neural activity may be involved, too. Much yet remains to do.

Feedback

Man is not a machine and a machine is not man, but as science and technology advance, man and machine seem to be becoming less and less distinguishable from each other.

If you analyse what it is that makes a man, one of the first thoughts that strikes you is that, more than any other living organism, he is a self-regulating system. He is capable of controlling not only himself but also his environment. He copes with changes in the environment, not by yielding but by reacting according to his own desires and standards. Let us see how close a machine can come to this ability.

About the simplest form of self-regulating mechanical device is the controlled valve. Simple versions were devised as early as A.D. 50 by Hero of Alexandria, who used one in a device to dispense liquid automatically. A very elementary version of a safety valve is exemplified in a pressure cooker invented by Denis Papin in 1679. To keep the lid on against the steam pressure, he placed a weight on it, but he used a weight light enough so that the lid could fly off before the pressure rose to the point where the pot would explode. The present-day household pressure cooker or steam boiler has more sophisticated devices for this purpose (such as a plug that will melt when the temperature gets too high); but the principle is the same.

Of course, this is a 'one-shot' sort of regulation. But it is easy to think of examples of continuous regulation. A primitive type was a device patented in 1745 by an Englishman, Edmund Lee, to keep a windmill facing squarely to the wind. He devised a 'fan-tail' with small vanes that caught the wind whenever the wind shifted direction; the turning of these vanes operated a set of gears that rotated the windmill itself so that its main vanes were again head-on to the wind in the new quarter. In that position the fantail vanes remained motionless; they turned only when the windmill was not facing the wind.

But the archetype of modern mechanical self-regulators is the 'governor' invented by James Watt for his steam engine. To keep the steam output of his engine steady, Watt conceived a device consisting of a vertical shaft with two weights attached to it laterally by hinged rods, allowing the weights to move up and down. The pressure of the steam whirled the shaft. When the steam pressure rose, the shaft whirled faster and the centrifugal force drove the weights upwards. In moving up they partly closed a valve, choking off the flow of steam. As the steam pressure fell, the shaft whirled less rapidly, gravity pulled the weights down, and the valve opened. Thus the governor kept the shaft speed, and hence the power delivered, at a uniform level. Each departure from that level set in train a series of events that corrected the deviation. This is called 'feedback': the error itself continually sends back information and serves as the measure of the correction required.

A very familiar example of a feedback device is the 'thermo-stat', first used in very crude form by the Dutch inventor

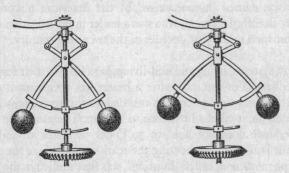

Watt's governor.

Cornelis Drebble in the early seventeenth century. A more sophisticated version, still used today, was invented in principle by a Scottish chemist named Andrew Ure in 1830. Its essential component consists of two strips of different metals laid against each other and soldered together. Since the two metals expand and contract at different rates with changes in temperature, the strip bends. The thermostat is set, say, at 70°F. When the room temperature falls below that, the thermocouple bends in such a fashion as to make a contact that closes an electric circuit and turns on the heating system. When the temperature rises above 70°, the thermocouple bends back enough to break the contact. Thus the heater regulates its own operation through feedback.

It is feedback that similarly controls the workings of the human body. To take one example of many, the glucose level in the blood is controlled by the insulin-producing pancreas, just as the temperature of a house is controlled by the heater. And just as the working of the heater is regulated by the departure of the temperature from the norm, so the secretion of insulin is regulated by the departure of the glucose concentration from the norm. A too-high glucose level turns on the insulin, just as a too-low temperature turns on the heater. Likewise, as a thermostat can be turned up to higher temperature, so an internal change in the body such as the secretion of adrenalin can raise the operation of the human body to a new norm, so to speak.

Self-regulation by living organisms to maintain a constant norm was named 'homoeostasis' by the American physiologist Walter Bradford Cannon, who was a leader in investigation of the phenomenon in the first decades of the twentieth century.

Most systems, living and non-living, lag a little in their response to feedback. For instance, after a heater has been turned off, it continues for a time to emit its residual heat; conversely, when it is turned on, it takes a little time to heat up. Therefore, the room temperature does not hold to 70°F but oscillates around that level; it is always overshooting the mark on one side or the other. This phenomenon, called 'hunting', was first studied in the 1830s by George Airy, the Astronomer Royal of England, in connection with devices he had designed to turn telescopes automatically with the motion of the earth.

Hunting is characteristic of most living processes, from control of the glucose level in the blood to conscious behaviour. When you reach to pick up an object, the motion of your hand is not a single movement but a series of movements continually adjusted in both speed and direction, with the muscles correcting departures from the proper line of motion, those departures being judged by the eye. The corrections are so automatic that you are not aware of them. But watch an infant, not yet practised in visual feedback, try to pick up something. It overshoots and undershoots because the muscular corrections are not precise enough. And victims of nerve damage that interferes with the ability to utilize visual feedback go into pathetic oscillations, or wild hunting, whenever they attempt a coordinated muscular movement.

The normal, practised hand goes smoothly to its target and stops at the right moment because the control centre looks ahead and makes corrections in advance. Thus, when you drive a car around a corner you begin to release the steering wheel before you have completed the turn, so that the wheels will be straight by the time you have rounded the corner. In other words, the correction is applied in time to avoid overshooting the mark to any significant degree.

It is the chief role of the cerebellum, evidently, to take care of this adjustment of motion by feedback. It looks ahead, and predicts the position of the arm a few instants ahead, organizing motion accordingly. It keeps the large muscles of the torso in constantly varying tensions to keep you in balance and upright if you are standing. It is hard work to stand and 'do nothing'; we all know how tiring just standing can be.

Now this principle can be applied to a machine. Matters can be arranged so that as the system approaches the desired condition, the shrinking margin between its actual state and the desired state will automatically shut off the corrective force before it overshoots. In 1868, a French engineer, Léon Farcot, used this principle to invent an automatic control for a steam-operated ship's rudder. As the rudder approached the desired position, his device automatically closed down the steam valve; by the time the rudder reached the specified position, the steam pressure had been shut off. When the rudder moved away from this position, its motion opened the appropriate valve so that it was pushed back. Farcot called his device a 'servomechanism', and in a sense it ushered in the era of 'automation' (a term introduced in 1946 by the American engineer D. S. Harder).

Servomechanisms did not really come into their own until the arrival of electronics. The application of electronics made it possible to endow the mechanisms with a sensitivity and swiftness of response even beyond the capabilities of a living organism. Furthermore, radio extended their sphere of action over a considerable distance. The German buzz bomb of the Second World War was essentially a flying servomechanism, and it introduced the possibility not only of guided missiles but also of self-operated or remotely operated vehicles of all sorts, from underground trains to space ships. Because the military establishments had the keenest interest in these devices, and the most abundant supply of funds, servomechanisms have reached perhaps their highest development in aiming-and-firing mechanisms for guns and rockets. These systems can detect a swiftly moving target hundreds of miles away, instantly calculate its course (taking into account the target's speed of motion, the wind, the temperatures of the various layers of air, and numerous other conditions), and hit the target with pinpoint accuracy, all without any human guidance.

Automation found an ardent theoretician and advocate in the mathematician Norbert Wiener. In the 1940s he and his group at the Massachusetts Institute of Technology worked out some of the fundamental mathematical relationships governing the handling of feedback. He named this branch of study 'cybernetics', from the Greek word for 'helmsman', which seems appropriate, since the first use of servomechanisms was in connection with a helmsman. (Cybernetics also harks back to Watt's centrifugal governor, for 'governor' comes from the Latin word for helmsman.)

Since the Second World War, automation has progressed fairly rapidly, especially in the United States and the Soviet Union. Oil refineries and factories manufacturing objects such as radio sets and aluminium pistons have been set up on an almost completely automatic basis, taking in raw materials at one end and pouring out finished products at the other, with all the processes handled by self-regulating machines. Automation has even invaded the farm: engineers at the University of Wisconsin announced in 1960 an automated pig-feeding system in which machines would feed each pig the correct amount of the correct type of feed at the correct time.

In a sense, automation marks the beginning of a new Industrial Revolution. Like the first revolution, it may bring great pains of adjustment, not only for workers but also for the economy in general. There is a story of a car executive who conducted a union official on a tour of a new automatic factory and observed: 'You won't be able to collect union dues from these machines, I'm afraid.' The union official shot back: 'And you won't be able to sell them cars, either.'

Automation will not make the human worker obsolete, any more than the steam engine or electricity did. But it will certainly mean a great shift in emphasis. The first Industrial Revolution made it no longer necessary for a man to be a muscle-straining workhorse. The second will make it no longer necessary for him to be a mind-dulled automaton.

Naturally, feedback and servomechanisms have stirred up as much interest among biologists as among engineers. Self-regulating machines can serve as simplified models for studying the workings of the nervous system.

A generation ago the imagination of men was excited – and disturbed – by Karel Čapek's play *R. U. R.* (for *Rossem's Universal Robots*, robot coming from the Czech word meaning 'to work'). In recent years scientists have begun to experiment with various forms of robots (sometimes called 'automata') not as mere mechanical substitutes for men but as tools to explore the nature of living organisms. For instance, L. D. Harmon of the Bell

Telephone Laboratories devised a transistorized circuit that, like a neuron, fires electrical pulses when stimulated. Such circuits can be assembled into devices that mimic some of the functions of the eye and the ear. In England the biologist W. Ross Ashby formed a system of circuits that exhibits simple reflex responses. He calls his creature a 'homoeostat', because it tends to maintain itself in a stable state.

The British neurologist William Grey Walter built a more elaborate system that explores and reacts to its surroundings. His turtle-like object, which he calls a 'testudo' (Latin for 'tortoise'), has a photoelectric cell for an eye, a sensing device to detect touch, and two motors, one to move forwards or backwards and the other to turn round. In the dark, it crawls about, circling in a wide arc. When it touches an obstacle, it backs off a bit, turns slightly and moves forwards again; it will do this until it gets around the obstacle. When its photoelectric eye sees a light, the turning motor shuts off and the testudo advances straight towards the light. But its phototropism is under control; as it gets close to the light the increase in brightness causes it to back away, so that it avoids the mistake of the moth. When its batteries run down, however, the now 'hungry' testudo can crawl close enough to the light to make contact with a recharger placed near the light bulb. Once recharged, it is again sensitive enough to back away from the bright area around the light.

The subject of robots brings up the thought of machines that, in general, mimic living systems. In a sense, human beings, as tool-makers, have always mimicked what they have seen in nature about them. The knife is an artificial tusk; the lever is an artificial arm; the wheel developed from the roller, which in turn was inspired by the tree-trunk; and so on.

It is only very recently, however, that the full resources of science have been turned upon the effort to analyse the functioning of living tissues and organs, in order that the manner in which they perform – worked out hit-and-miss over thousands of millions of years of evolution – might be imitated in man-made machines. This study is called 'bionics', a term – suggested by '*bio*logical electro*nics*' but much broader in scope – coined by the American engineer Jack Steele in 1960.

As one example of what bionics might do, consider the structure of dolphin skin. Dolphins swim at speeds that would require 2·6 horsepower if the water about them were as turbulent as it would be about a vessel of the same size. For some reason, water flows past the dolphin without turbulence, and therefore little power is consumed overcoming water resistance. Apparently this happens because of the nature of dolphin skin. If we can reproduce that effect in vessel walls, the speed of an ocean liner could be increased and its fuel consumption decreased – simultaneously.

Then, too, the American biophysicist Jerome Lettvin studied the frog's retina in detail by inserting tiny platinum electrodes into its optic nerve. It turned out that the retina did not merely transmit a melange of light and dark dots to the brain and leave it to the brain to do all the interpretation. Rather, there were five different types of cells in the retina, each designed for a particular job. One reacted to edges – that is, to sudden changes in the nature of illumination, as at the edge of a tree marked off against the sky. A second reacted to dark curved objects (the insects eaten by the frog). A third reacted to anything moving rapidly (a dangerous creature that might better be avoided). A fourth reacted to dimming light, and a fifth to the watery blue of a pond. In other words, the retinal message went to the brain already analysed to a considerable degree. If man-made sensors made use of the tricks of the frog's retina, they could be made far more sensitive and versatile than they now are.

If, however, we are to build a machine that will imitate some living device, the most attractive possibility is the imitation of that unique device that interests us most profoundly – the human brain.

Thinking Machines

Can we build a machine that thinks? To try to answer that question we must first define 'thinking'.

Certainly we can take mathematics as a representation of one form of thinking. And it is a particularly good example for our purposes. For one thing, it is distinctly a human attribute. Some lower organisms are able to distinguish between three objects and four, say, but no species except *Homo sapiens* can perform the simple operation of dividing $\frac{3}{4}$ by $\frac{7}{8}$. Second, mathematics involves a type of reasoning that operates by fixed rules and includes (ideally) no undefined terms or procedures. It can be analysed in a more definite and more precise way than can the kind of thinking that goes into, say, literary composition or high finance or industrial management or military strategy. As early as 1936, the English mathematician Alan Mathison Turing showed that any problem could be solved mechanically if it could be expressed in the form of a finite number of manipulations that could be performed by the machine. So let us consider machines in relation to mathematics.

Tools to aid mathematical reasoning are undoubtedly as old as mathematics itself. The first tools for the purpose must have been man's own fingers. Mathematics began when man used his fingers to represent numbers and combinations of numbers. It is no accident that the word 'digit' stands both for a finger (or toe) and for a numerical integer.

From that, another step leads to the use of other objects in place of fingers – small pebbles, perhaps. There are more pebbles than fingers, and intermediate results can be preserved for future reference in the course of solving the problem. Again, it is no accident that the word 'calculate' comes from the Latin word for pebble.

Pebbles or beads lined up in slots or strung on wires formed the 'abacus', the first really versatile mathematical tool. With this device it became easy to represent units, tens, hundreds, thousands, and so on. By manipulating the pebbles, or counters, of an abacus, one could quickly carry through an addition such as $576+289$. Furthermore, any instrument that can add can also multiply, for multiplication is only repeated addition. And multiplication makes raising to a power possible, because this is only repeated multiplication (e.g., 4^5 is shorthand for $4\times4\times4\times4\times4$). Finally, running the instrument backwards, so to speak, makes possible the operations of subtraction, division, and extracting a root.

The abacus can be considered the second 'digital computer'. (The first, of course, was the fingers.)

For thousands of years the abacus remained the most advanced form of calculating tool. It actually dropped out of use in the West after the end of the Roman Empire and was reintroduced by Pope Sylvester II about A.D. 1000, probably from Moorish Spain, where its use had lingered. It was greeted on its return as an eastern novelty, its western ancestry forgotten.

The abacus was not replaced until a numerical notation was introduced which imitated the workings of the abacus. (This notation, the one familiar to us nowadays as 'Arabic numerals', was originated in India some time about A.D. 800, was picked up by the Arabs, and finally introduced to the West about A.D. 1200 by the Italian mathematician Leonardo of Pisa.)

In the new notation, the nine different pebbles in the units row of the abacus were represented by nine different symbols, and those same nine symbols were used for the tens row, hundreds row, and thousands row. Counters differing only in position were replaced by symbols differing only in position, so that in the written number 222, for instance, the first 2 represents 200, the second 20, and the third represents two itself; that is, $200+20+2 = 222$.

This 'positional notation' was made possible by recognition of an all-important fact which the ancient users of the abacus had overlooked. Although there are only nine counters in each row of the abacus, there are actually ten possible arrangements. Besides using any number of counters from one to nine in a row, it is also possible to use *no* counter – that is, to leave the place at the counting position empty. This escaped all the great Greek mathematicians and was not recognized until the ninth century, when some unnamed Hindu thought of representing the tenth alternative by a special symbol which the Arabs called 'sifr' ('empty') and which has come down to us, in consequence, as 'cipher' or, in more corrupt form, 'zero'. The importance of the zero is recorded in the fact that the manipulation of numbers is still sometimes called 'ciphering', and that to solve any hard problem is to 'decipher' it.

Another powerful tool grew out of the use of the exponents to

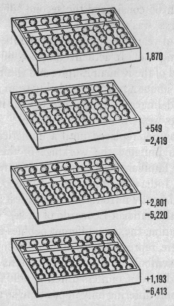

1,870

+549
=2,419

+2,801
=5,220

+1,193
=6,413

Adding with an abacus. Each counter below the bar counts 1; each counter above the bar counts 5. A counter registers when it is pushed to the bar. Thus in the top setting here the right-hand column reads 0; the one to the left of that reads 7 or $(5+2)$; the next left reads 8 or $(5+3)$; and the next left reads 1: the number shown, then, is 1,870. When 549 is added to this, the right column becomes 9 or $(9+0)$; the next addition $(4+7)$ becomes 1 with 1 to carry, which means that one counter is pushed up in the next column; the third addition is $9+5$, or 4 with 1 to carry; and the fourth addition is $1+1$ or 2; the addition gives 2,419, as the abacus shows. The simple manoeuvre of carrying 1 by pushing up a counter in the next column makes it possible to calculate very rapidly; a skilled operator can add faster than an adding machine can, as was shown by an actual test in 1946.

express powers of numbers. To express 100 as 10^2, 1,000 as 10^3, 100,000 as 10^5, and so on, is a great convenience in several respects; not only does it simplify the writing of large numbers but it reduces multiplication and division to simple addition or subtraction of the exponents (e.g., $10^2 \times 10^3 = 10^5$) and makes raising to a power or extraction of a root a simple matter of multiplying or dividing exponents (e.g., the cube root of 1,000,000 is $10^{6/3} = 10^2$). Now this is all very well, but very few numbers can be put into simple exponential form. What could be done with a number such as 111? The answer to that question led to the tables of logarithms.

The first to deal with this problem was the seventeenth-century Scottish mathematician John Napier. Obviously, expressing a number such as 111 as a power of 10 involves assigning a fractional exponent to 10 (the exponent is between 2 and 3). In more general terms, the exponent will be fractional whenever the number in question is not a multiple of the base number. Napier worked out a method of calculating the fractional exponents of numbers, and he named these exponents 'logarithms'. Shortly afterwards the English mathematician Henry Briggs simplified the technique and worked out logarithms with 10 as the base. The 'Briggsian logarithms' are less convenient in calculus, but they are the more popular for ordinary computations.

All non-integral exponents are irrational, that is, they cannot be expressed in the form of an ordinary fraction. They can be expressed only as an indefinitely long decimal expression lacking a repeating pattern. Such a decimal can be calculated, however, to as many places as necessary for the desired precision.

For instance, let us say we wish to multiply 111 by 254. The Briggsian logarithm of 111 to five decimal places is 2·04532, and for 254 it is 2·40483. Adding these logarithms, we get $10^{2·04532} \times 10^{2·40483} = 10^{4·45015}$. That number is approximately 28,194, the actual product of 111×254. If we want to get still closer accuracy, we can use the logarithms to six or more decimal places.

Tables of logarithms simplified computation enormously. In 1622 an English mathematician named William Oughtred made things still easier by devising a 'slide rule'. Two rulers are marked with a 'logarithmic scale', in which the distances between numbers get shorter as the numbers get larger: for example, the first division holds the numbers from 1 to 10; the second division, of the same length, holds the numbers from 10 to 100; the third from 100 to 1,000, and so on. By sliding one rule along the other to an appropriate position, one can read off the result of an operation involving multiplication or division. The slide rule makes computations as easy as addition and subtraction on the abacus, though in both cases, to be sure, one must be skilled in the use of the instrument.

The first step towards a truly automatic calculating machine was taken in 1642 by the French mathematician Blaise Pascal. He invented an adding machine that did away with the need to move the counters separately in each row of the abacus. His machine consisted of a set of wheels connected by gears. When the first wheel – the units wheel – was turned ten notches to its 0 mark, the second wheel turned one notch to the number 1, so that the two wheels together showed the number 10. When the tens wheel reached its 0, the third wheel turned a notch, showing 100, and so on. (The principle is the same as that of the mileage indicator in a car.) Pascal is supposed to have had more than fifty such machines constructed; at least five are still in existence.

Pascal's device could add and subtract. In 1674 the German mathematician Gottfried Wilhelm von Leibnitz went a step further and arranged the wheels and gears so that multiplication and division were as automatic and easy as addition and subtraction. In 1850 a United States inventor named D. D. Parmalee patented an important advance which added greatly to the calculator's convenience: in place of moving the wheels by hand, he introduced a set of keys – pushing down a marked key with the finger turned the wheels to the correct number. This is the mechanism of what is now familiar to us as the old-fashioned cash register.

It remained only to electrify the machine (so that motors did the work dictated by the touching of keys), and the Pascal–Leibnitz device graduated into the modern desk computer.

The desk computer, however, represents a dead end, not the wave of the future. The computer that we have in mind when we consider thinking machines is an entirely different affair. Its granddaddy is an idea conceived early in the nineteenth century by an English mathematician named Charles Babbage.

A genius far in advance of his time, Babbage imagined an analytical machine that would be able to perform any mathematical operation, be instructed by punch cards, store numbers in a memory device, compare the results of operations, and so on. He worked for thirty-seven years on his ideas, spending a fortune, both his own and the government's money, putting together awesomely elaborate structures of wheels, cams, levers, and wires, in an age when every part had to be hand-fitted. In the end he failed, and died a bitterly disappointed man, because what he was seeking to achieve could not be accomplished with mere mechanical devices.

The machine had to wait a century for the development of electronics. And electronics in turn suggested the use of a mathematical language much easier for a machine to handle than the decimal system of numerals. It is called the 'binary system' and was invented by Leibnitz. To understand the modern computer we must acquaint ourselves with this system.

The binary notation uses only two digits: 0 and 1. It expresses all numbers in terms of powers of 2. Thus, the number one is 2^0, the number two is 2^1, three is 2^1+2^0, four is 2^2, and so on. As in the decimal system, the power is indicated by the position of the symbol. For instance, the number four is represented by 100, read thus: $(1\times2^2)+(0\times2^1)+(0\times2^0)$, or $4+0+0=4$ in the decimal system.

As an illustration, let us consider the number 6,413. In the decimal system it can be written $(6\times10^3)+(4\times10^2)+(1\times10^1)+(3\times10^0)$; remember that any number to the power of nought equals 1. Now in the binary system we add numbers in powers of 2, instead of powers of 10, to compose a number. The highest power of 2 that leaves us short of 6,413 is 12; 2^{12} is 4,096. If we now add 2^{11}, or 2,048, we have 6,144, which is 269 short of 6,413. Next, 2^8 adds 256 more, leaving 13; we can then add 2^3, or 8, leaving 5; then 2^2, or 4, leaving 1; and 2^1 is 1. Thus we might write the number 6,413 as $(1\times2^{12})+(1\times2^{11})+(1\times2^8)+(1\times2^3)+(1\times2^2)+(1\times2^0)$. But as in the decimal system, each digit in a number, reading from the left, must represent the next smaller power. Just as in the decimal system we represent the additions of the third, second, first, and zero powers of 10 in stating the number 6,413, so in the binary system we must represent the additions of the powers of 2 from 12 down to 0. In the form of a table this would read:

$$
\begin{aligned}
1 \times 2^{12} &= 4,096 \\
1 \times 2^{11} &= 2,048 \\
0 \times 2^{10} &= 0 \\
0 \times 2^{9} &= 0 \\
1 \times 2^{8} &= 256 \\
0 \times 2^{7} &= 0 \\
0 \times 2^{6} &= 0 \\
0 \times 2^{5} &= 0 \\
0 \times 2^{4} &= 0 \\
1 \times 2^{3} &= 8 \\
1 \times 2^{2} &= 4 \\
0 \times 2^{1} &= 0 \\
1 \times 2^{0} &= 1 \\
\hline
&\ 6,413
\end{aligned}
$$

Taking the successive multipliers in the column at the left (as we take 6, 4, 1, and 3 as the successive multipliers in the decimal system), we write the number in the binary system as 1100100001101.

This looks pretty cumbersome. It takes 13 digits to write the number 6,413, whereas in the decimal system we need only four. But for a computing machine the system is just about the simplest imaginable. Since there are only two different digits, any operation can be carried out in terms of just the two states of an electrical circuit – on and off. On (i.e., the circuit closed) can represent 1; off (the circuit open) can represent 0. With proper circuitry and the use of diodes the machine can carry out all sorts of mathematical manipulations. For instance, with two switches in parallel it can perform addition: $0 + 0 = 0$ (off plus off equals off), and $0 + 1 = 1$ (off plus on equals on). Similarly, two switches in series can perform multiplication: $0 \times 0 = 0$; $0 \times 1 = 0$; $1 \times 1 = 1$.

So much for the binary system. As for electronics, its contribution to the computing machine is incredible speed. It can perform an arithmetical operation almost instantaneously: some of the electronic computers carry out thousands of millions of operations per second! Calculations that would take a man with a pencil a lifetime can be completed by a computer in a matter of days. To cite a typical case, before the coming of the computer an English mathematician named William Shanks spent fifteen years calculating the value of π, the ratio of the circumference of a circle to its diameter, and carried it to 707 decimal places (getting the last hundred-odd places wrong). Some years ago, an electronic computer carried out the calculation accurately to 10,000 decimal places, taking only a few days for the job.

In 1925, the American electrical engineer Vannevar Bush and his colleagues constructed a machine capable of solving differential equations. This was the first modern computer, though it still made use of mechanical switches, and represented a successful version of the sort of thing Babbage had tackled a century earlier.

For complete success, however, the switches had to be electronic. This would greatly increase the speed of computers for a flow of electrons can be started, shifted, or stopped in millionths of a second – far faster than mechanical switches, however delicate, could be manipulated.

The first large electronic computer, containing 19,000 vacuum tubes, was built at the University of Pennsylvania by John Presper Eckert and John William Mauchly during the Second World War. It was called ENIAC, for 'Electronic Numerical Integrator and Computer'. ENIAC ceased operation in 1955 and was dismantled in 1957, a hopelessly outmoded dotard at twelve years of age, but it left behind an amazingly numerous and sophisticated progeny. Whereas ENIAC weighed 30 tons and took up 1,500 square feet of floor space, the equivalent computer today – using switching units far smaller, faster, and more reliable than the old vacuum tubes – could be built into an object the size of a refrigerator. Modern computers contain half a million switching components and 10 million high-speed memory-elements.

So fast was progress that by 1948, small electronic computers were being produced in quantity; within five years, 2,000 were in use; by 1961, the number was 10,000. By 1970, the number had passed the 100,000 mark.

Computers working directly with numbers are 'digital computers'. They can work to any desired accuracy but can only answer the specific problem asked. An abacus, as I said earlier, is a primitive example. A computer may work not with numbers but with current-strengths made to vary in fashion analogous to the variables being considered. This is an 'analog computer'; a famous example is Remington Rand's 'Universal Analog Computer' (UNIVAC), the first large commercial computer, produced by Eckert and Mauchly in 1950 and put to use on television in 1952 to analyse the results of the presidential election as they came in. The slide rule is actually a very simple analogue computer. Analogue computers are of limited accuracy, but can supply answers, at one stroke, to a whole related family of questions.

Is ENIAC, or any of its more sophisticated descendants, truly a 'thinking machine'? Hardly. Essentially it is no more than a very rapid abacus. It slavishly follows instructions that are fed into it.

The basic tool for instructing most computers is the punch card, still much like those Babbage once planned to use. A hole punched in a definite position on such a card can signify a number; a combination of two or more holes in specific positions can

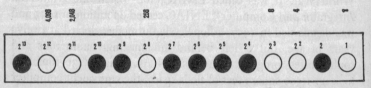

The number 6,413 represented by lights on a computer panel. The unshaded circles are lamps turned on.

signify a letter of the alphabet, an operation of mathematics, a quality – anything one chooses to let it represent. Thus the card can record a person's name, the colour of his eyes and hair, his position or income, his marital status, his special talents or educational qualifications. When the card is passed between electric poles, the contacts made through the holes set up a specific electrical pattern. By scanning cards in this way and selecting only those that set up a particular pattern, the machine can pick out of a large population just those individuals, say, who are over six feet tall, have blue eyes, and speak Russian.

Similarly, a punch card carrying a 'programme' for a computer sets up electrical circuits that cause the machine to put certain numbers through certain operations. In all except the simplest programmes, the computer has to store some numbers or data until it is ready to use them in the course of its series of operations. This means it must have a memory device. The memory in a modern computer usually is of the magnetic type: the information is stored in the form of magnetized spots on a spinning drum or, more recently, on magnetic tape. A particular spot may represent the digit 1 when it is magnetized and 0 when it is not magnetized. Each such item of information is a 'bit' (for 'binary digit'). The spots are magnetized by an electric current in recording the information and also are read by the production of an induced voltage.

It is quite apparent, now, however, that the useful punch card has had its day. Systems are in use whereby instructions can be fed computers (such as UNIVAC III, built in 1960) by means of typed English words. The computer is designed to react properly to certain words and combinations of words, which can be used quite flexibly but according to a definite set of rules. Such 'computer languages' have proliferated rapidly; one of the better known ones is FORTRAN (short for 'formula translation').

The computer is, of course, the crowning instrument that makes full automation possible. Servomechanisms can only carry out the slave aspects of a task. In any complex process, such as the aiming and firing of a gun or the operation of an automatic factory, computers are necessary to calculate, interpret, coordinate, and supervise – in other words, to serve as mechanical minds to direct the mechanical muscles. And the computer is by no means limited to this sort of function. It makes out your electric bills; it handles reservations for airlines; it keeps the records of cheques and banking accounts; it deals with company payrolls, maintains a running record of inventories, in short, handles a great deal of the business of large companies today.

The speed of a computer, and its immunity to tedium, make it possible to perform tasks too long-enduring (though not necessarily too difficult in principle) for the human brain. The computer can deal with purely random processes like the mutual collisions of thousands of particles by actually conducting mathematical 'games' that simulate such processes and following thousands of cases in very short periods, in order to set up the probabilities of various possible events. This is called the 'Monte Carlo method', since it is the same principle as that used by human gamblers to try to bring some sort of system into the workings of a roulette wheel.

The mathematicians, physicists, and engineers investigating the possible uses of computers are sure that all this is only a beginning. They have programmed computers to play chess, as an exercise in storing and applying information. (Of course, the machine plays only a mediocre game, but very likely it will eventually be possible to programme a computer to learn from its opponent's moves.) In laboratory experiments computers have been programmed to digest articles from scientific journals, to index such articles, and to translate from one language into another. For instance, with a vocabulary of Russian words and their English equivalents stored in its memory, a computer can scan Russian print, recognize key words, and render a crude translation. A great deal of research is being done on this problem of translating by machine, because the flood of technical literature the world over is so vast that human translators cannot possibly keep up with it.

Not only can a computer 'read'; it can also listen. Workers at the Bell Telephone Laboratories have built a machine called 'Audrey' which can distinguish words of ordinary speech. Presumably this creates the possibility that computers will some day be able to 'understand' spoken instructions. It is already possible with machines to programme speech and even song.

In 1965, systems were devised to enable people or machines to communicate with the computer from remote terminals – via a typewriter, for instance. This is called 'time-sharing', because many people may be sharing the computer's time, though response is so quick that each user is scarcely aware of the others.

These developments suggest still further possibilities. Computers are fast and far more efficient than a man in performing routine mathematical operations, but they do not begin to approach the flexibility of the human mind. Whereas even a child learning to read has no difficulty in recognizing that a capital B, a small b, and an italic *b* in various type sizes and styles all stand for the same letter, this is far beyond any present machine. To attain any such flexibility in a machine would call for prohibitive size and complexity.

The human brain weighs a little over 3 pounds and works on a practically negligible input of energy. In contrast, ENIAC, which was probably less than one millionth as complex as the brain, weighed 30 tons, and required 150 kilowatts of energy. To be sure, devices such as the transistor and the cryotron, a kind of switch employing super-conducting wires at low temperatures, have miniaturized computers and reduced the power requirements. The use of ferrite components to store bits of information has increased the capacity of computer memory as the components have been made smaller. Ferrite components as small as a full-stop on this page now exist.

Tiny tunnel sandwiches, operating under super-conductive conditions and called 'neuristors', bear the promise of being put together with the compactness and complexity of cells in the human brain. So perhaps we are in sight of duplicating the amazing qualities of that organ, which, in its three pounds, contains some 10,000 million nerve cells together with 90 million auxiliary cells.

As computers grow more intricate, subtle, and complex at a

breakneck rate, the question naturally arises: Even if computers today cannot truly think, will the day come when one will? They can, already, calculate, remember, associate, compare, recognize. Will one someday reason? The answer, many think (as, for instance, the American mathematician Marvin L. Minsky), is: Yes.

In 1938, a young American mathematician and engineer, Claude Elwood Shannon, pointed out in his master's thesis that deductive logic, in the form known as Boolean algebra, could be handled by means of the binary system. Boolean algebra refers to a system of 'symbolic logic' suggested in 1854 by the English mathematician George Boole in a book entitled *An Investigation of the Laws of Thought*. Boole observed that the types of statement employed in deductive logic could be represented by mathematical symbols, and he went on to show how such symbols could be manipulated according to fixed rules to yield appropriate conclusions.

To take a very simple example, consider the following statement: 'Both A and B are true.' We are to determine the truth or falsity of this statement by a strictly logical exercise, assuming that we know whether A and B, respectively, are true or false. To handle the problem in binary terms, as Shannon suggested, let 0 represent 'false' and 1 represent 'true'. If A and B are both false, then the statement 'Both A and B are true' is false. In other words, 0 and 0 yield 0. If A is true but B is false (or vice versa), then the statement again is false. That is, 1 and 0 (or 0 and 1) yield 0. If A is true and B is true, then the statement 'Both A and B are true' is true. Symbolically, 1 and 1 yield 1.

Now these three alternatives correspond to the three possible multiplications in the binary system, namely: $0 \times 0 = 0$, $1 \times 0 = 0$, and $1 \times 1 = 1$. Thus the problem in logic posed by the statement 'Both A and B are true' can be manipulated by multiplication. A computer (properly programmed) therefore can handle this logical problem as easily, and in the same way, as it handles ordinary calculations.

In the case of the statement 'Either A or B is true', the problem is handled by addition instead of by multiplication. If neither A nor B is true, then this statement is false. In other words, $0+0 = 0$. If A is true and B false, or vice versa, the statement is true; in these cases $1+0 = 1$ and $0+1 = 1$. If both A and B are true, the statement is certainly true, and $1+1 = 10$. (The significant digit in the 10 is the 1; the fact that it is moved over one position is immaterial. In the binary system 10 represents $(1 \times 2^1)+(0 \times 2^0)$, which is equivalent to 2 in the decimal system.)

Boolean algebra has become important in the engineering of communications, and it forms part of what is now known as 'information theory'.

What remains of thinking, then, that cannot be acquired by the machine? We are left finally with creativity and the ability of the human mind to cope with the unknown: its intuition, judgement, weighing of a situation and the possible consequences – call it what you will. This, too, was put in mathematical form, after a fashion, by the mathematician John von Neumann, who with the economist Oskar Morgenstern wrote *The Theory of Games and Economic Behaviour* in the early 1940s.

Von Neumann took certain simple games, such as matching coins and poker, as models for analysis of the typical situation in which one tries to find a winning strategy against an opponent who is himself selecting the best possible course of action for his own purposes. Military campaigns, business competition, many matters of great moment involve decisions of this kind. Even scientific research can be viewed as a game of man against nature, and game theory can be helpful in selecting the optimal strategy of research, assuming that nature stacks the cards in the way that will hamper man most (which, in fact, it often seems to do).

And as machines threaten to become recognizably human, humans become more mechanical. Mechanical organs can replace organic hearts and kidneys. Electronic devices can make up for organic failures so that artificial pacemakers can be implanted in bodies to keep hearts going that would otherwise stop. Artificial hands have been devised that can be controlled by amplified nerve-impulses in the arms to which they are attached. The word 'cyborg' ('*cyb*ernetic *org*anism') was coined in 1960 to refer to mechanically amplified men.

Attempts to mimic the mind of man are as yet in their infancy. The road, however, is open, and it conjures up thoughts which are exciting but also in some ways frightening. What if man eventually were to produce a mechanical creature, with or without organic parts, equal or superior to himself in all respects, including intelligence and creativity? Would it replace man, as the superior organisms of the earth have replaced or subordinated the less well adapted in the long history of evolution?

It is a queasy thought: that we represent, for the first time in the history of life on the earth, a species capable of bringing about its own possible replacement. Of course, we have it in our power to prevent such a regrettable *dénouement* by refusing to build machines that are too intelligent. But it is tempting to build them nevertheless. What achievement could be grander than the creation of an object that surpasses the creator? How could we consummate the victory of intelligence over nature more gloriously than by passing on our heritage, in triumph, to a greater intelligence – of our own making?

Bibliography

A guide to biological science would be incomplete without a guide to more reading. I am setting down here a brief selection of books. The list is a miscellaneous one and does not pretend to be a comprehensive collection of the best modern books about biological science, but I have read most or all of each of them myself and can highly recommend all of them, even my own.

Chapter 1 – *The Molecule*

FIESER, L. F., and M., *Organic Chemistry*, D. C. Heath, Boston, 1956.

GIBBS, F. W., *Organic Chemistry Today*, Penguin Books, Harmondsworth, 1961.

HUTTON, KENNETH, *Chemistry*, Penguin Books, Harmondsworth, 1957.

PAULING, LINUS, *The Nature of the Chemical Bond* (3rd edn), Cornell University Press, Ithaca, N.Y., 1960.

PAULING, LINUS, and HAYWARD, R., *The Architecture of Molecules*, W. H. Freeman, San Francisco and London, 1964.

Chapter 2 – *The Proteins*

ASIMOV, ISAAC, *Photosynthesis*, Allen & Unwin, London, 1970.

BALDWIN, ERNEST, *Dynamic Aspects of Biochemistry* (5th edn), Cambridge University Press, Cambridge, 1967.

BALDWIN, ERNEST, *The Nature of Biochemistry*, Cambridge University Press, Cambridge, 1967.

HARPER, HAROLD A., *Review of Physiological Chemistry* (12th edn), Blackwell, Oxford, 1969.

KAMEN, MARTIN D., *Isotope Tracers in Biology*, Academic Press, New York, 1957.

KARLSON, P., *Introduction to Modern Biochemistry*, Academic Press, New York and London, 1963.
LEHNINGER, A. L., *Bioenergetics*, Benjamin Company, New York, 1965.
SCIENTIFIC AMERICAN (editors), *The Physics and Chemistry of Life*, Simon & Schuster, New York, 1955.

Chapter 3 – *The Cell*

ANFINSEN, CHRISTIAN B., *The Molecular Basis of Evolution*, John Wiley, New York, 1959.
ASIMOV, ISAAC, *The Genetic Code*, Murray, London, 1964.
ASIMOV, ISAAC, *A Short History of Biology*, Nelson, London, 1965.
BOYD, WILLIAM C., *Genetics and the Races of Man*, Blackwell, Oxford, 1950.
BUTLER, J. A. V., *Inside the Living Cell*, Allen & Unwin, London, 1959.
DOLE, STEPHEN H., *Planets for Man*, Methuen, London, 1965.
HARTMAN, P. E., and SUSKIND, S. R., *Gene Action*, Prentice-Hall, Englewood Cliffs, 1965.
HUGHES, ARTHUR, *A History of Cytology*, Abelard-Schuman, New York and London, 1959.
NEEL, J. V., and SCHULL, W. J., *Human Heredity*, University of Chicago Press, Chicago, 1954.
OPARIN, A. I., *The Origin of Life on the Earth*, Oliver & Boyd, Edinburgh, 1957.
SINGER, CHARLES, *A Short History of Anatomy and Physiology from the Greeks to Harvey*, Dover Publications, New York, 1957.
SULLIVAN, WALTER, *We Are Not Alone*, Hodder & Stoughton, London, 1965.
TAYLOR, GORDON R., *The Science of Life*, Thames & Hudson, London, 1963.
WALKER, KENNETH, *Human Physiology*, Penguin Books, Harmondsworth, 1956.

Chapter 4 – *The Micro-organisms*

BURNET, F. M., *Viruses and Man* (2nd edn), Penguin Books, Harmondsworth, 1955.
DE KRUIF, PAUL, *Microbe Hunters*, Cape, London, 1963.
DUBOS, RENÉ, *Louis Pasteur*, Gollancz, London, 1951.
LUDOVICI, L. J., *The World of the Infinitely Small*, Phoenix House, London, 1958.
MCGRADY, PAT, *The Savage Cell*, Basic Books, New York, 1964.
RIEDMAN, SARAH R., *Shots without Guns*, Rand, McNally, Chicago, 1960.

SMITH, KENNETH M., *Beyond the Microscope*, Penguin Books,
   Harmondsworth, 1957.
WILLIAMS, GREER, *Virus Hunters*, Hutchinson, London, 1960.
ZINSSER, HANS, *Rats, Lice and History*, Little, Brown, Boston, 1935.

Chapter 5 – *The Body*

ASIMOV, ISAAC, *The Human Body*, Nelson, London, 1965.
CARLSON, ANTON J., and JOHNSON, VICTOR, *The Machinery of the Body*,
   University of Chicago Press, Chicago and London, 1961.
CHANEY, MARGARET S., *Nutrition*, Houghton Mifflin, Boston, 1954.
MCCOLLUM, ELMER VERNER, *A History of Nutrition*, Houghton Mifflin,
   Boston, 1957.
WILLIAMS, ROGER J., *Nutrition in a Nutshell*, Doubleday, New York, 1962.

Chapter 6 – *The Species*

ASIMOV, ISAAC, *Wellsprings of Life*, Abelard-Schuman, London, 1960.
BOULE, M., and VALLOIS, H. V., *Fossil Men*, Thames & Hudson, London,
   1957.
CARRINGTON, RICHARD, *A Biography of the Sea*, Chatto & Windus,
   London, 1960.
DARWIN, FRANCIS (editor), *The Life and Letters of Charles Darwin* (2 vols.),
   Basic Books, New York, 1959.
DE BELL, G., *The Environmental Handbook*, Ballantine Books, New York,
   1970.
HANRAHAN, JAMES S., and BUSHNELL, DAVID, *Space Biology*, Thames &
   Hudson, London, 1960.
HARRISON, R. J., *Man, the Peculiar Animal*, Penguin Books,
   Harmondsworth, 1958.
HOWELLS, WILLIAM, *Mankind in the Making*, Secker & Warburg, London,
   1960.
HUXLEY, T. H., *Man's Place in Nature*, University of Michigan Press, Ann
   Arbor, 1959.
MEDAWAR, P. B., *The Future of Man*, Methuen, London, 1960.
MILNE, L. J. and M. J., *The Biotic World and Man*, Prentice-Hall, New
   York, 1958.
MONTAGU, ASHLEY, *The Science of Man*, Odyssey Press, New York, 1964.
MOORE, RUTH, *Man, Time, and Fossils*, Cape, London, 1962.
ROMER, A. S., *Man and the Vertebrates* (2 vols.), Penguin Books,
   Harmondsworth, 1954.
ROSTAND, JEAN, *Can Man Be Modified?*, Secker & Warburg, London,
   1959.
SAX, KARL, *Standing Room Only*, Beacon Press, Boston, 1955.

SIMPSON, GEORGE G., PITTENDRIGH, C. S., and TIFFANY, L. H., *Life: An Introduction to College Biology* (2nd edn), Harcourt, Brace, New York, 1965.

TINBERGEN, NIKO, *Curious Naturalists*, Country Life, London, 1958.

UBBELOHDE, A. R., *Man and Energy*, Penguin Books, Harmondsworth, 1963.

Chapter 7 – *The Mind*

ANSBACHER, H. and R. (editors), *The Individual Psychology of Alfred Adler*, Allen & Unwin, London, 1958.

ARIETI, SILVANO (editor), *American Handbook of Psychiatry* (3 vols.), Basic Books, New York, 1959, 1966.

ASIMOV, ISAAC, *The Human Brain*, Nelson, London, 1965.

BERKELEY, EDMUND C., *Symbolic Logic and Intelligent Machines*, Reinhold, New York, 1959.

FREUD, SIGMUND, *Collected Papers* (5 vols.), International Psycho-Analytical Society, London, 1924–50.

JONES, ERNEST, *The Life and Work of Sigmund Freud* (3 vols.), Hogarth Press, London, 1953–7.

LASSEK, A. M., *The Human Brain*, Charles C. Thomas, Springfield, Ill., 1957.

MENNINGER, KARL, *Theory of Psychoanalytic Technique*, Imago, London, 1958.

MURPHY, GARDNER, *Human Potentialities*, Allen & Unwin, London, 1960.

RAWCLIFFE, D. H., *Illusions and Delusions of the Supernatural and Occult*, Dover Publications, New York, 1959.

SCIENTIFIC AMERICAN (editors), *Automatic Control*, Simon & Schuster, New York, 1955.

SCOTT, JOHN PAUL, *Animal Behavior*, University of Chicago Press, Chicago and London, 1957.

THOMPSON, CLARA, and MULLAHY, PATRICK, *Psychoanalysis: Evolution and Development*, Allen & Unwin, London, 1952.

General

ASIMOV, ISAAC, *Asimov's Biographical Encyclopedia of Science and Technology*, Allen & Unwin, London, 1966.

ASIMOV, ISAAC, *The Words of Science*, Houghton Mifflin, Boston, 1959.

HUTCHINGS, EDWARD, JR (editor), *Frontiers in Science*, Allen & Unwin, London, 1960.

SHAPLEY, HARLOW, RAPPORT, SAMUEL, and WRIGHT, HELEN (editors), *A Treasury of Science* (3rd edn), Angus & Robertson, London, 1954.

WATSON, JANE WERNER, *The World of Science*, Adprint, London, 1959.

Name Index

Abbe, Ernst Karl, 193
Abderhalden, Emil, 68
Abel, Frederick Augustus, 44
Abelson, Philip, 177
Addison, Thomas, 288
Adrian, Edgar Douglas, 395
Agamemnon, 340
Airy, George Biddell, 192, 418
Albert (Prince), 21
Alcmaeon, 123
Alder, Kurt, 25
Alhazen, 191
Appert, François, 253
Aquinas, Thomas, 173
Aristotle, 173, 304, 315, 380
Armstrong, Neil A., 367
Arnon, Daniel I., 117
Arrhenius, Svante August, 175, 176
Aschheim, Selmar, 287
Ashby, W. Ross, 422
Ashurbanipal, 339, 340
Astbury, William Thomas, 72
Avery, Oswald Theodore, 161

Babbage, Charles, 428, 429, 432
Babinski, Joseph F. F., 392
Bacon, Roger, 191
Baekeland, Leo Hendrik, 48, 72
Baer, Karl Ernst von, 129, 310
Baeyer, Aldolf von, 23, 48, 118
Baker, B. L., 182
Balfour, Francis Maitland, 310
Bang, Olaf, 243
Banting, Frederick Grant, 283
Barghoorn, Elso Sterrenberg, 327

Barnard, Christiaan, 235
Bateson, William, 138
Baumann, Eugen, 276
Bawden, Frederick Charles, 217
Bayliss, William Maddock, 279, 282
Beadle, George Wells, 148ff.
Beckwith, Jonathan, 142
Behring, Emil von, 228, 229
Beijerinck, Martinus Willem, 215–16
Benda, C., 104
Benedict, Francis Gano, 285
Berger, Hans, 397
Bergmann, Max, 68, 72
Bernard, Claude, 283
Berthelot, Pierre E. M., 19, 20
Berzelius, Jöns Jakob, 2, 3, 4, 5, 40,
    63, 87
Best, Charles Herbert, 283
Bijvoet, J. M., 14
Binet, Alfred, 380
Biot, Jean Baptiste, 9, 10, 11
Bittner, John Joseph, 243, 244
Blaiberg, Philip, 235
Bleuler, Eugen, 410
Bloch, Felix, 75
Bloch, Konrad Emil, 111–12
Blout, Elkan Rogers, 75
Blumenbach, Johann Friedrich, 351–2
Bonaparte, Napoleon, 338
Bonnet, Charles, 316
Boole, George, 436
Bordet, Jules, 229
Borman, Frank, 367
Borrel, Amédée, 243
Bosch, Karl, 46

443

King, Charles Glen, 259, 262
Kipling, Rudyard, 44
Kipping, Frederic Stanley, 53
Kirchhoff, Gottlieb Sigismund, 36, 87
Klebs, Edwin, 199
Knoop, Franz, 107, 108
Koch, Robert, 198–9, 228
Kocher, Emil Theodor, 276
Koenigswald, Gustav H. R. von, 347, 351
Köhler, Wolfgang, 404–5
Kolbe, Adolph W. H., 3, 19
Koller, Carl, 28
Komarov, Vladimir M., 366, 367
Korff, Serge, 344
Kornberg, Arthur, 164–5
Kossel, Aibrecht, 157, 158
Krebs, Hans Adolf, 103, 106
Kruse, Walther, 216
Kuhn, Richard, 266, 267, 287
Kühne, Wilhelm, 92
Kylstra, Johannes A., 365–6

Laennec, René T. H., 195
Lafayette, Marquis de, 408
Lamarck, Jean Baptiste de, 318, 319
Land, Edwin Herbert, 9
Landsteiner, Karl, 145, 146, 233
Langerhans, Paul, 282
Langley, John Newport, 390
Langmuir, Irving, 88, 94
Larsen, John Augustus, 396
Lartet, Édouard Armand, 342
Laveran, Charles L. A., 210
Lavoisier, Antoine Laurent, 87, 408
Lawes, John Bennet, 250
Lazear, Jesse William, 211
Leakey, Louis S. B.. 349
Leakey, Mary, 349
Lebedev, Peter Nicolaevich, 175
Le Bel, Joseph Achille, 11, 13
LeCanu, L. R., 32
Lee, Edmund, 416–17
Leeuwenhoek, Anton van, 90, 129, 192, 193
Leibnitz, Gottfried Wilhelm von, 428, 429
Lejeune, Jérôme Jean, 134
Leonardo of Pisa, 425

Leonov, Aleksei A, 367
Leopold (Prince), 139
Lettvin, Jerome, 423
Levene, Phoebus A. T., 158, 159
Lewis, G. Edward, 349
Li, Cho Hao, 84, 293
Libby, Willard Frank, 344, 345
Liebig, Justus von, 4, 5, 20, 21, 65, 78, 249, 270, 273
Lilly, John C., 379
Lincoln, Abraham, 296
Lind, James, 254
Lindbergh, Charles Augustus, 230
Lindbergh, Jon, 365
Linnaeus, Carolus, 305, 306, 314, 315
Lipmann, Fritz Albert, 102ff.
Lippershey, Hans, 192
Lister, Joseph, 198
Lister, Joseph Jackson, 193
Loewi, Otto, 397
Löffler, Friedrich August, 216
Lohmann, K., 267
Long, Crawford Williamson, 28
Lorenz, Konrad Zacharias, 393
Lowell, Percival, 185
Lucretius, 341
Lunge, George, 88, 94
Luzzi, Mondino de, 124
Lwoff, Andre Michael, 170
Lyell, Charles, 318, 322, 326, 332
Lynen, Feodor, 105, 112
Lysenko, Trofim Denisovich, 319

Macintosh, Charles, 39
Macleod, Colin Munro, 161
MacLeod, John J. R., 283
Macquer, Pierre Joseph, 63
Magellan, Ferdinand, 253
Magendie, François, 250
Malpighi, Marcello, 127, 192
Malthus, Thomas Robert, 320, 322, 357
Malus, Étienne Louis, 9
Mann, Thaddeus, 96, 271
Manson, Patrick, 210
Marcus Aurelius, 196
Marine, David, 276
Mariner, Ruth, 178
Martin, Archer J. P., 76ff.
Matthaei, J. Heinrich, 170–71

Subject Index

abacus, 424ff.
abnormal haemoglobin, 153–4
absorption spectrum, 77
acceleration, 371
acetate ion, 111
acetic acid, 3
acetylation, 105
acetylcholine, 397
acetylene, 7
achromatic microscope, 193
Achromycin, 205
Acrilan, 58
acromegaly, 293
acrylonitrile, 58
ACTH, 98, 291, 292
activated carbon, 89
actomyosin, 86
Addison's disease, 288, 290
adenine, 157
adenine monophosphate, 295
adenosine triphosphate (ATP), 102, 178
adenyl cyclase, 295
adipic acid, 57
adolescence, 287
adrenal glands, 280, 288
adrenalin, 280; nerve action and, 396
adrenocortical hormones, 111
adrenocorticotropic hormone (ACTH), 98, 291
adsorption, 89
afferent nerve cells, 375
agammaglobulinaemia, 237, 238
agar, 199

Agnatha, 310
air conditioning, 51
alanine, 96, 177
alanine–transfer–RNA, 169
albino, 148
albumen, 63
albuminous substances, 63
alcoholism, 31
aldosterone, 290
algae, 307; blue-green 182
aliphatic compounds, 16
alizarin, 22; synthetic, 23
alkaloids, 26
alleles, 135
allergic reactions, 235, 237
alpha-amino acids, 65
Alpha Centauri, 189
alpha waves, 398
Alpines, 354
amethopterin, 247
amine group, 65, 256
amine oxidase, 413
amino acids, 14, 56, 65; combinations of, 67; dietary requirement for, 251–2; essential, 252; fossils and, 333; haemoglobin and, 154; order of, 8off.; polymerization of, 72; primordial formation of, 177ff.; separation of, 76; three-dimensional structure of, 73–4
ammonia, 6; nitrogenous wastes and, 333; synthetic, 46
ammonium cyanate, 3
Amphibia, 310–11, 332
amphioxus, 311

453

More about Penguins and Pelicans

Penguinews, which appears every month, contains details of all the new books issued by Penguins as they are published. From time to time it is supplemented by *Penguins in Print*, which is a complete list of all titles available. (There are some five thousand of these.)

A specimen copy of *Penguinews* will be sent to you free on request. For a year's issues (including the complete lists) please send 50p if you live in the British Isles, or 75p if you live elsewhere. Just write to Dept EP, Penguin Books Ltd, Harmondsworth, Middlesex, enclosing a cheque or postal order, and your name will be added to the mailing list.

In the U.S.A.: For a complete list of books available from Penguin in the United States write to Dept CS, Penguin Books Inc., 7110 Ambassador Road, Baltimore, Maryland 21207.

In Canada: For a complete list of books available from Penguin in Canada write to Penguin Books Canada Ltd, 41 Steelcase Road West, Markham, Ontario.

The Chemistry of Life

Steven Rose

The molecular structure of a protein (insulin) was described in detail for the first time in 1956: today such procedures are routine. Not only has the pace of biochemistry accelerated in recent years: with the perfection of the electron microscope and the development of cybernetics, the science has also widened and grown more complex.

The Chemistry of Life outlines the scope and achievement of a science which began as the study of the chemical constituents of living matter. Dealing successively with the chemical analysis of the living animal cell, the conversions induced between chemicals by the enzymes acting as catalysts, and the self-regulating nature of cells, Dr Rose explains how the design of particular cells influences their functions within the living organism as a whole.

Biochemistry is a difficult subject. But it is presented here as simply as accuracy will permit by a young research biochemist who conveys much of the adventure of discovery implicit in a science which may one day answer the eternal question: 'What is life?'

Radio Astronomy

F. Graham Smith

Fourth Edition

Radio astronomy, as an organized body of science, only started after the Second World War. Ten years ago nobody could have predicted the discovery of pulsars or the bewildering behaviour of quasars.

In such a science each new edition of a Pelican must virtually be a new book, and a publisher is lucky to find an active radio astronomer with the time to review the history, the developing techniques and the main achievement of his science. The account Graham Smith gives us is exciting: even if no messages have been intercepted from the 'little green men' of space, the universe he reveals contains other wonders – astral generators, flashing lighthouses, transmitting stations and a wealth of celestial phenomena to which optical astronomy is 'blind'.

Today a worldwide network of radio telescopes of various types can probe the unimaginable depths of space, keep count of observable changes in the universe and shed new light on astrophysics. This latest edition of *Radio Astronomy* admirably suggests the boundless potential of a young science which has tapped the *silence éternel* that terrified Pascal.